Biodiversity and Spatial Distribution of Fishes

Biodiversity and Spatial Distribution of Fishes

Guest Editors

Heng Zhang
Ying Xue
Yunzhi Yan
Luoliang Xu

MDPI

Basel • Beijing • Wuhan • Barcelona • Belgrade • Novi Sad • Cluj • Manchester

Guest Editors

Heng Zhang
East China Sea Fisheries
Research Institute
Chinese Academy of
Fishery Sciences
Shanghai
China

Ying Xue
College of Fisheries
Ocean University of China
Qingdao
China

Yunzhi Yan
School of Ecology and
Environment
Anhui Normal Univesity
Wuhu
China

Luoliang Xu
Center for limnology
University of Wisconsin-Madison
Madison, WI
United States

Editorial Office
MDPI AG
Grosspeteranlage 5
4052 Basel, Switzerland

This is a reprint of the Special Issue, published open access by the journal *Fishes* (ISSN 2410-3888), freely accessible at: www.mdpi.com/journal/fishes/special_issues/4WN7CH6VLU.

For citation purposes, cite each article independently as indicated on the article page online and using the guide below:

Lastname, A.A.; Lastname, B.B. Article Title. *Journal Name* **Year**, *Volume Number*, Page Range.

ISBN 978-3-7258-2806-7 (Hbk)
ISBN 978-3-7258-2805-0 (PDF)
https://doi.org/10.3390/books978-3-7258-2805-0

Contents

Preface

With the increasing impact of human activities, the diversity and distribution of fish across various aquatic ecosystems, ranging from freshwater to marine environments, have undergone profound changes. These changes are characterized by a decline in biodiversity, the simplification of community structures, and notable shifts in species distribution ranges. Gaining a timely and comprehensive understanding of the distribution patterns and key influencing factors driving these changes, whether due to environmental shifts or anthropogenic activities, is crucial for the effective conservation of fish populations and their habitats.

In recent years, alongside traditional research approaches in fishery ecology, emerging methodologies such as environmental DNA (eDNA) analysis and deep learning models have significantly advanced our understanding of fish diversity and distribution mechanisms. While these approaches have yielded promising results, there remains considerable potential for further methodological refinement and application.

This Special Issue focuses on monitoring, identifying, and refining methodologies to assess biodiversity and the spatial–temporal distribution of fish across diverse aquatic systems. Particular emphasis is placed on evaluating habitat-induced changes using advanced tools such as species distribution models and deep learning techniques.

To enrich this Special Issue, we invited four distinguished experts, both domestic and international, to serve as guest editors. We also extended invitations to research institutions and scholars from fisheries-related universities to contribute their findings. This endeavor received substantial support, resulting in 23 submissions, of which 13 high-quality papers were published following rigorous peer review. The selected studies span a wide range of topics, including species distribution, biodiversity dynamics, and the environmental factors influencing fish and fisheries across varied regions such as China's nearshore and estuarine areas, coastal island waters, the Pacific high seas, Poyang Lake, and streams in Europe and the Americas. These contributions offer valuable insights into the significant changes occurring in fish populations across different water bodies and provide a foundation for understanding natural change patterns, ecosystem functions, and structural dynamics. Importantly, the findings offer practical implications for the sustainable management and conservation of fishery resources.

Finally, we extend our heartfelt gratitude to all reviewers, editors, and contributors for their invaluable efforts and dedication. Your commitment and expertise were instrumental in the success of this Special Issue.

Heng Zhang, Ying Xue, Yunzhi Yan, and Luoliang Xu
Guest Editors

fishes

MDPI

Article

Distribution, Occupancy, and Habitat of the Endangered Carolina Madtom: Implications for Recovery of an Endemic Stream Fish

W. Robert Cope [1,†,‡], Thomas J. Kwak [2,§], Tyler R. Black [3,||], Krishna Pacifici [4], Jennifer M. Archambault [5] and W. Gregory Cope [6,*]

1 North Carolina Cooperative Fish and Wildlife Research Unit, Department of Applied Ecology, Campus Box 7617, North Carolina State University, Raleigh, NC 27695, USA; wrcope@iastate.edu
2 U.S. Geological Survey, North Carolina Cooperative Fish and Wildlife Research Unit, Department of Applied Ecology, Campus Box 7617, North Carolina State University, Raleigh, NC 27695, USA
3 North Carolina Wildlife Resources Commission, 2430 Turner Road, Mebane, NC 27032, USA; tblack@rkk.com
4 Department of Forestry and Environmental Resources, North Carolina State University, 3120 Jordan Hall, Raleigh, NC 27607, USA; jkpacifi@ncsu.edu
5 U.S. Fish and Wildlife Service, Ecological Services Field Office, Raleigh, NC 27636, USA; jennifer_archambault@fws.gov
6 Department of Applied Ecology, Campus Box 7617, North Carolina State University, Raleigh, NC 27695, USA
* Correspondence: greg_cope@ncsu.edu; Tel.: +1-919-515-5296
† Current address: Department of Natural Resource Ecology & Management, 339 Science II, Iowa State University, Ames, IA 50011, USA.
‡ This work is part of the Master of Science thesis of the first author W. Robert Cope. Master of Science Program at North Carolina State University, Raleigh, NC 27695, USA.
§ Deceased.
|| Current address: RK&K, 8601 Six Forks Road, Forum 1, Suite 700, Raleigh, NC 27615, USA.

Abstract: Endemic fish are important components of freshwater ecosystems because they contribute to biodiversity and provide vital ecological functions. The Carolina Madtom, *Noturus furiosus*, is a small catfish endemic to the Neuse and Tar river basins of North Carolina, USA. Three previous surveys over the past 60 years have shown declining occurrence and abundance in the basins, and as such, the species was listed as federally endangered in 2021. To provide critical information to guide Carolina Madtom conservation and recovery strategies, we surveyed 36 sites (75 locations) in both basins to (1) determine the current distribution of the Carolina Madtom, (2) develop occupancy models to estimate probability of detection and occurrence of the species throughout its range, and (3) determine its instream habitat use and suitability. We collected 59 Carolina Madtom during snorkel surveys in the Tar River basin and none from the Neuse River basin, indicating that Carolina Madtom populations are still declining in both occurrence and abundance throughout their historical and recent range, especially in the Neuse River basin. Occupancy modeling estimated low occupancy probability (0.35), while detection probability was high (0.81). Carolina Madtom occupied slow-to-moderate velocity water over sand and gravel substrate, using cobble and woody debris as cover. Habitat suitability distributions quantified the most suitable ranges of microhabitat parameters for Carolina Madtom occupancy. A comparison of available suitable habitat in the two river basins revealed that adequate suitable habitat was available in both basins, suggesting that other factors such as pollution or predation from the non-native Flathead Catfish *Pylodictis olivaris*, may be contributing to population declines. The application of our results will aid in management and recovery actions for the species.

Keywords: endangered species; conservation; microhabitat; biodiversity

Key Contribution: Gained new knowledge about the endangered Carolina Madtom's populations and habitat use, which is helping to guide informed recovery and protective management decisions.

check for updates

Citation: Cope, W.R.; Kwak, T.J.; Black, T.R.; Pacifici, K.; Archambault, J.M.; Cope, W.G. Distribution, Occupancy, and Habitat of the Endangered Carolina Madtom: Implications for Recovery of an Endemic Stream Fish. *Fishes* **2024**, *9*, 454. https://doi.org/10.3390/fishes9110454

Academic Editors: Heng Zhang, Ying Xue, Yunzhi Yan and Luoliang Xu

Received: 26 September 2024
Revised: 31 October 2024
Accepted: 1 November 2024
Published: 7 November 2024

1. Introduction

North America supports the greatest temperate freshwater biodiversity globally, and the southeastern United States is especially rich with freshwater fauna [1,2]. Sixty-two percent of United States fishes (493 species) occur in the southeast, and 91% (269 species) of the Nation's freshwater mussels are found in the southeastern United States [3,4]. The southeastern United States also has the greatest number of imperiled freshwater species in North America [1]. Through anthropogenic factors such as habitat degradation, fragmentation and loss, flow modification, and pollution, many southeastern fishes are experiencing population declines [5–7]. Many of these imperiled fishes are endemic species and critical to freshwater ecosystems because they contribute to biodiversity and provide important ecological functions; however, many are in need of conservation and are understudied and data deficient [7,8]. Moreover, endemic species are limited in range or suitable habitat, typically exhibit small population sizes, and are at increased risk of extinction due to human-induced habitat modifications [5,9]. Thus, many endemic species remain poorly understood and minimally managed globally [10,11].

The Carolina Madtom *Noturus furiosus*, is a small catfish (less than 200 mm total length) endemic to the Neuse and Tar river basins of North Carolina [12]. The species occurs in free-flowing streams in riffles, runs, or pools in shallow areas [12,13]. Carolina Madtom are benthic-associated organisms, and as such, most commonly occupy areas with substrates consisting of a mixture of sand or gravel, with leaf litter and small cobble included for cover [13]. The Carolina Madtom spawning season begins in May, and reproduction may last through July [12]. The species seeks out spawning areas with low water velocity and adequate cover to nest. Males guard nests consisting of natural cavities, such as leaf litter packs, large woody debris, under small rocks, inside empty native unionid mussel shells, or in discarded beverage containers [12,13]. Artificially constructed and deployed cover units may enhance Carolina Madtom detections [14], and they were found to select such artificial cover units over natural cover items in laboratory choice tests [15].

The Carolina Madtom has experienced substantial declines in distribution since the 1960s [16,17], and are currently protected with federal endangered species status [18]. Carolina Madtom populations have been relatively stable in the Tar River basin but have declined precipitously in the Neuse River basin [19,20]. The Neuse River is a degraded river basin, and the North Carolina Department of Environmental Quality has determined that 14% of the total stream distance in the basin is impaired, according to their water quality thresholds. Impacts from urban wastewater, fertilizer, residential and industrial development, and commercial animal operations contribute to pollution and eutrophication of the system [21].

The Carolina Madtom has not been widely studied and was until recently, considered data deficient [22]. In the past 60 years, only three intensive sampling efforts have been conducted for the species. In the 1960s, Smith and Bayless performed basin-wide rotenone sampling and found that the Carolina Madtom was common in both basins [23,24]. In the 1980s, Burr et al. [12] re-sampled the Neuse and Tar river basins and found slight declines in the species' distributions in both basins. Wood and Nichols [19,20] then sampled Carolina Madtom in 2007 and found substantial decreases in their distribution in the Neuse River, while Tar River distributions were stable, relative to those sampled in the 1980s. Population distribution research had not occurred since 2007 on the Carolina Madtom until this study in 2016–2017.

Our study was designed to provide an evaluation of the status and trends of Carolina Madtom populations and their habitat, prior to an impending endangered species status assessment and to build upon the aforementioned studies. Our specific objectives were to (1) determine the current range and distribution of the Carolina Madtom in the Neuse and Tar river basins, (2) develop occupancy models to estimate probability of detection and occurrence of the species throughout its range, and (3) determine its current instream habitat use and suitability and compare findings to previous studies.

2. Materials and Methods

2.1. Study Area

This research was conducted in the Neuse and Tar river basins of the Piedmont and Coastal Plain physiographic provinces of North Carolina, USA. The Neuse River flows approximately 325 km through North Carolina from its headwaters originating in the Piedmont at the confluence of the Eno and Flat rivers to its mouth at Pamlico Sound near the city of New Bern [25]. The basin covers an area of 10,034 km^2 and spans 18 counties. Approximately 2.5 million people currently live in the Neuse River basin, and human-associated activities in the watershed adversely impact the habitat and water quality of the Neuse River [25]. Approximately 13% of the basin is considered urban, 45% forested, and 29% cropland and pastureland [26]. Non-point source pollution from a variety of intensive land uses has degraded water quality and habitats throughout the basin (e.g., urbanization, agriculture, and forestry without adequate riparian protections or proper erosion and sediment management). Commercial farming inputs, such as animal waste and fertilizers, contribute 60% of the nitrates and phosphates in the system [26]. Due to the dense human population in the basin, many municipalities have constructed dams and withdraw water from the created impoundments for human use, affecting river flow. Aquatic habitat loss, especially in the Neuse River basin, is an issue of importance, as an increasing human population results in the loss of natural areas and increases sedimentation and polluted runoff from impervious surfaces [25].

The Tar River runs through North Carolina from its origin in Person County to the town of Washington, where it becomes the Pamlico River and flows 65 more kilometers to its mouth at Pamlico Sound [27]. The basin covers 8755 km^2 and spans 16 counties. The Tar-Pamlico basin is more rural and less impacted by human activities than the Neuse River basin, with a human population of only 415,000. Approximately 55% percent of the basin is classified as forest and wetland, 28% cropland and pastureland, and 7% urban [28]. The primary habitat problems affecting the basin are erosion and sedimentation [26]. Future urbanization in this area of the southeastern Piedmont of North Carolina [29] is a concern for water quality and instream habitat degradation.

2.2. Study Design and Conditions

In 2016, we studied 11 sites in the Tar River basin and 9 sites in the Neuse River basin for a total of 20 sites to encompass previously extant Carolina Madtom population distributions. These sites included the main waterways of each basin, including the mainstem Tar, Fishing Creek, and Swift Creek in the Tar River basin and Contentnea Creek and Little River in the Neuse River basin. The surveyed sites included areas of historical occurrence, recent occurrence (post-2006), and exploratory areas where Carolina Madtom had not been detected. Each site was 150-m in length and followed GPS coordinates corresponding to the same 150-m site sampled by Wood and Nichols [19]. In 2017, we expanded our study site length from 150-m to 4-km to improve detection of this rare species within waterways containing a patchwork of suitable habitat. This modification was made due to low catch-rates in 2016 and the need to establish a flexible sampling protocol for long-term monitoring efforts (i.e., ability to adjust sampling sites within dynamic river systems). Surveys in 2017 included three to four 30-m locations within the expanded site length, which focused on areas with, or adjacent to, recent madtom occurrence (post-2006). Survey sites were established at areas deemed to have suitable Carolina Madtom habitat, with a random survey (30-m) added to reduce positive sampling bias. In total, we sampled twelve 4-km sites in the Tar River basin and four 4-km sites in the Neuse River basin for a total of sixteen sites consisting of fifty-five 30-m locations in 2017.

We surveyed stream sites for Carolina Madtom from May through October in 2016 and 2017 [30]. Each survey was conducted following a snorkeling protocol similar to that of Wood and Nichols [19]. Based on concerns over impaired visibility and sampling efficiency [31–33], our snorkel sampling was not performed when Secchi depth was $\leq$1 m. In 2016, surveys were conducted at the 150-m sites, with two to five snorkelers participating, and in 2017, surveys were conducted at the 30-m locations within the 4-km site, with two to four snorkelers. Each snorkeler covered no more than a 5-m wetted width to standardize effort and ensure thorough searching. If the river width (in meters) was five times greater than the total number of snorkelers, then each snorkeler sampled more than one 5-m section of the stream width. Surveys were conducted during daylight hours between 0800 and 1800. Each snorkeler started at the downstream portion of the site or location and slowly surveyed upstream, upturning any leaf litter, woody debris, cobble, or litter present in the river to maximize detection potential. Detected Carolina Madtom were promptly collected in a dip net and placed into buckets along the shore for further processing upon conclusion of the snorkel survey. We noted only three Carolina Madtom that were detected but escaped capture during surveys; all occurred in the Tar River basin and these observations were not included in our analyses. We weighed the captured fish to the nearest 0.1 g (Ohaus Scout Model SPX8200, Ohaus Corporation, Parsippany, NJ, USA) and measured total length in millimeters. A right-side pelvic fin clip was collected for genetic analysis [34,35]. The point-of-capture was designated by a weighted marker for the data collection of microhabitat use. After data collection was complete, each fish was released alive at their point-of-capture. We also recorded the presence of co-occurring native Margined Madtom *Noturus insignis*, two co-occurring non-native ictalurid fish species (Channel Catfish *Ictalurus punctatus*, and Flathead Catfish *Pylodictis olivaris*), and the endemic Neuse River Waterdog *Necturus lewisi*, a federally threatened aquatic salamander [17,36,37] at the surveyed sites. These co-occurring species may serve as bioindicators of similar habitat quality needs or offer insight into possible interactions (positive or negative) with the Carolina Madtom.

Microhabitat use data were collected during both the 2016 and 2017 seasons at all sites with Carolina Madtom occurrence. Depth (m), bottom velocity (m/s), mean-column velocity (m/s), dominant substrate composition, subdominant substrate composition, distance to nearest cover, and cover type were measured at the point-of-capture of each Carolina Madtom. Depth, bottom velocity, and mean-column velocity were measured using a top-set wading rod and a Marsh-McBirney Model 2000 flow meter. Bottom velocity was measured directly on the substrate surface and mean-column velocity was measured at 60% of the total depth from the water surface. Dominant and subdominant substrate compositions were determined by greatest and second greatest percentage of substrate type at the location, according to a modified Wentworth particle scale [38]. Distance to nearest cover was measured as the distance from point-of-capture to the nearest cover object (zero if the fish was directly associated with cover). Cover type was classified as small woody debris, large woody debris, cobble, leaf litter, artificial, and none.

Available instream microhabitat surveys were performed at all sites during 2016 and 2017 under base-flow conditions. Instream microhabitat data were collected by sampling cross-sectional transects. For the 150-m sites in 2016, 10 cross-sectional transects were spaced 15 m apart, starting from a random point within the first 15 m and moving upstream through the reach. The same transect spacing was used in 2017; however, each location encompassed two cross-sectional transects for a total of six to eight transects per location. Each transect was sampled at a minimum of 10 evenly spaced points, usually 1–3 m apart, depending on stream width. At each transect point, we measured depth, bottom velocity, mean column velocity, dominant substrate, subdominant substrate, distance to nearest cover, and cover type using the methods previously described.

2.3. Occupancy Modeling Data Analysis

Snorkel survey results from 2017 were modeled to estimate Carolina Madtom occupancy and detection probabilities using a single-species, single-season occupancy model [39] in the software program PRESENCE version 2.13.47 ([40], USGS Patuxent Wildlife Research Center, Laurel, MD, USA, https://www.mbr-pwrc.usgs.gov/software/presence.html; accessed on 5 November 2024). Given the large potential habitat range of the Carolina Madtom and the large time allotment for each sampling event, repeat visits to each individual surveying location were not logistically feasible during the field season. Instead, we implemented a spatial replication approach with each 4-km site serving as the single sampling unit relevant to occupancy and each 30-m location within the site serving as a repeat visit [41]. In total, 16 sites were snorkel surveyed, each with three to four 30-m locations (spatial replicates) within each site. This resulted in 16 occupancy sites with three to four repeat visits serving as the detection histories for each site. Prior to developing occupancy models, we assessed our 10 microhabitat covariates and placed them into the following two groups: variables expected to influence occupancy and those expected to influence detection. Using information from the most commonly used microhabitats by madtom species [42,43], we selected the following seven microhabitat covariates as most likely to influence Carolina Madtom occupancy at a site: depth, mean-column velocity, dominant substrate, subdominant substrate, distance to nearest cover; and cover type (large woody debris; cobble). We selected the following three variables likely to influence detection of Carolina Madtom: depth, mean-column velocity, and distance to nearest cover. Due to the relatively low number of sites surveyed and Carolina Madtom detected, we restricted the complexity of the fitted models by investigating models with only one covariate influencing occupancy and one influencing detection. We compiled all models and ranked them following an information-theoretic approach (Akaike information criterion corrected for small sample size, AIC_c) [44]. Individual estimates of occupancy (Ψ) and detection (p) probabilities were provided for all fitted models. Models within 10% of the AIC weight (W_i) of the top-performing model were included as confidence models, as described by Ruiz and Peterson [45].

2.4. Habitat Data Analysis

Using the cross-sectional transect and point-of-capture microhabitat data collected from sites surveyed in 2016 and 2017, we analyzed the available instream microhabitat and microhabitat use data to determine current habitat use, selectivity, and suitability distributions for Carolina Madtom. To determine Carolina Madtom microhabitat use, we calculated the mean depth, bottom velocity, and mean-column velocity measurements of occupied microhabitats and the mode of substrate and cover type. Habitat selectivity was assessed using the percentage of habitat used by Carolina Madtom compared to the available instream habitat from our surveyed sites. To determine these percentages, we developed separate frequency distributions for Carolina Madtom microhabitat use and available instream microhabitat for each variable. The two frequency distributions were then compared using a Kolmogorov–Smirnov (K-S) two-sample test for the continuous variables (depth, bottom velocity, mean column velocity, dominant substrate, subdominant substrate) and a log-likelihood ratio G-test for independence for the categorical variable (cover type), with p-values < 0.05 (α) indicating significant differences between available instream habitat and Carolina Madtom habitat use (i.e., non-random habitat use). Habitat suitability was calculated as the proportion of habitat use divided by available habitat for each habitat variable range in corresponding frequency distributions. The resulting values were standardized to a maximum of 1.0, and the range of the habitat variable with a suitability value of 1.0 was deemed the most suitable. If multiple ranges had values approximate to 1.0, the entire combined range was deemed most suitable. Habitat suitability analysis was based on fish locations from the Tar River, as no Carolina Madtom were detected in the Neuse River basin during our snorkeling surveys.

We also compared the available habitat between the Neuse and Tar river basins to determine if the drastic decrease in Neuse River basin populations may be attributed to a lack of suitable instream physical habitat. All available instream habitat measurements were pooled over both seasons separately for the Neuse and Tar river basins. Available instream habitat frequency distributions were then developed for each microhabitat variable, and the two distributions were compared using the K-S two-sample test and G-test as described previously. Using the ranges of the most suitable Carolina Madtom habitat from our calculations, we determined if there was available habitat in the suitable ranges in the Neuse River basin and compared the proportion of each basin's available instream habitat to determine if there were differences between the amount of suitable habitat in the Neuse and Tar river basins.

3. Results

3.1. Snorkel Surveys

We captured a total of 59 Carolina Madtom in standardized snorkeling surveys during 2016–2017. During the 2016 season, we captured only 15 fish, which all occurred at only one of the 20 surveyed 150-m sites (5%). The one site with observed specimens was Swift Creek in the Tar River basin (site 9; Figure 1). No individuals were found in the Neuse River basin in 2016.

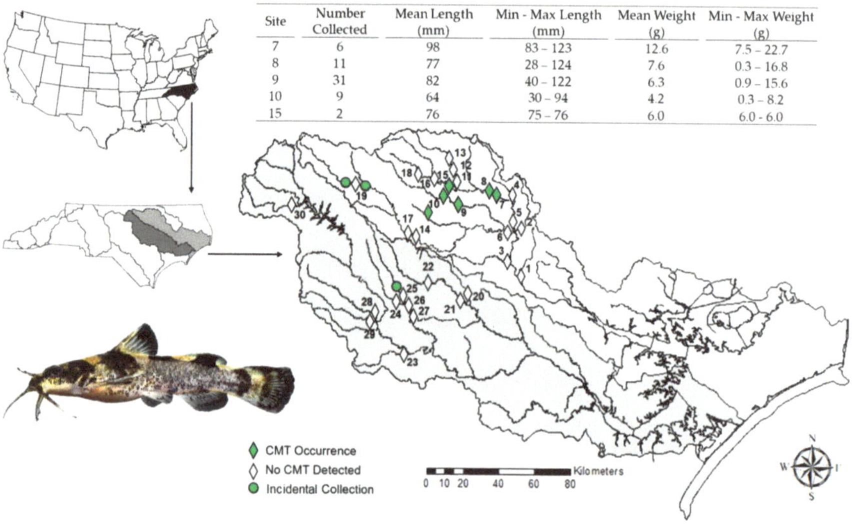

Site	Number Collected	Mean Length (mm)	Min - Max Length (mm)	Mean Weight (g)	Min - Max Weight (g)
7	6	98	83 – 123	12.6	7.5 – 22.7
8	11	77	28 – 124	7.6	0.3 – 16.8
9	31	82	40 – 122	6.3	0.9 – 15.6
10	9	64	30 – 94	4.2	0.3 – 8.2
15	2	76	75 – 76	6.0	6.0 - 6.0

Figure 1. A map showing the United States of America, the State of North Carolina, and Carolina Madtom (CMT) collections from the Tar River basin (white) and Neuse River basin (gray) during 2016 and 2017 sampling. The sites with Carolina Madtom occurrence and incidental collections are denoted by green diamonds and circles, respectively. The white diamonds denote sites with no occurrence. The incidental collections (green circles) from the mainstem Tar River and Little River (Neuse River basin) were not from standardized snorkel surveys. A summary of the CMT collection numbers with the associated length (mm) and weight (g) data are provided in the upper right panel. The bottom left image is of the study species, the Carolina Madtom (*Noturus furiosus*); photo credit Tracy et al. [46].

During the 2017 season, we captured 44 Carolina Madtom. They were found at five of the 16 surveyed 4-km sites (31%), or 14 of the 55 total 30-m survey locations (25%; Figure 1). Carolina Madtom were found at three sites on Fishing Creek and two sites on Swift Creek in the Tar River basin. Among the five sites where Carolina Madtom were captured, their overall mean total length was 79 mm (range 64–98 mm) and their overall mean total weight was 7.3 g (range 4.2–12.6 g; Figure 1). Although not collected during our standardized snorkeling surveys in the Neuse River basin in 2017, we captured two Carolina Madtom in the Little River while assisting state agency biologists performing freshwater mussel surveys. These two fish were not included in our analyses or capture total but reported herein as incidental collections (Figure 1) because of their importance to future conservation and management. Likewise, we report the incidental collection of six Carolina Madtom in 2017 from two sites in the Tar River basin (Figure 1) while assisting with mussel or salamander surveys. These six fish were not included in our capture total, but their data were used in the habitat suitability analysis.

While conducting snorkel surveys over the two sampling seasons, we also found 206 Margined Madtom, a native congeneric and potential competitor to the Carolina Madtom (Table 1). We also observed 143 Channel Catfish, a non-native competitor and possible predator, and 16 Flathead Catfish, a voracious non-native predator [47–49]. Finally, we counted 87 occurrences of the Neuse River Waterdog salamander (Table 1).

Table 1. Number (#) of Carolina Madtom, co-occurring native Margined Madtom, two co-occurring non-native ictalurid fish species (Channel Catfish and Flathead Catfish), and the endemic salamander, Neuse River Waterdog, observed during Carolina Madtom surveys in 2016 and 2017.

	2016		**2017**	
Species	**# Observed**	**# Sites**	**# Observed**	**# Sites (Locations)**
Carolina Madtom	15	1	44	5 (14)
Margined Madtom	60	9	146	12 (31)
Channel Catfish	49	11	94	7 (16)
Flathead Catfish	12	1	4	1 (3)
Neuse River Waterdog	29	8	58	11 (23)

3.2. Occupancy Modeling

Our naïve occupancy value was $\Psi = 0.31$, which is based on our survey results of five of 16 sites with Carolina Madtom occurrences. We ran a suite of 31 models and ranked them by lowest AIC_c value and highest W_i (Table 2). Because of constraints in model fitting due to small sample size, we were unable to run a fully parameterized (global) model, but considered models with one covariate for Ψ and one covariate for p. Our most plausible occupancy model was Ψ (dominant substrate) and p (mean-column velocity), with dominant substrate having a negative effect on occupancy ($\beta = -0.09$, 95% CI = −0.23–0.04) and mean-column velocity having a negative effect on detection probability ($\beta = -33.00$, 95% CI = −63.00–−3.01) (Table 3). Goodness of fit (χ^2) testing determined that the model adequately fit the observed data ($p = 0.47$) and there was no evidence of overdispersion ($\hat{c} = 0.99$). Site-specific occupancy ranged from <0.01 (95% CI = <0.01–0.98, site 26) to 0.68 (95% CI = 0.25–0.93, site 7), and site-specific detection probability ranged from 0.07 (95% CI = <0.01–0.66, site 5) to 0.99 (95% CI = 0.72–>0.99, site 9). Because of limitations in the occupancy modeling dataset and inherent uncertainty in fish modeling [45], eight other occupancy models were selected as confidence models to help determine which microhabitat covariates influence Carolina Madtom occupancy and detection (Table 2). All confidence models had W_i greater than 10% of our selected model (i.e., $\geq$0.02), and covariates included in the confidence models for occupancy were subdominant substrate, depth, large woody debris, and cobble. Covariates included in confidence models for detection probability were depth and distance to nearest cover. Confidence models within two ΔAIC_c yielded similar beta values and confidence intervals for covariates (Table 3), which resulted in similar occupancy and detection probability estimates.

Table 2. The occupancy models estimating occupancy probability ($\hat{\Psi}$) and detection probability ($\hat{p}$) for Carolina Madtom in the Tar River basin, including microhabitat covariates (in parentheses). The models presented are within 10% AIC weight (W_i) of the top performing model. Dominant = dominant substrate; mean velocity = mean-column water velocity; depth = water depth; subdominant = subdominant substrate; distance to cover = distance to nearest cover object; large woody = large woody debris as cover; cobble = cobble as cover; (.) = constant (null).

Model	**AIC_c**	**ΔAIC_c**	**W_i**
Ψ(dominant), p(mean velocity)	37.64	0.00	0.20
Ψ(.), p(mean velocity)	39.03	1.39	0.10
Ψ(dominant), p(.)	39.36	1.72	0.08
Ψ(dominant), p(depth)	39.98	2.34	0.06

Table 2. *Cont.*

Model	AIC_c	ΔAIC_c	W_i
Ψ(subdominant), *p*(mean velocity)	40.06	2.42	0.06
Ψ(dominant), *p*(distance to cover)	40.39	2.75	0.05
Ψ(.), *p*(.)	40.60	2.96	0.04
Ψ(.), *p*(distance to cover)	40.89	3.25	0.04
Ψ(depth), *p*(mean velocity)	40.89	3.25	0.04
Ψ(.), *p*(depth)	40.91	3.27	0.04
Ψ(large woody), *p*(mean velocity)	41.03	3.39	0.04
Ψ(cobble), *p*(mean velocity)	41.19	3.55	0.03
Ψ(subdominant), *p*(.)	41.61	3.97	0.03
Ψ(large woody), *p*(.)	42.14	4.50	0.02

Table 3. Top performing occupancy model and models within two ΔAIC_c of top performing model for Carolina Madtom across its extant range. Beta values and 95% confidence intervals for Ψ and *p* covariates are reported; dominant = dominant substrate; mean velocity = mean-column velocity. (.) indicates no covariate used for given parameter, so intercept beta and 95% CI are reported.

Model	β(Ψ)	95% CI(Ψ)	β(*p*)	95% CI(*p*)
Ψ(dominant), *p*(mean velocity)	−0.09	−0.23–0.04	−33.00	−63.00–−3.01
Ψ(.), *p*(mean velocity)	0.67	−0.43–1.76	−31.69	−63.73–0.34
Ψ(dominant), *p*(.)	−0.08	−0.20–0.04	1.53	0.26–2.79

3.3. Microhabitat Use, Availability, and Suitability

From our collections of Carolina Madtom (59 from the standardized snorkeling surveys plus the six incidental from the Tar basin in 2017), we found that Carolina Madtom occupied areas with a mean water depth of 0.47 m, a mean bottom velocity of 0.10 m/s, and a mean-column velocity of 0.22 m/s (Table 4). The most commonly occupied substrate was sand, and the predominant cover was cobble. All microhabitat use and availability data came from Carolina Madtom collections in the Tar River basin in our study because there were zero standardized collections in the Neuse River basin and only two incidental collections, which were not appropriate to use for this analysis.

Table 4. Comparison of Carolina Madtom microhabitat use between Midway et al. [13] and the current study.

	Midway et al. [13]			Current Study		
Variable	Mean/Mode	95% CI	Range	Mean/Mode	95% CI	Range
Depth (m)	0.42	0.39–0.45	0.01–0.92	0.47	0.42–0.52	0.18–1.00
Bottom velocity (m/s)	0.14	0.12–0.16	0.00–0.43	0.1	0.08–0.12	−0.03–0.39
Mean-column velocity (m/s)	0.22	0.20–0.24	0.00–0.58	0.22	0.18–0.26	−0.02–0.77
Substrate	Sand	-	-	Sand	-	-
Cover type	Cobble	-	-	Cobble	-	-

By comparing the available instream habitat with the habitat use of Carolina Madtom, we found that the species non-randomly selected instream habitat. For all six habitat parameters that we measured, Carolina Madtom inhabited a specific subset of the total available microhabitat in the basin (K-S two-sample test, $p < 0.05$). A wide range of depths were available in the Tar River basin from extremely shallow to over 1 m deep. However, Carolina Madtom microhabitat use followed a distribution with peak occurrences between 0.30 and 0.49 m deep (Figure 2). The distribution for bottom velocity availability was positively skewed, with a majority of the velocity values negative (i.e., upstream flow) or near 0.00 m/s. Carolina Madtom occupied some of the same slow-moving areas; however, bottom velocity use was recorded up to 0.39 m/s, indicating their selection for slightly faster water than the slow, stagnant velocities that dominated the available microhabitat. The distribution of available mean-column velocities followed the same pattern as bottom velocities with a positively skewed distribution and low velocities prevalent. The mean-column velocity use was approximately normal in distribution, and again, Carolina Madtom were found occupying areas with slightly faster moving water than the slow, stagnant velocities that were common throughout the surveyed reaches. The dominant substrate composition was predominantly sand and silt, with lower occurrences of gravel and cobble. Carolina Madtom most frequently occupied habitats over sand with gravel used almost as frequently, and cobble occupied less frequently (Figure 2). Carolina Madtom were never observed over silt substrate, even though silt was prevalent in over 16% of the total stream area surveyed. The subdominant substrate habitat associations were similar to the dominant substrate, with sand and gravel over the majority of the available habitat, and Carolina Madtom occupying substrates of sand, gravel, and cobble, with a strong selection for gravel subdominant substrates. The available cover type in the basin was equally distributed among cobble, small woody debris, and large woody debris. Artificial habitat (i.e., beverage containers or other litter) and leaf packs were much less commonly available cover. Nearly 40% of the surveyed available habitat in the Tar River basin had little to no adequate cover for Carolina Madtom to inhabit. Carolina Madtom were found most frequently under cobble cover, although both large and small woody debris were used as well, and 20% of our captures came from Carolina Madtom occupying artificial cover (Figure 2).

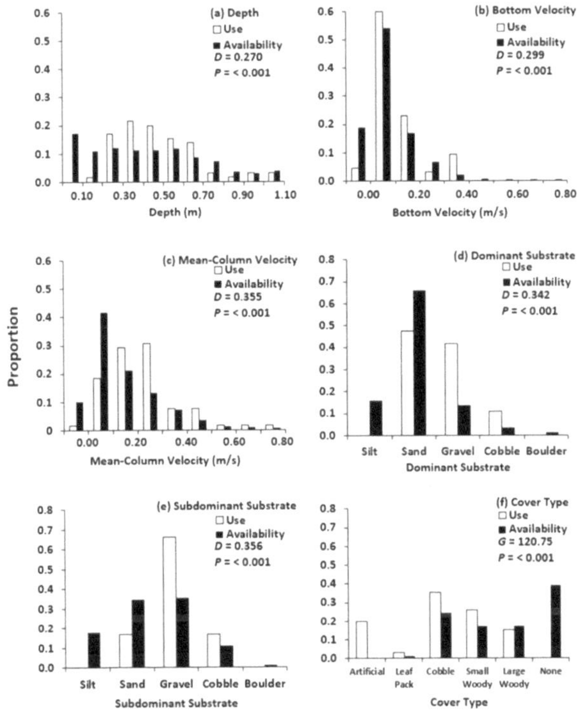

Figure 2. Frequency distributions of Carolina Madtom microhabitat use and availability for (**a**) depth, (**b**) bottom velocity, (**c**) mean-column velocity, (**d**) dominant substrate, (**e**) subdominant substrate, and (**f**) cover type for the Tar River basin. Significant differences between distributions were tested using a Kolmogorov–Smirnov two-sample test for continuous variables (depth, bottom velocity, mean column velocity, dominant substrate, subdominant substrate) and a log-likelihood ratio G-test for the categorical variable (cover type).

We developed Carolina Madtom habitat suitability distributions, based on microhabitat use and availability data, to determine the most suitable ranges of microhabitat for each of the six microhabitat parameters. The most suitable range of water depth was between 0.30 and 0.49 m (Figure 3). The most suitable bottom velocity was 0.30–0.39 m/s. Interestingly, six Carolina Madtom were collected at a site with particularly fast velocity, as compared to relatively slow water elsewhere in the basin. Mean column velocity had multiple suitable ranges with modal peaks at 0.20–0.29, 0.40–0.49, and 0.70–0.79 m/s. These multiple peaks were likely influenced by the six Carolina Madtom captures at the site with particularly high water velocity. For both the dominant and subdominant substrate composition, gravel and cobble were the most suitable substrate types (Figure 3).

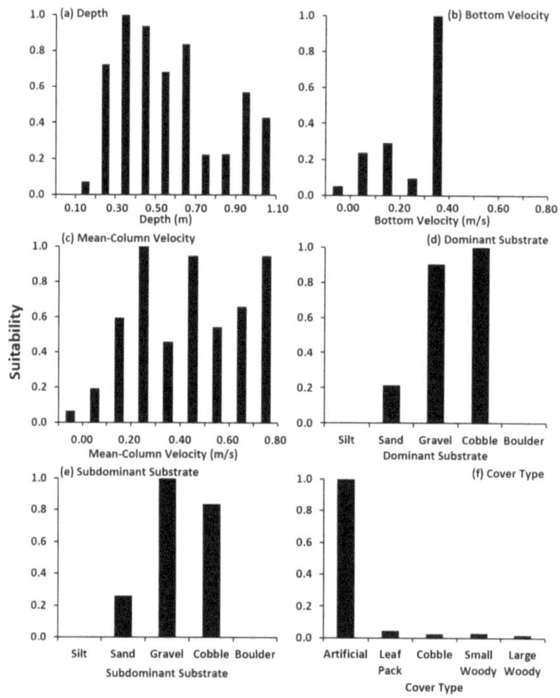

Figure 3. Carolina Madtom microhabitat suitability for (**a**) depth, (**b**) bottom velocity, (**c**) mean-column velocity, (**d**) dominant substrate, (**e**) subdominant substrate, and (**f**) cover type based on use and availability data collected from the Tar River basin.

The most suitable cover type was artificial. This is because we placed constructed artificial cover units in the stream for another component of the study [14,30] and treated them as existing instream cover if they were in place when we conducted our snorkel surveys. Carolina Madtom occurrence was high inside the cover units; however, artificial habitat availability, either placed purposefully by us or occurring as instream litter, was extremely low throughout the basin, leading to high suitability values. Availability of the suitable habitat ranges was generally low (Table 5). Most suitable ranges accounted for less than 15% of the available instream habitat for each parameter measured.

We also compared the instream habitat between the Tar River basin and Neuse River basin to assess whether the decline in Carolina Madtom populations in the Neuse River basin might be attributed to a loss of suitable instream habitat. Comparisons of the two basins showed significant differences in the availability of habitat for each parameter (Figure 4). The Tar River basin was shallower than the Neuse, as the depth availability in the Neuse peaked at 0.20–0.29 m and had higher proportion percentages of 0.70 m and above. The Tar River basin reaches were also faster moving in general. Although both basins had right-skewed distributions, the Neuse River had slower velocities and exhibited lower proportions of available habitat at velocities of 0.20 m/s and above. Substrate composition was similar between the basins, with the Neuse River basin having more cobble and boulders as the dominant substrate, whereas the Tar River basin was composed of a sand and gravel mixture.

Table 5. Comparison of the percentage by area of available microhabitat at surveyed sites, suitable for Carolina Madtom occupancy in the Neuse and Tar river basins.

		Neuse Basin	Tar Basin
Variable	**Suitable Range**	**% Available**	**% Available**
Depth	0.30–0.49 m	23.80	22.22
Bottom velocity	0.30–0.39 m/s	2.71	1.97
Mean-column velocity	0.20–0.29 m/s	10.96	13.25
	0.40–0.49 m/s	2.91	3.50
	0.70–0.79 m/s	0.42	0.70
Dominant substrate	Cobble	7.91	3.24
Subdominant substrate	Gravel	33.66	35.43
Cover type	Artificial	0.97	0.39

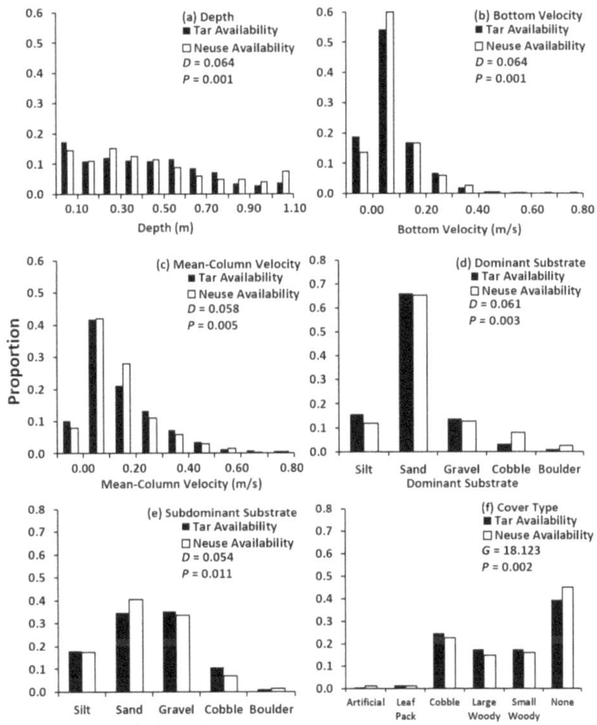

Figure 4. A frequency distribution comparison of the available microhabitat in the Neuse and Tar river basins for (**a**) depth, (**b**) bottom velocity, (**c**) mean-column velocity, (**d**) dominant substrate, (**e**) subdominant substrate, and (**f**) cover type from the Tar River basin. The significant differences between distributions were tested using a Kolmogorov–Smirnov two-sample test for the continuous variables (depth, bottom velocity, mean column velocity, dominant substrate, subdominant substrate) and a log-likelihood ratio G-test for the categorical variable (cover type).

The Tar River basin also had a higher amount of total available cover for all cover types except for artificial cover, which could be explained by the more degraded, urban nature of the Neuse River basin that may be more likely to contain artificial litter. The suitability analysis between the two basins revealed variability in the amount of available suitable habitat. Using the ranges calculated above, the Neuse River basin generally had more habitat available in the suitable ranges for Carolina Madtom (Table 5). The Neuse River basin had higher percentages of available habitat in the suitable ranges for depth, bottom velocity, dominant substrate, and cover type, while the Tar River basin had higher percentages of suitable habitat for mean column velocity and subdominant substrate. As such, the lack of suitable instream physical habitat for Carolina Madtom cannot wholly explain the lack of Carolina Madtom occurrences in the Neuse River basin.

4. Discussion

Carolina Madtom populations have decreased steadily and substantially over time in both occurrence and abundance throughout their historical and current range, and our results revealed these continued downward trends between the study of Wood and Nichols [19] and ours in 2016–2017. The Neuse River population may be approaching extirpation, as we only captured two individuals, which were not part of our standardized sampling, from a single site in the Neuse River basin. The Tar River population is experiencing less drastic population losses; however, we documented a decrease in the current range of populations in the Tar River basin. Along with a decline in occurrence, we also observed a large decrease in captured individuals from the previous studies, suggesting that Carolina Madtom populations have also decreased in abundance. We captured a total of 67 fish (59 from standardized surveys, six from incidental collections in the Tar, and two from incidental collections in the Neuse) from only five sites during the 2016 and 2017 field seasons, even with substantial sampling efforts among 36 sites (75 sampling locations). Previous surveys have also shown these steady declines in Carolina Madtom populations over the previous 30 years of sampling; our research confirmed the continued trend of precipitous declines. For example, Wood and Nichols [19] found only two sites in the Neuse River basin that supported Carolina Madtom, accounting for a 92% decrease in population occurrence since the 1960s. Their corresponding capture abundance was 208 individuals during the 2007 field season [19], and a temporally overlapping study by [13] observed 274 Carolina Madtom over their 2007 and 2008 field seasons. Even with different sampling efforts and coverage between these two studies, their relatively consistent capture numbers suggest that Carolina Madtom population abundances have decreased when compared to our recent findings. Although occurrences and capture abundances were low in our surveys, our model results suggest that the visual snorkel survey technique, which was also employed by Wood and Nichols [19] and Midway et al. [13], remains an efficient means to study Carolina Madtom populations.

The detection probability of Carolina Madtom during our snorkel surveys was quite high but varied among sites ($\hat{p}$ = 0.07–0.99), validating the snorkel technique as an effective method to capture Carolina Madtom. Even though Carolina Madtom occur in and under cover, employing a thorough search in snorkel surveys allowed us to effectively find and capture fish. Wood and Nichols [19] also estimated high detection probabilities ($\hat{p}$ = 0.94) in their snorkel surveys for Carolina Madtom, and although occupancy models were not developed by Midway et al. [13], they reported Carolina Madtom densities similar to Wood and Nichols [19], further validating the snorkel sampling method. In wadable streams, snorkeling is a common and effective visual technique to sample fish with minimal mortality. Although snorkel surveying sometimes raises concerns over sampling bias and impaired visibility [31–33], studies on related madtom species and other benthic fishes have shown that visual detection through snorkeling is as effective as traditional methods such as electrofishing or seining [50–52]. Snorkel surveying has also been recommended as a non-lethal method to sample rare species, as in our study of the Carolina Madtom [53,54].

Our modeled estimate of Carolina Madtom occupancy probability was low, but variable among sites ($\hat{\Psi}$ = <0.01–0.68). Low occupancy with high detection demonstrates the patchy spatial distribution of the remaining Carolina Madtom populations. If an area encompassing a small patch population is sampled, it is likely that Carolina Madtom will be detected in nearly every survey of the sampling area. Low occupancy is expected among rare, endemic species sampling, and occupancy estimates below 0.60 are common [55,56], primarily because of both their rare or imperilment status and their fragmented distributions. Our modeling analysis also found that microhabitat covariates influenced occupancy and detection probability estimates. The top performing model included the influence of dominant substrate on occupancy and mean-column velocity on detection. Substrate composition is an important microhabitat variable for all madtom species, as most are found in riffles with sand and gravel substrates [42,43], including the Carolina Madtom. Our site-specific occupancy estimates showed that occupancy was greatest at sites with a predominantly sand substrate composition. The mean-column velocity detection results were similar, confirming that madtom species are found in free-flowing streams with adequate velocity to flush fine sediments and provide oxygenated water [42,43].

The microhabitat use observed during our study was in accord with all known documented sources of Carolina Madtom habitat occupancy [12,13,19]. Substrate, flow, and instream cover are important microhabitat features for madtom occupancy; however, they are generally among the first variables affected in impacted streams [9]. Interestingly, we found that almost 40% of the surveyed available habitat in the Tar River basin had no adequate cover for Carolina Madtom to inhabit. Carolina Madtom were found most frequently under cobble cover, although both large and small woody debris were used as well. In the more impacted Neuse River basin, we are uncertain whether the available cover percentage is similar to the Tar because of the lack of fish detections. Nonetheless, this is a potentially important parameter to evaluate in the future.

Among the threats to the continued existence of Carolina Madtom, increased urbanization and related land development are among them [18,29]. These activities are a major cause of sedimentation, and increased fine sediment and silt load in a river can be detrimental to aquatic life, including Carolina Madtom [18,21,57]. Suspended sediment decreases the ability of light to penetrate the water, leading to decreases in growth and biomass of primary producers and macrophytes, and adversely affects the visibility, respiratory, and feeding function of fishes and invertebrates [58,59]. Deposited sediment associated with land development also has adverse effects on freshwater organisms. Fine sediment deposition fills in interstitial spaces between larger substrate particles and increases their embeddedness [57]. This not only reduces abundance and biomass of primary fish food sources, such as invertebrate larvae, but also decreases suitable habitat and areas of cover for benthic fishes such as the Carolina Madtom, which loses spawning habitat as available cavities become filled with sediment. Sedimentation problems from bridge construction, mining operations, and clearcutting riparian vegetation have been well-documented for the Ouachita Madtom *N. lachneri* [60]; Frecklebelly Madtom *N. munitus* [61]; and Neosho Madtom *N. placidus* [62,63]. Sedimentation is likewise among the threats to the Carolina Madtom [13,18,64].

Dams are another threat to many freshwater species. Reduced river continuity and altered flows from dams have been shown to negatively affect the intensively studied Neosho Madtom [62,65]. There are multiple dams causing habitat and population fragmentation along both the Neuse and Tar river basins [26] that are likely affecting the Carolina Madtom. Although habitat degradation is common throughout most areas inhabited by madtom species, they continue to inhabit specific microhabitats. Whether due to reproductive constraints or obligate cavity nesting common to all madtom species [43], we found that Carolina Madtom non-randomly select microhabitat. These findings support those of Midway et al. [13] and identify important specific stream microhabitats and reaches, especially relating to substrate, flow, and cover, that support madtom occurrence. These findings may assist with guidance for maintaining or restoring these habitats that are necessary to conserve Carolina Madtom populations in the future.

We developed suitability functions to determine current suitable habitat parameter ranges for the Carolina Madtom. Monitoring instream flow and habitat to develop models or management scenarios to maximize suitable habitat generally rely upon either single- or multi-species suitability functions, in which suitable or preferred microhabitat ranges are determined for the target species [66,67]. Determination of suitable habitat is especially important for madtom species because their obligate cavity nesting and benthic association does not allow them to adapt to rapidly changing instream conditions compared to other more generalist species. Habitat use has been widely studied for many madtom species [42,43], and habitat suitability criteria have been developed for the Freckled Madtom *N. nocturnus* [68], and Orangefin Madtom *N. gilberti* [69]. Habitat suitability criteria are especially useful in impacted or flow-regulated basins, such as the Neuse River basin, where flow and substrate composition can change rapidly and drastically. Midway et al. [13] originally determined some limited suitable habitat ranges for Carolina Madtom, and the suitable habitat ranges that we developed were similar to theirs, indicating that microhabitat use and availability distributions had changed little over the 8–10 years between the studies, even though the Neuse River basin has likely continued to experience habitat change and degradation [21]. These findings suggest that the species' habitat affinities have remained relatively stable, but the area and number of sites where the species occur have decreased over time.

Given that we and Midway et al. [13] have both shown that suitable instream physical habitat seems to exist in both the Neuse and Tar river basins, another threat that may be contributing to the decline in Carolina Madtom populations is possible predation by the non-native Flathead Catfish. Between the two river basins, Flathead Catfish were first detected in the Neuse basin and more recently, the Tar basin [48]. Flathead Catfish are common in the mainstem and larger tributaries, where historical Carolina Madtom populations no longer exist. This study is the first to document the co-occurrence of Carolina Madtom and Flathead Catfish at the same site. We found both species occupying a site on the Little River in the Neuse River basin (site Neuse 25; incidental collection) and a site on Fishing Creek in the Tar River basin (Tar 7). Both of these instances are highly concerning findings because Flathead Catfish are voracious predators that have been shown to suppress native fish populations by up to 50% and positively select for ictalurid prey, including native bullhead catfish and madtom species in North Carolina rivers. [47,49,70]. Future studies focusing on Flathead Catfish range overlap and diet analysis may be warranted if they continue expansion into smaller tributaries that have previously been refugia for Carolina Madtom.

Another potential threat influencing Carolina Madtom decline is pollution. Although physical habitat degradation has been widely studied in madtom species, water quality and toxicology studies are less common. However, metals and other aquatic contaminants have been identified as potential factors leading to population decrease in the Neosho Madtom [62,63,71]. Such pollutants may also affect the Neuse and Tar river basins and their biota. Human population growth, as well as increases in intensive animal agriculture in the basins, may be contributing to greater point and non-point source pollutants entering the ecosystem (e.g., via sewage, industrial effluent, road and field runoff). Future research may include water quality analysis throughout both basins in relation to Carolina Madtom occurrence. Moreover, little is known of Carolina Madtom chemical sensitivity; however, given their restricted range and low fecundity (hundreds of eggs) [43], water quality impacts are of concern and toxicological studies of pollutants on reproduction and survival are warranted.

The Carolina Madtom is listed as Endangered in the United States by the U.S. Fish and Wildlife Service, and our findings herein on occurrence, abundance, and habitat should help inform future recovery actions. Reintroduction efforts are common among federally listed species as viable population recovery goals [72]. Concerns have been raised over reintroduction efforts, as species are often reintroduced into habitat that retains problematic factors that caused the original population collapse [73]. However, multiple madtom species have been successfully reintroduced in the southeastern United States. The Smoky Madtom *N. baileyi*, and Yellowfin Madtom *N. flavipinnis*, have both been successfully propagated in captivity, returned to stream habitat in their native ranges in Tennessee, and successful post-reintroduction reproduction and population increases in the wild have been documented [74–76]. Successful captive breeding, augmentation of extant wild populations, and reintroduction of Carolina Madtom into historically occupied habitats are integral to preventing the species' extinction and promoting its resiliency and recovery [18,64]. In 2018, the U.S. Fish and Wildlife Service and North Carolina Wildlife Resources Commission began partnering with Conservation Fisheries, Inc., in Knoxville, TN, USA to develop and refine successful captive spawning and husbandry protocols for the species. Since then, these partners have collected adults, eggs, and larvae from the wild, and Conservation Fisheries, Inc., has been successful in developing controlled propagation protocols. Partners released the first cohort of about 260 juvenile Carolina Madtom in 2021, augmenting the Fishing and Swift Creek populations in the Tar River basin. Progeny have been released annually at these locations since then, thus far contributing about 700 fish to improve the resilience of Carolina Madtom in those watersheds. Captive propagation will also support reintroduction into the species' historical range under a new Safe Harbor Agreement, carried out through an Endangered Species Act Section 10 Enhancement of Survival Permit (ESPER0041144) issued to the North Carolina Wildlife Resources Commission by the U.S. Fish and Wildlife Service. However, identifying and ameliorating the drivers of Carolina Madtom decline will be essential before expanding species restoration efforts beyond populations in suitable habitats and protected watersheds (e.g., State Game Lands and conservation holdings).

5. Conclusions

The Carolina Madtom is small but mighty and iconic among ictalurid species; along with the Tar River Spinymussel (*Parvaspina steinstansana*) and Neuse River Waterdog, this triplet of endemic species highlights the special nature of the Neuse and Tar Rivers—a biodiversity trait broadly characteristic of many southeastern United States freshwater ecosystems. Swift and Fishing Creeks in the Tar River basin have been called the most biologically diverse waterways in the state, and Swift Creek may be the most important lotic system remaining on the Atlantic Seaboard [77,78]. Conservation of these fauna and ecosystems is vital [79], not only to maintain a unique, endemic species, but to also maintain diversity and overall health of these waterways. Although we found a relatively high detection probability of Carolina Madtom at our surveyed locations, their patchy distribution along with the fact that not all possible locations in the Neuse and Tar river basins were sampled, indicate that our findings may underestimate the actual number of fish still residing in the systems. However, this research adds to and reinforces the body of knowledge about the Carolina Madtom's populations and habitat use. Our results will help guide natural resource managers toward informed recovery and protective management decisions to improve the viability of this important endemic species.

Author Contributions: Conceptualization, W.R.C., T.J.K., T.R.B. and K.P.; methodology, W.R.C. and T.R.B.; validation, W.R.C. and T.J.K.; formal analysis, W.R.C., T.J.K. and K.P.; investigation, W.R.C. and T.R.B.; data curation, W.R.C.; writing—original draft preparation, W.R.C.; writing—review and editing, W.R.C., T.J.K., T.R.B., K.P., J.M.A. and W.G.C.; visualization, W.R.C.; supervision, T.J.K.; project administration, T.J.K.; funding acquisition, T.J.K. All authors have read and agreed to the published version of the manuscript.

Funding: This research was funded by the North Carolina Wildlife Resources Commission State Wildlife Grant Program through Agreement Number SWG-NCSU-15-16-3 to the U.S. Geological Survey, North Carolina Cooperative Fish and Wildlife Research Unit at North Carolina State University.

Institutional Review Board Statement: Fish collection and processing procedures were approved by the North Carolina State University Institutional Animal Care and Use Committee under protocol no. 15-042.

Informed Consent Statement: Not applicable.

Data Availability Statement: The data presented in this study are available on request from the corresponding author. The data are not publicly available due to lack of mandate.

Acknowledgments: Grant administration was facilitated by Todd Ewing and Ruby Valeton. We thank William Wood, Joseph McIver, Spencer Gardner, Tom Fox, Zoe Nichols, and Mike Walter for sampling support. This article is dedicated to the memory of Tom Kwak, who died unexpectedly in November 2021. He was the major advisor of W. Robert (Bobby) Cope and contributed greatly to this study and many others of his students and collaborators in the fisheries and aquatic science fields over his 28 year career. The North Carolina Cooperative Fish and Wildlife Research Unit is jointly supported by North Carolina State University, North Carolina Wildlife Resources Commission, U.S. Geological Survey, U.S. Fish and Wildlife Service, and Wildlife Management Institute. The findings and conclusions in this article are those of the authors and do not necessarily represent the views of the U.S. Fish and Wildlife Service. Any use of trade, firm, or product names is for descriptive purposes only and does not imply endorsement by the U.S. Government.

Conflicts of Interest: The authors declare no conflicts of interest.

References

1. Jelks, H.L.; Walsh, S.J.; Burkhead, N.M.; Contreras-Balderas, S.; Diaz-Pardo, E.; Hendrickson, D.A.; Lyons, J.; Mandrak, N.E.; McCormick, F.; Nelson, J.S.; et al. Conservation status of imperiled North American freshwater and diadromous fishes. *Fisheries* **2008**, *33*, 372–407. [CrossRef]
2. Ross, S.T.; Matthews, W.J. Evolution and ecology of North American freshwater fish assemblages. In *Freshwater Fishes of North America: Petromyzontidae to Catostomidae*; Warren, M.L., Burr, B.M., Eds.; Johns Hopkins University Press: Baltimore, MD, USA, 2014; pp. 1–49.
3. Neves, R.J.; Bogan, A.E.; Williams, J.D.; Ahlstedt, S.A.; Hartfield, P.W. Status of aquatic mollusks in the southeastern United States: A downward spiral of diversity. In *Aquatic Fauna in Peril: The Southeastern Perspective*; Benz, G.W., Collins, D.E., Eds.; Southeast Aquatic Research Institute, Lenz Design and Communications: Decatur, GA, USA, 1997; pp. 43–85.
4. Warren, M.L.; Angermeier, P.L.; Burr, B.M.; Haag, W.R. Decline of a diverse fish fauna: Patterns of imperilment and protection in the southeastern United States. In *Aquatic Fauna in Peril: The Southeastern Perspective*; Benz, G.W., Collins, D.E., Eds.; Southeast Aquatic Research Institute, Lenz Design and Communications: Decatur, GA, USA, 1997; pp. 105–164.
5. Burkhead, N.M.; Walsh, S.J.; Freeman, B.J.; Williams, J.D. Status and restoration of the Etowah River, and imperiled southern Appalachian ecosystem. In *Aquatic Fauna in Peril: The Southeastern Perspective*; Benz, G.W., Collins, D.E., Eds.; Southeast Aquatic Research Institute, Lenz Design and Communications: Decatur, GA, USA, 1997; pp. 375–444.
6. Etnier, D.A. Jeopardized southeastern freshwater fishes: A search for causes. In *Aquatic Fauna in Peril: The Southeastern Perspective*; Benz, G.W., Collins, D.E., Eds.; Southeast Aquatic Research Institute, Lenz Design and Communications: Decatur, GA, USA, 1997; pp. 87–104.
7. Warren, M.L.; Burr, B.M.; Walsh, S.J.; Bart, H.L.; Cashner, R.C.; Etnier, D.A.; Freeman, B.J.; Kuhajda, B.R.; Mayden, R.L.; Robison, H.W.; et al. Diversity, distribution, and conservation status of the native freshwater fishes of the Southern United States. *Fisheries* **2000**, *25*, 7–31. [CrossRef]
8. Master, L.L.; Flack, S.R.; Stein, B.A. *Rivers of Life: Critical Watersheds for Protecting Freshwater Biodiversity*; The Nature Conservancy: Arlington, VA, USA, 1998.
9. Angermeier, P.L. Ecological attributes of extinction-prone species: Loss of freshwater fishes of Virginia. *Cons. Biol.* **1995**, *9*, 143–158. [CrossRef]
10. Dudgeon, D.A.; Arthington, H.; Gessner, M.O.; Kawabata, Z.; Knowler, D.J.; Lévêque, C.; Naiman, R.J.; Prieur-Richard, A.; Soto, D.; Stiassny, M.L.; et al. Freshwater biodiversity: Importance, threats, status and conservation challenges. *Biol. Rev.* **2006**, *81*, 163–182. [CrossRef]
11. Kwak, T.J.; Porath, M.T.; Michaletz, P.H.; Travnichek, V.H. Catfish science: Status and trends in the 21st century. In *Conservation, Ecology, and Management of Catfish: The Second International Symposium*; Michaletz, P.H., Travnichek, V.H., Eds.; American Fisheries Society Symposium 77: Bethesda, MD, USA, 2011; pp. 755–780.
12. Burr, B.M.; Kuhajda, B.R.; Dimmick, W.W.; Grady, J.M. Distribution, biology, and conservation status of the Carolina Madtom, *Noturus furiosus*, an endemic North Carolina catfish. *Brimleyana* **1989**, *15*, 57–86.
13. Midway, S.R.; Kwak, T.J.; Aday, D.D. Habitat suitability of the Carolina Madtom, an imperiled, endemic stream fish. *Trans. Am. Fish. Soc.* **2010**, *139*, 325–338. [CrossRef]
14. Cope, W.R.; Kwak, T.J.; Black, T.R.; Pacifici, K. Evaluation of artificial cover units as a sampling technique and habitat enhancement for madtoms in rivers. *N. Am. J. Fish. Manag.* **2019**, *39*, 778–787. [CrossRef]

15. Midway, S.R.; Aday, D.D.; Kwak, T.J.; Gross, K. Cover preference of the Carolina Madtom (*Noturus furiosus*), an imperiled endemic southeastern stream fish. *J. Freshwat. Ecol.* **2010**, *25*, 151–154. [CrossRef]
16. LeGrand, H.E.; McRae, S.E.; Hall, S.P.; Finnegan, J.T. *Natural Heritage Program List of the Rare Animal Species of North Carolina*; North Carolina Natural Heritage Program, Division of Natural Resource Planning and Conservation, North Carolina Department of Environment and Natural Resources: Raleigh, NC, USA, 2008.
17. North Carolina Natural Heritage Program. *Natural Heritage Program List of Rare Animal Species of North Carolina 2016*; North Carolina Natural Heritage Program: Raleigh, NC, USA, 2016.
18. U.S. Fish and Wildlife Service. Endangered and threatened wildlife and plants; threatened species status with section 4(d) rule for Neuse River Waterdog, and endangered species status for Carolina Madtom, and designations of critical habitat. *Fed. Regist.* **2021**, *86*, 30688.
19. Wood, C.J.; Nichols, R.B. *Status Assessment of the Carolina Madtom Final Report*; North Carolina Wildlife Resources Commission: Raleigh, NC, USA, 2008.
20. Wood, C.J.; Nichols, R.B. Status assessment of the Carolina Madtom: A rare North Carolina endemic. In *Conservation, Ecology, and Management of Catfish: The Second International Symposium*; Michaletz, P.H., Travnichek, V.H., Eds.; American Fisheries Society Symposium 77: Bethesda, MD, USA, 2011; pp. 295–303.
21. North Carolina Department of Environmental Quality. *Neuse River Basinwide Water Quality Plan*; North Carolina Department of Environmental Quality: Raleigh, NC, USA, 2009.
22. NatureServe. *Noturus furiosus*. The IUCN Red List of Threatened Species 2014: E.T14901A19033935. 2014. Available online: https://www.iucnredlist.org/species/14901/19033935 (accessed on 10 May 2016).
23. Bayless, J.D.; Smith, W.B. *Survey and Classification of the Neuse River and Tributaries, North Carolina*; North Carolina Wildlife Resources Commission: Raleigh, NC, USA, 1962.
24. Smith, W.B.; Bayless, J.D. *Survey and Classification of the Tar River and Tributaries, North Carolina*; North Carolina Wildlife Resources Commission: Raleigh, NC, USA, 1964.
25. North Carolina Division of Water Resources. *Neuse River Basin Water Resources Plan*; North Carolina Department of Environment and Natural Resources: Raleigh, NC, USA, 2010.
26. North Carolina Wildlife Resources Commission. *North Carolina Wildlife Action Plan*; North Carolina Wildlife Resources Commission: Raleigh, NC, USA, 2015.
27. North Carolina Division of Water Resources. *State Water Supply Plan*; North Carolina Department of Environment and Natural Resources: Raleigh, NC, USA, 2001.
28. North Carolina Department of Environmental Quality. *Tar-Pamlico River Basin 2010 Water Quality Plan*; North Carolina Department of Environmental Quality: Raleigh, NC, USA, 2010.
29. Terando, A.J.; Costanza, J.; Belyea, C.; Dunn, R.R.; McKerrow, A.; Collazo, J.A. The southern megalopolis: Using the past to predict the future of urban sprawl in the southeast U.S. *PLoS ONE* **2014**, *9*, e102261. [CrossRef]
30. Cope, W.R. Status, Trends, Habitat, and Genetics of the Endemic Carolina Madtom. Master's Thesis, North Carolina State University, Raleigh, NC, USA, 2018.
31. Ensign, W.E.; Angermeier, P.L.; Dolloff, C.A. Use of line transect methods to estimate abundance of benthic stream fishes. *Can. J. Fish. Aquat. Sci.* **1995**, *52*, 213–222. [CrossRef]
32. Thompson, W.L. Hankin and Reeves' approach to estimating fish abundance in small streams: Limitations and alternatives. *Trans. Am. Fish. Soc.* **2003**, *132*, 69–75. [CrossRef]
33. Weaver, D.M.; Kwak, T.J.; Pollock, K.H. Sampling characteristics and calibration of snorkel counts to estimate stream fish populations. *N. Am. J. Fish. Manag.* **2014**, *34*, 1159–1166. [CrossRef]
34. Moyer, G.R.; Williams, A. Assessment of genetic diversity of American shad in the Santee-Cooper River basin of South Carolina before hatchery augmentation. *Mar. Coast. Fish.* **2012**, *4*, 312–326. [CrossRef]
35. Cope, W.R.; Kwak, T.J.; Black, T.R.; Pacifici, K.; Harris, S.C.; Miller, C.M.; Raley, M.E.; Hallerman, E.M. Genetic structure and diversity of the endemic Carolina Madtom and conservation implications. *N. Am. J. Fish. Manag.* **2021**, *41*, S27–S41. [CrossRef]
36. Braswell, A.L.; Ashton, R.E., Jr. Distribution, ecology, and feeding habits of *Necturus lewisi* (Brimley). *Brimleyana* **1985**, *10*, 13–35.
37. Teitsworth, E.W. Multi-Scale Assessment of the Drivers of Neuse River Waterdog (*Necturus lewisi*) Distribution and Recruitment. Ph.D. Thesis, North Carolina State University, Raleigh, NC, USA, 2024.
38. Bovee, K.D.; Milhous, R.T. *Hydraulic Simulation in Instream Flow Studies: Theory and Technique*; Instream Flow Information Paper 5; U.S. Fish and Wildlife Service: Fort Collins, CO, USA, 1978.
39. MacKenzie, D.I.; Nichols, J.D.; Lachman, G.B.; Droege, S.; Royle, J.A.; Langtimm, C.A. Estimating site occupancy rates when detection probabilities are less than one. *Ecology* **2002**, *83*, 2248–2255. [CrossRef]
40. Hines, J.E. *PRESENCE2: Software to Estimate Patch Occupancy and Related Parameters*; U.S. Geological Survey; Patuxent Wildlife Research Center: Laurel, MD, USA, 2006.
41. Kendall, W.L.; White, G.C. A cautionary note on the substituting spatial subunits for repeated temporal sampling in studies of site occupancy. *J. Appl. Ecol.* **2009**, *46*, 1182–1188. [CrossRef]
42. Taylor, R.T. *A Revision of the Catfish Genus Noturus (Rafinesque), with an Analysis of Higher Groups in the Ictaluridae*; Smithsonian Institution Press: Washington, DC, USA, 1969.

43. Burr, B.M.; Stoeckel, J.N. The natural history of madtoms (genus *Noturus*), North America's diminutive catfishes. In *Catfish 2000: Proceedings of the International Ictalurid Symposium*; American Fisheries Society, Symposium 24; Irwin, E.R., Hubert, W.A., Rabeni, C.F., Schramm, H.L., Jr., Coon, T., Eds.; American Fisheries Society: Bethesda, MD, USA, 1999; pp. 51–101.
44. Akaike, H. Maximum likelihood identification of Gaussian autoregressive moving average models. *Biometrika* **1973**, *60*, 255–265. [CrossRef]
45. Ruiz, J.C.; Peterson, J.T. An evaluation of the relative influence of spatial, statistical, and biological factors on the accuracy of stream fish species presence models. *Trans. Am. Fish. Soc.* **2007**, *136*, 1640–1653. [CrossRef]
46. Tracy, B.H.; Rohde, F.C.; Smith, S.A.; Bissette, J.L.; Hogue, G.M. *A Guide to North Carolina's Freshwater Fishes*; The University of North Carolina Press: Chapel Hill, NC, USA, 2024.
47. Pine, W.E., III; Kwak, T.J.; Waters, D.S.; Rice, J.A. Diet selectivity of introduced Flathead Catfish in coastal rivers. *Trans. Am. Fish. Soc.* **2005**, *134*, 901–909. [CrossRef]
48. Kwak, T.J.; Pine, W.E., III; Waters, D.S. Age, growth, and mortality of introduced Flathead Catfish in Atlantic rivers and a review of other populations. *N. Am. J. Fish. Manag.* **2006**, *26*, 73–87. [CrossRef]
49. Baumann, J.R.; Kwak, T.J. Trophic relations of introduced Flathead Catfish in an Atlantic river. *Trans. Am. Fish. Soc.* **2011**, *140*, 1120–1134. [CrossRef]
50. Hankin, D.G.; Reeves, G.H. Estimating total fish abundance and habitat area in small streams based on visual estimation methods. *Can. J. Fish. Aquat. Sci.* **1988**, *45*, 834–844. [CrossRef]
51. Persinger, J.W.; Orth, D.J.; Newcomb, T.J. A comparison of snorkeling versus throwable anode electrofishing for evaluating stream fish habitat use. *J. Freshwat. Ecol.* **2004**, *19*, 547–557. [CrossRef]
52. Albanese, B.; Owers, K.A.; Weiler, D.W.; Pruitt, W. Estimating occupancy of rare fishes using visual surveys, with a comparison to backpack electrofishing. *Southeast. Nat.* **2011**, *10*, 423–442. [CrossRef]
53. Dunham, J.B.; Rosenberger, A.E.; Thurow, R.F.; Dolloff, C.A.; Howell, P.J. Coldwater fish in wadeable streams. In *Standard Methods for Sampling North American Freshwater Fishes*; Bonar, S.A., Hubert, W.A., Willis, D.W., Eds.; American Fisheries Society: Bethesda, MD, USA, 2009; pp. 1–20.
54. Thurow, R.F.; Dolloff, C.A.; Marsden, J.E. Visual observation of fishes and aquatic habitats. In *Fisheries Techniques*, 3rd ed.; Zale, A.V., Parrish, D.L., Sutton, T.M., Eds.; American Fisheries Society: Bethesda, MD, USA, 2012; pp. 781–817.
55. Collier, B.A.; Groce, J.E.; Morrison, M.L.; Newnam, J.C.; Campomizzi, A.J.; Farrell, S.L.; Mathewson, H.A.; Snelgrove, R.T.; Carroll, R.J.; Wilkins, R.N. Predicting patch occupancy in fragmented landscapes at the rangewide scale for an endangered species: An example of an American warbler. *Divers. Distrib.* **2012**, *18*, 158–167. [CrossRef]
56. Magoulick, D.D.; Lynch, D.T. Occupancy and abundance modeling of the endangered Yellowcheek Darter in Arkansas. *Copeia* **2015**, *103*, 433–439. [CrossRef]
57. Waters, T.F. *Sediment in Streams: Sources, Biological Effects, and Control*; American Fisheries Society Monograph 7: Bethesda, MD, USA, 1995.
58. Van Nieuwenhuyse, E.E.; LaPerriere, J.D. Effects of placer gold mining on primary production in subarctic streams of Alaska. *Water Resour. Bull.* **1986**, *22*, 91–99. [CrossRef]
59. Aldridge, D.W.; Payne, B.S.; Miller, A.C. The effects of intermittent exposure to suspended solids and turbulence on three species of freshwater mussels. *Environ. Pollut.* **1987**, *45*, 17–28. [CrossRef]
60. Robison, H.W.; Harp, G.L. Distribution, habitat and food of the Ouachita Madtom, *Noturus lachneri*, a Ouachita River drainage endemic. *Copeia* **1985**, *1985*, 216–220. [CrossRef]
61. Piller, K.R.; Bart, H.L., Jr.; Tipton, J.A. Decline of the Frecklebelly Madtom in the Pearl River based on contemporary and historical surveys. *Trans. Am. Fish. Soc.* **2004**, *133*, 1004–1013. [CrossRef]
62. Wildhaber, M.L.; Tabor, V.M.; Mulhern, D.W.; Powell, K.L.; Sowa, S.P. 2000. Natural and anthropogenic influences on the distribution of the threatened Neosho Madtom in a midwestern warmwater stream. *Trans. Am. Fish. Soc.* **2000**, *129*, 243–261. [CrossRef]
63. Boroughs, K.L.; Whitney, J.E.; King, A.D.; Holloway, J.A.; Clemens, A.N.; Thompson, A.D. Comparison of threatened Neosho Madtom (*Noturus placidus*) densities between riverscapes differing in anthropogenic stressors, with a particular focus on recovery from mining-derived metal pollution. *Environ. Biol. Fishes* **2024**, *107*, 59–73. [CrossRef]
64. U.S. Fish and Wildlife Service. *Recovery Outline for Carolina Madtom (Noturus furiosus)*; U.S. Fish and Wildlife Service: Atlanta, GA, USA, 2021.
65. Tiemann, J.S.; Gillette, D.P. Effects of lowhead dams on riffle-dwelling fishes and macroinvertebrates in a midwestern river. *Trans. Am. Fish. Soc.* **2004**, *133*, 705–717. [CrossRef]
66. Bovee, K.D. *Development and Evaluation of Habitat Suitability Criteria for Use in the Instream Flow Incremental Methodology*; U.S. Fish and Wildlife Service: Atlanta, GA, USA, 1986; Biological Report 86(7).
67. Annear, T.; Chisholm, I.; Beecher, H.; Locke, A.; Aarrestad, P.; Burkardt, N.; Coomer, C.; Estes, C.; Hunt, J.; Jacobson, R.; et al. *Instream Flows for Riverine Resource Stewardship*, revised ed.; Instream Flow Council: Cheyenne, WY, USA, 2004.
68. Orth, D.J.; Maughan, O.E. Evaluation of the incremental methodology for recommending instream flows for fishes. *Trans. Am. Fish. Soc.* **1982**, *111*, 413–445. [CrossRef]
69. Simonson, T.D.; Neves, R.J. Habitat suitability and reproductive traits of the Orangefin Madtom *Noturus gilberti* (Pisces: Ictaluridae). *Am. Midl. Nat.* **1992**, *127*, 115–124. [CrossRef]

70. Pine, W.E., III; Kwak, T.J.; Rice, J.A. Modeling management scenarios and the effects of an introduced apex predator on a coastal riverine fish community. *Trans. Am. Fish. Soc.* **2007**, *136*, 105–120. [CrossRef]
71. Allen, G.T.; Blackford, S.H.; Tabor, V.M.; Cringan, M.S. Metals, boron, and selenium in Neosho Madtom habitat in the Neosho River in Kansas, USA. *Environ. Monit. Assess.* **2001**, *66*, 1–21. [CrossRef]
72. Tear, T.H.; Scott, J.M.; Hayward, P.H.; Griffith, B. Status and prospects for success of the endangered species act: A look at recovery plans. *Science* **1993**, *262*, 976–977. [CrossRef]
73. Conant, S. Saving endangered species by translocation: Are we tinkering with evolution? *BioScience* **1988**, *38*, 254–257. [CrossRef]
74. Rakes, P.L.; Shute, J.R.; Shute, P.W. Reproductive behavior, captive breeding, and restoration ecology of endangered fishes. *Environ. Biol. Fishes* **1999**, *55*, 31–42. [CrossRef]
75. Shute, J.R.; Rakes, P.L.; Shute, P.W. Reintroduction of four imperiled fishes in Abrams Creek, Tennessee. *Southeast. Nat.* **2005**, *4*, 93–110. [CrossRef]
76. Throneberry, J.K. Reintroduction Success of Smoky Madtom *Noturus baileyi* and Yellowfin Madtom *Noturus flavipinnis* in Abrams Creek, Great Smoky Mountains National Park. Master's Thesis, Tennessee Tech University, Cookeville, TN, USA, 2009.
77. Alderman, J.M.; Braswell, A.L.; Hall, S.P.; Kelly, A.W.; McGrath, C. *Biological Inventory: Swift Creek Subbasin*; North Carolina Natural Heritage Program, North Carolina State Museum of Natural History, North Carolina Wildlife Resources Commission: Raleigh, NC, USA, 1993.
78. North Carolina Natural Heritage Program. *The Upper Tar River Basin: Swift Creek and Fishing Creek Subbasins*; Biological Conservation Database: Raleigh, NC, USA, 1997.
79. North Carolina Wildlife Resources Commission. *Conservation Plan for Five Rare Aquatic Species Restricted to the Neuse and Tar-Pamlico River Basins in North Carolina*; North Carolina Wildlife Resources Commission: Raleigh, NC, USA, 2020.

fishes

MDPI

Article

Distribution Characteristics of *Trichiurus japonicus* and Their Relationships with Environmental Factors in the East China Sea and South-Central Yellow Sea

Xinyu Shi [1,2], Zhanhui Lu [2,3], Zhongming Wang [2,3], Jianxiong Li [1,2], Xin Gao [1,2], Zhuang Kong [1,2] and Wenbin Zhu [1,3,*]

1 Marine and Fishery Institute, Zhejiang Ocean University, Zhoushan 316021, China; shixinyu@zjou.edu.cn (X.S.); lijianxiong@zjou.zjou.edu.cn (J.L.); gaoxin@zjou.zjou.edu.cn (X.G.); kongzhuang@zjou.edu.cn (Z.K.)

2 Zhejiang Marine Fisheries Research Institute, Zhoushan 316021, China; lzh4286@zjou.edu.cn (Z.L.); wangzhongming@zjou.edu.cn (Z.W.)

3 Key Laboratory of Sustainable Utilization of Technology Research for Fishery Resource of Zhejiang Province, Zhoushan 316021, China

* Correspondence: wenbinzhu@zjou.edu.cn

Abstract: The largehead hairtail (*Trichiurus japonicus*) is the most productive fish caught in China. In order to understand the seasonal distribution of *T. japonicus* in the East China Sea and the central and southern parts of the Yellow Sea, three species distribution models were used in this study, namely the random-forest model, K-nearest-neighbor algorithm, and gradient-ascending decision-tree model, based on the data of trawling surveys in the East China Sea and central and southern parts of the Yellow Sea from 2008 to 2009. Combined with a variance inflation factor and cross-check, a distribution model of *T. japonicus* was screened and constructed to analyze the influence of environmental factors on the distribution of *T. japonicus* in the East China Sea and central and southern parts of the Yellow Sea. The results showed that the random-forest model had the advantages of fitting effect and prediction ability among the three models. The analysis of this model showed that the water depth, bottom water temperature, and surface salinity had a great influence on the habitat distribution of *T. japonicus*. The relative resources of *T. japonicus* increased with the increase of bottom water temperature, reached the maximum at 23.8 °C, and first increased and then decreased with the increase of water depth and surface salinity, reaching the maximum when water depth is 72 m and surface salinity is 31.2%. This study also used the random-forest model to predict the spatial distribution of *T. japonicus* in the central and southern waters of the East China Sea and south-central Yellow Sea from 2008 to 2009, and the results showed that the predicted results were close to the actual situation. The research results can provide a reference for the exploitation and protection of *T. japonicus* resources in the East China Sea and the south-central Yellow Sea.

Keywords: East China Sea; south-central Yellow Sea; *T. japonicus*; species distribution model; spatial distribution; environmental factor

Key Contribution: In this study, we used a random-forest model to analyze the effects of environmental factors on the distribution of *T. japonicus* in the East China Sea and the south-central Yellow Sea and made distribution predictions. The results of this study can help to comprehensively understand the distribution of *T. japonicus* fisheries in the East China Sea and the south-central Yellow Sea and provide valuable theoretical support for their rational development and utilization.

Citation: Shi, X.; Lu, Z.; Wang, Z.; Li, J.; Gao, X.; Kong, Z.; Zhu, W. Distribution Characteristics of *Trichiurus japonicus* and Their Relationships with Environmental Factors in the East China Sea and South-Central Yellow Sea. *Fishes* **2024**, *9*, 439. https://doi.org/10.3390/fishes9110439

Academic Editors: Heng Zhang, Yunzhi Yan, Luoliang Xu and Ying Xue

Received: 14 September 2024
Revised: 9 October 2024
Accepted: 12 October 2024
Published: 29 October 2024

1. Introduction

The largehead hairtail (*Trichiurus japonicus*) (Temminck and Schlegel, 1844) belongs to the order Perciformes, family Trichiuridae, and genus *Trichiurus*. It is a warm–temperate species that typically forms schools near the seafloor. The catch of this species exceeded 1 million tons in 1995 [1], making it one of the few marine species in China with over one million tons in landing. Currently, the main fishing methods in the East China Sea and Yellow Sea include bottom trawling and seine netting. *T. japonicus* resources in the East China Sea have been exploited since the 1950s, with catches often ranking first among various species since the late 1950s. Consequently, *T. japonicus* is a key species in domestic fisheries research and management. Many resource management systems in the East China Sea, including fishing bans and protected areas, are based on research findings related to *T. japonicus* resources, primarily focusing on conserving traditional economic fish species, with *T. japonicus* being the primary target [2].

One of the hot topics in fishery ecology is the spatial distribution characteristics of species and their relationships with environmental factors [3]. The spatial distribution of fish populations is influenced by a variety of control factors, both external and internal, of which the external control, also known as environmental control, includes hydrological conditions, substrate types, etc., and is generally considered to be the main factor affecting the spatial distribution of fish populations [4]. On the other hand, population size, age structure, fish condition, diversity, behavior, etc., are internal control factors that can also regulate the spatial distribution of fish populations through density-dependent, age-dependent habitat preference, migration ability differences, etc. [5]. The adaptability and limitation of fish to marine environments are one of the key factors determining their migration, distribution, and movement, and the study of the influence of environmental factors on the spatial distribution of fish populations is of great reference value for fishery analysis, fishing ground exploration, and rational use of fishery resources [6].

A species distribution model (SDM) is a mathematical model that uses environmental data to predict the spatial distribution of species according to their survival conditions and has become one of the important methods in the application of conservation biology and ecology [7]. Widely used species distribution models in fisheries include generalized additive models and generalized linear models [8,9], with relatively fewer applications of machine learning methods. As automation and intelligence advance, machine learning algorithms increasingly predict fish abundance and distribution [10], identify populations [11], standardize catch per unit effort (CPUE) [12], and explore relationships between fishery resources and environmental factors [13,14], showing distinct advantages.

For instance, Chen [15] developed a forecasting model for Indian Ocean yellowfin tuna fisheries using a random-forest model, enhancing the forecasting capabilities of distant offshore fisheries. Hou [16] researched the modeling and forecasting of South Pacific yellowfin tuna fisheries using six ensemble learning models, improving the accuracy of their predictions. Gao [17] constructed a forecasting model for mackerel in the East and Yellow Seas, employing gradient boosting decision trees and playing a crucial role in managing and protecting mackerel resources. Song [18] built a forecasting model for bigeye tuna in the Atlantic tropical waters using K-nearest-neighbors and gradient-boosting decision trees, enhancing the accuracy of their model predictions. Currently, research utilizing species distribution models to examine the habitat distribution of ribbonfish remains scarce.

Based on the trawl survey data in the central and southern waters of the East China Sea and the Yellow Sea from 2008 to 2009, this study used random-forest model, K-proximity algorithm, and gradient-lifting decision tree to analyze the distribution characteristics of *T. japonicus* and their relationships with environmental factors and then compared and analyzed the fitting effect and prediction ability of the models. The habitat index was used to predict the distribution of *T. japonicus* in the East China Sea and the south-central Yellow Sea, so as to provide a basis for the rational utilization and scientific conservation of its resources and provide a reference for fishery policy management.

2. Materials and Methods

2.1. Data Sources

The samples of belt fish in this study were collected from the fixed bottom trawl survey of the national science and technology support program "Investigation and Assessment of important fishery Resources in the Main fishing grounds of the East China Sea" conducted in May (spring), August (summer), and December (autumn) of 2008 and February (winter) of 2009 in the waters of the East China Sea and the south-central Yellow Sea. The sea area covered 121–126.5° E and 26–35° N (Figure 1), with 119 stations. The survey ship used a 6 × 80 m target net; the width of the network port was 48 m, the mesh size of the bag net was 30 mm, the towing speed of the survey ship was 2.0 kn, and the towing time of each station was 1 h. The relative catch Y (g/h) was obtained by using the trawling time of 1 h and trawling speed of 2 kn.

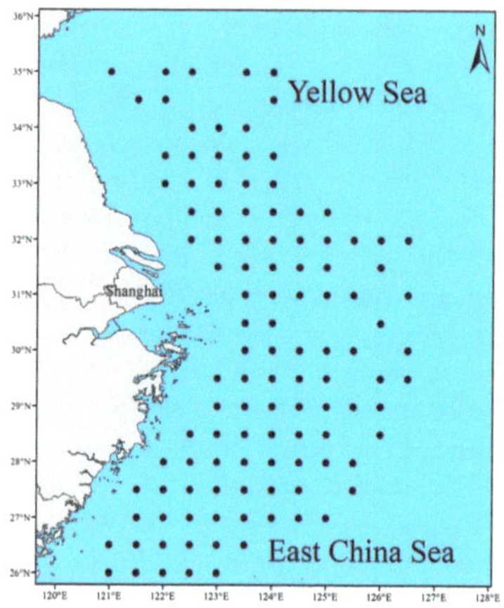

Figure 1. Survey stations.

Sample processing and environmental factor measurements adhered to the "Ocean Survey Standards" [19]. Environmental data were collected using a shipborne synchronized CTD instrument, which measured sea water depth (SWD), sea surface temperature (SST), sea bottom temperature (SBT), sea surface salinity (SSS), and sea bottom salinity (SBS).

2.2. Model Construction

Random forest (RF), proposed by Breiman [20], is an ensemble-learning method based on the classification and regression tree algorithm. This approach improves the predictive performance of models by combining multiple decision trees. Specifically, random forest achieves this through several steps: first, it randomly extracts multiple samples from the original data set, known as bootstrap samples; next, it models decision trees for each bootstrap sample; finally, it aggregates the predictions from each decision tree, arriving at the final prediction through voting or averaging. This method exhibits high tolerance to noise and outliers, achieves high classification accuracy and predictive precision, shows a lower probability of overfitting, and possesses strong generalization capabilities [21,22].

The expression for a random forest is as follows:

$$\mathrm{f}(x) = \sum_{m=1}^{M} c_m I(x \in R_m), I = \begin{cases} 1, X \in R_m \\ 0, x \notin R_m \end{cases} \quad (1)$$

In the formula, x is the independent variable, y is the dependent variable, R_m is the feature space partition unit, and c_m is the fixed output value on each unit.

The K-nearest-neighbor (KNN) algorithm serves as a widely adopted classification method. The steps for classification are straightforward: First, compute the distance between an object, whose category is unknown, and every sample in the training set. Next, select the K-most-similar (nearest) samples within the feature space. Then, determine which category most of these K samples belong to. Finally, if the majority of samples fall into a specific category, classify the object into that category as well [23]. The fundamental concept behind the KNN algorithm is clear: if most of the K-nearest samples reside in a particular category, the sample should be assigned to that category too [24]. The KNN algorithm can facilitate both regression and classification by evaluating distances between various feature values. For any N-dimensional input vector, which correlates to a point in the feature space, the outcome is the category label or predicted value associated with that feature vector. While the concept remains simple and intuitive, the algorithm boasts significant maturity and stability [25–27].

The expression for KNN is as follows:

$$d(x, y_i) = \|x - yi\|_2 = \left(\sum_{k=1}^{n} |x_k - y_{ik}|^2\right)^{1/2} \tag{2}$$

In the formula, x is the sample to be classified, y is a known category sample, n represents the data dimension, and i, k is the sample number.

A gradient boosting decision tree (GBDT) is an enhanced ensemble learning model based on the classification and regression trees (CART) algorithm [28]. It is one of the important algorithms in the field of machine learning. This model combines multiple weak classifiers into strong classifiers by iterating continuously. In each iteration, based on the previous iteration, the loss function is calculated to obtain the pseudo-residual and the iteration is obtained, and then, a new decision tree is constructed. Then, all the generated decision trees are weighted and fused according to the weight of the decision tree through gradient descent [29]. The model can deal with nonlinear relations effectively, has good generalization performance and accuracy in many prediction studies, and can identify and correct errors in the modeling process. However, GBDT is sensitive to outliers, and in multiple iterations, GBDT models will try to fit outliers, which may lead to overfitting. Therefore, when applying this model, hyperparameter tuning should be performed on the imported data to obtain the optimal solution of parameters and reduce the risk of overfitting [30].

The expression for GBDT is as follows:

$$F_i(x) = \sum_{t=1}^{m} r^{(t)} f^{(t)}(x_i) \tag{3}$$

In the formula, $Fi\ (x)$ represents the final prediction result for the i-th sample, and $Fi\ (x)$ is the predicted value of the i-th sample in the t-th tree for the observed sample x.

2.3. Factor Screening and Model Fitting

ln(Y+1) was obtained by natural logarithm conversion of the relative resource amount (Y) of *T. japonicus* as the response variable and SWD, SST, SBT, SSS, and SBS were selected as the explanatory variables. A significant correlation between two or more explanatory variables in a multicollinearity representation model can negatively affect the final result. In order to avoid such influence, a variance inflation factor (VIF) [31] is used in this study to test the multicollinearity of the above five factors and screen out the factors to be added to the model. In general, $\sqrt{V_{IF}} < 2$ indicates that there is no multicollinearity, and explanatory variables that exceed the threshold need to be removed.

$$V_{IF} = \frac{1}{1 - R^2} \tag{4}$$

In the formula, R^2 represents the goodness of fit of the relationship between the independent variable and other independent variables through a simple linear regression model.

2.4. Evaluation of Model Prediction Ability

We compared the fitting effects and predictive capabilities of three models to select the optimal one. We analyzed the relationship between the distribution characteristics of *T. japonicus* in the East China Sea and south-central Yellow Sea and environmental factors. Subsequently, we predicted their distribution.

The prediction ability of each model was tested by the 50-fold cross-validation method. The total data set was randomly and equally divided into 5 sub-data sets. Each time, 4 sub-data sets were randomly selected as the training set, and the other one was used as the validation set for the accuracy evaluation of the model prediction. The calculation was repeated 100 times, and the average effect was taken for the accuracy evaluation of each model. According to the mean squared error (MSE) and coefficient of determination (R^2) obtained, the prediction ability of each model was determined.

The MSE is the ratio of the square sum of the deviation between the predicted value and the true value and the number of observations n, which can reflect the degree of dispersion of the data set [32]. The smaller the MSE value, the higher the accuracy of the model prediction and the more accurate the description of the test data. R^2 is the proportion of the sum of squares caused by the independent variable X in the total sum of squares of the dependent variable Y [33], which can be used to evaluate the fitting degree of the prediction model. The closer R^2 is to 1, the higher the reference value of the model, which can well describe the trend and rule of the data set. The closer R^2 is to 0, the lower the reference value of the model, and the trend and rule of the data set cannot be well described [34].

The formulas for calculating the MSE and R^2 are as follows:

$$MSE(y,p) = \frac{1}{n}\sum_{i=1}^{n}(y_i - p_i)_2^2 \quad (5)$$

$$R^2(y,p) = 1 - \frac{\sum_{i=0}^{n}(y_i - p_i)^2}{\sum_{i=0}^{n}(y_i - \overline{y})^2} \quad (6)$$

In the formula, y represents the original value, p stands for the predicted value, and n denotes the sample size.

2.5. Mapping Habitat Distribution Prediction

The habitat suitability index (HSI) was initially proposed in the 1980s [35] and is primarily utilized for assessing habitat quality, providing a more comprehensive depiction of the adaptation process of marine organisms to their environment. Currently, it has gained widespread application in the fields of biological spatial distribution and fishing-ground forecasting [36–38]. In this study, after conducting comparisons, we selected the species distribution model with superior predictive performance. Subsequently, HSI values were calculated for each station and the ArcGIS 10.2 software's spatial analysis module was employed to generate habitat distribution maps for *T. japonicus* during different seasons using Kriging interpolation based on an exponential semi-variance function [39].

3. Results and Analysis

3.1. Impact Factor Screening

Five factors (SWD, SST, SBT, SSS, and SBS) were tested with multicollinearity using V_{IF}, and their values were 1.78, 1.32, 1.51, 1.53 and 1.27, respectively. The results show that there is no multicollinearity between the factors and they all can be added to the model.

3.2. Model Performance Evaluation

As can be seen from Table 1, after model fitting, the MSE of the random-forest model is 0.348, which is smaller than values for the KNN and GBDT models, and the R^2 is 0.919, which is higher and closer to one than the values for the KNN and GBDT models. Therefore, this model has the best fitting effect. The mean values of the MSE and R^2 of 100 model predictions and observations were obtained by cross-validation. The results show that the MSE of the random-forest model is 2.566 ± 1.734, is smaller than the values for the KNN and GBDT models, and its R^2 is 0.373 ± 0.563, which is higher than the values for the KNN and GBDT models. The difference between the prediction results of random-forest model and the observed values is smaller, so the prediction ability is the best. Therefore, the random-forest model is superior to KNN and GBDT in all aspects, so the random-forest model is adopted for follow-up research.

Table 1. Cross-validation comparison between three models.

Inspection Method	Statistical Parameters	RF	KNN	GBDT
Model fitting	MSE	0.348	2.120	2.445
	R^2	0.919	0.506	0.431
Cross validation	MSE	2.566 ± 1.734	3.295 ± 2.161	3.004 ± 1.264
	R^2	0.373 ± 0.563	0.203 ± 0.385	0.275 ± 0.255

3.3. Importance Ranking of Impact Factors

In RF, the contribution rate of a feature is usually calculated based on the number of node splits of the feature in the decision tree and the information gain obtained by splitting. The random-forest model was constructed, and the input variables were SWD, SST, SBT, SSS, and SBS, and the output variable was resource density. The results show that in the random-forest model, the contribution rates of each impact factor to resource density in different months are shown in Figures 2 and 3.

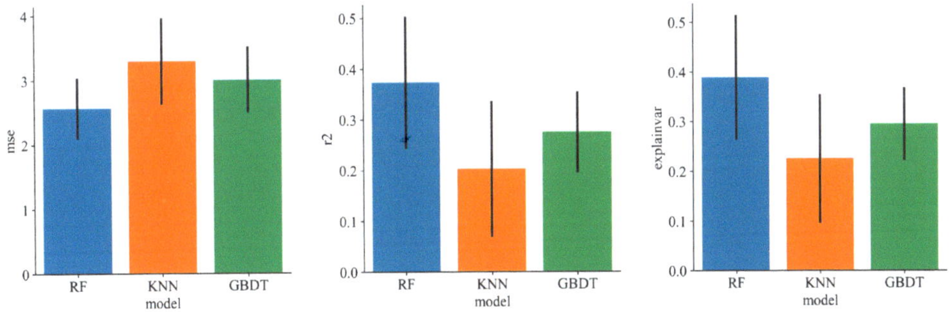

Figure 2. Performance comparison of the three machine learning methods.

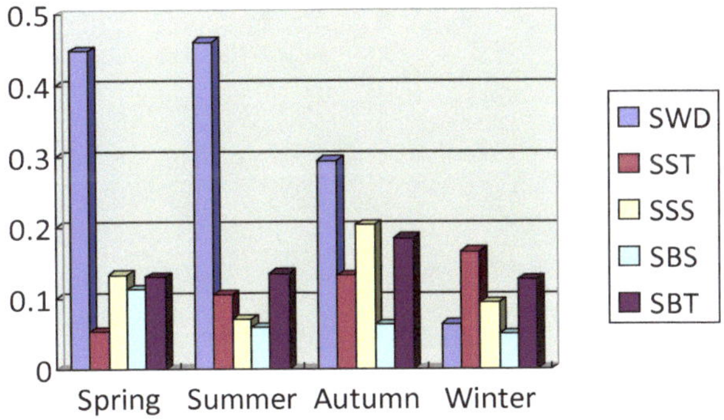

Figure 3. Importance ranking of factors affecting the density distribution of *T. japonicus* in the study area.

The results show that SWD is the most important in May (spring), followed by SSS, SBT, SBS, and SST; SWD is the most important in August (summer), followed by SBT, SST, SSS, and SBS; SWD is the most important in November (autumn), followed by SSS, SBT, SST, and SBS; and SST is the most important in February (winter), followed by SBT, SSS, SWD, and SBS. It can be seen that among the five factors, SWD, SBT, and SSS are relatively more important.

3.4. Relationship between the T. japonicus Distribution and Explanatory Variables

The influences of various factors on the relative resources for *T. japonicus* are shown in Figure 4. The relative resources of belt fish increased slowly when the SST was less than 24.8 °C, fluctuated after 24.8 °C, and became stable after 27 °C. The relative resources of *T. japonicus* increased slowly when the SBT was less than 22.2 °C and the increase rate increased after 22.2 °C and reached the maximum at 23.8 °C, showing an overall increasing trend. The relative resources of *T. japonicus* increased when the SSS was less than 31.2%, reached the maximum at 31.2%, and decreased when the SSS was more than 31.2%. The relative resource amount of *T. japonicus* showed a higher level when the SBS was less than 33.3%, and a lower level when the SBS was more than 33.3%, showing a decreasing trend in general. The relative resources of belt fish increased when the SWD was less than 72 m, reached the maximum at 72 m, and decreased after 72 m.

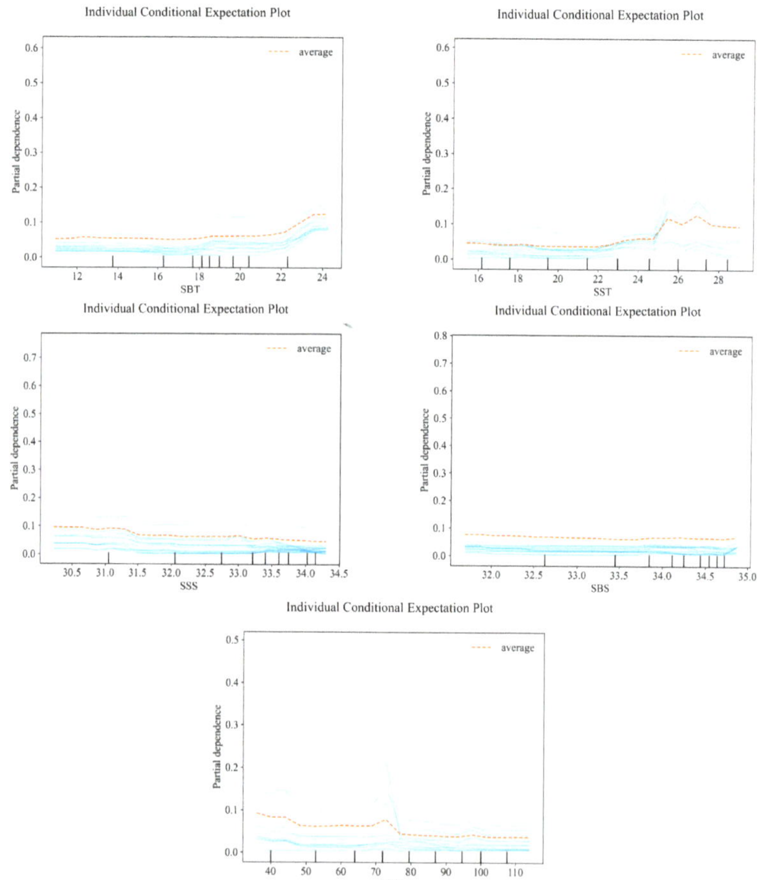

Figure 4. Impact of environmental factors on the relative resource of *T. japonicus*. From left to right and top to bottom are SBT SST, SSS, SBS, and SWD.

3.5. Prediction of Habitat Distribution of *T. japonicus* in the East China Sea and South-Central Yellow Sea

The prediction performance of the three models was compared, and it was found that the RF model had the best prediction performance. The environmental data simulated by the HSI were added to the random-forest model for prediction, and the spatial distribution map was drawn. It was found that the abundance distribution of *T. japonicus* in May is high in the southwest, low in the northeast, and mainly distributed in the sea areas of 25.5–31.5° N and 119.5–124° E. In August, the abundance of *T. japonicus* is mainly concentrated in the northwest waters, mainly distributed in the waters of 30.5–33° N and 121–125° E. In November, *T. japonicus* resources were mainly concentrated in the middle of the sea area, mainly distributed in the sea areas of 29–33° N and 122–126° E. In February, the distribution of *T. japonicus* abundance showed the characteristics of being high in the southwest sea area, low in the northeast sea area, and mainly distributed in the sea areas of 27–30° N and 121–124.5° E (Figure 5). These results are consistent with the characteristics that *T. japonicus* likes to cluster in the warm environment near the bottom. In Figure 5, the predicted results are compared with the actual results, and it is found that the predicted results are close to the actual results, which shows that the predicted results have a certain accuracy.

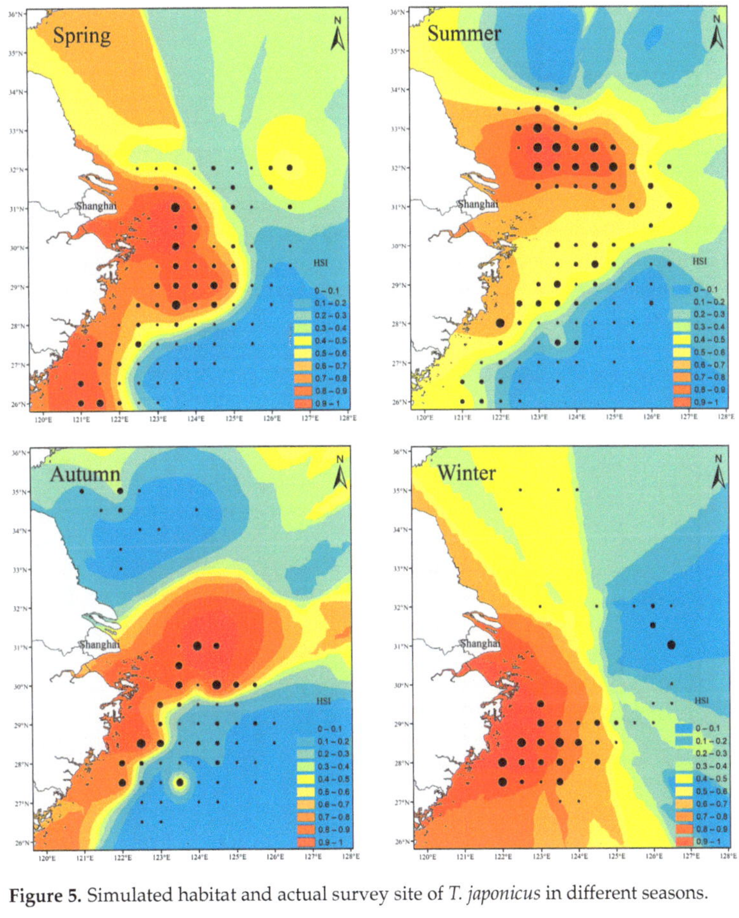

Figure 5. Simulated habitat and actual survey site of *T. japonicus* in different seasons.

4. Discussion

4.1. Model Analysis

At present, there are few studies on the distribution of *T. japonicus* using a species distribution model. Zhang [40] used a GAM model to study the distribution characteristics of *T. japonicus* in the Beibu Gulf from 2006 to 2018 and the relationship between its resource density and environmental factors. The results showed that the distribution of *T. japonicus* fishing grounds in the Beibu Gulf was southwest–northeast, and the center of gravity of *T. japonicus* resources moved to southwest–northeast and south–north in summer and autumn, respectively. Chlorophyll A affected the resource density and spatial distribution of *T. japonicus*, and abnormal values of water depth and longitude and sea surface temperature affected the resource density of *T. japonicus* but did not affect its spatial distribution. Liu [41] predicted the potential distribution areas of *T. japonicus* in the coastal waters of China in 2040–2050 and 2090–2100 by using nine species distribution models such as a random-forest model based on the survey data of fishery resources from 1998 to 2000, and the prediction results showed that the distribution hotspots of *T. japonicus* tended to move to high latitudes, and the investigation stations and research simulation in that study overlapped with those in this study.

In this study, three machine learning models, the RF, KNN and GLDT models, were compared in order to choose the most suitable model to analyze the habitat distribution characteristics of *T. japonicus* in the East China Sea and the central and southern parts of Yellow Sea and their relationships with environmental factors. The results showed that the random-forest model has a good fitting effect and cross-validation result, which may be attributed to its advantages in data processing ability and algorithm. Firstly, a random-forest model introduces the concept of randomness, and randomly selects training samples and feature subsets, thus effectively enhancing the classification ability and anti-noise ability and reducing the possibility of over-fitting of random forest. At the same time, random forest can effectively deal with the situation of less data, lost features, and unbalanced data sets and has a high tolerance for outliers. In addition, the random-forest model has the characteristics of integrated learning, and the accuracy of the results can be improved by constructing many different regression trees, avoiding the weak generalization ability of a single decision tree [21,22].

The utilization of the random-forest model in the fisheries domain has been progressively increasing in recent years. In comparison to conventional species, distribution models, and other machine learning techniques, the random-forest model can effectively capture the interaction between environmental variables through constructing a randomized decision tree [42]. Furthermore, it demonstrates greater robustness against outliers and random interference [20] during regression analysis. Luan [3] employed a GLM, a GAM, and the random-forest model to evaluate the spatial distribution of Portunid crab across different seasons in Haizhou Bay in 2011. Liu [42] utilized both the random-forest model and a GAM model to analyze the relationship between krill catch per unit fishing effort and environmental factors in Antarctica. Cui [13] employed artificial-neural-network models, random-forest models, and generalized-enhanced-regression models to predict and compare habitat distributions for *Tetragnatha tetragnatha* in Haizhou Bay. All the above results show that RF has good fitting effect and forecasting ability and has certain advantages, which is similar to the results in this study. However, since random forests are composed of a large number of decision trees containing only partial feature variables, the prediction results are dependent on the mode of the output category of the decision tree. As a result, the interpretation of the results obtained from random forests is challenging. Therefore, when employing random forests for fishery forecasting, it is essential to complement this approach with other analytical techniques to gain a deeper understanding of the underlying dynamics shaping the fishery. This will help to mitigate the limitations of this model in fishery forecasting.

4.2. The Influence of Environmental Factors on the Distribution of *T. japonicus*

The temporal and spatial differences of environmental factors are one of the main reasons for the temporal and spatial changes of fish resources; environmental factors influence the growth of individual fish and their age at maturity [43] and fish usually distribute along the distribution characteristics according to environmental gradients [44]. In this study, the distribution characteristics of *T. japonicus* and their relationships with environmental factors were analyzed by a random-forest model. It was found that environmental factors have different effects on the distribution of *T. japonicus* in different seasons, and the SWD, SSS and SBT are relatively important.

Water temperature is a very important environmental factor, which can directly affect the growth, development, reproduction, metabolism, migration, and distribution of marine organisms and other ecological processes [45]. As a warm–temperate fish clustered near the bottom, the growth and reproduction of *T. japonicus* are directly affected by water temperature, so its distribution area will also be affected by water temperature [46–48]. Wang [46,47] found that the increase of water temperature is not only beneficial to the gonad development and maturity of *T. japonicus*, but also can increase the feed supply of *T. japonicus*. The fluctuation of *T. japonicus* catch in the East China Sea is significantly related to the sea surface temperature; Yuan [48] found that the hot spots of *T. japonicus* in the East China Sea will move adaptively with the first mode change of sea surface temperature, and they are all close to the left waters of the northern branch of the Kuroshio. You [49] found that the bottom water temperature of the central fishing ground of Zhoushan fishing ground in summer flood was between 16 °C and 22 °C, and the temperature at that time might be lower than now, earlier in this research year. In addition, this research only studied *T. japonicus* in summer flood, so the result was different from this research. It was found that the relative resources of *T. japonicus* were small when the SST was less than 24.8 °C and obviously fluctuated and rose above 24.8 °C, reaching the maximum at 27 °C. The SBT is considered of low resource density when it is less than 22.2 °C, gradually increases when it is higher than 22.2 °C, and reaches the maximum at 23.8 °C, which indicates that *T. japonicus* has certain requirements on water temperature and is suitable for its survival within a certain temperature range. The relative resources of *T. japonicus* reached the maximum when the SBT was 23.8 °C, which indicated that the water temperature was suitable for *T. japonicus* survival. In the SBT range of 22.2–24.2 °C, the relative resources of *T. japonicus* are relatively large, and this temperature range is the suitable water temperature range for its life.

As one of the important environmental factors affecting the spatial and temporal distribution of marine life, salinity can affect the spatial distribution of marine life to a certain extent [50]. Previous studies have found that the fishing season of *T. japonicus* is directly affected by salinity [49,51]. You [49] found that the central fishing ground of *T. japonicus* in the Zhoushan fishing ground is located near the 34% isosalinity line. Zhu [51] and others found that the zonal fluctuation of the 34% isosalinity line was obviously related to the fishing ground in central Zhejiang during winter flood. Wang [52] found that sea surface salinity has a significant influence on the change of *T. japonicus* catch in the Zhejiang sea area, and the *T. japonicus* catch shows a linear upward trend with the increase of sea surface salinity. The above research is earlier, which may lead to a gap between those results and the results of this study. This study found that the relative resources of *T. japonicus* were higher when the SSS was less than 31.2%, and lower after the SSS was more than 31.2%. It is higher when the SBS is less than 33.3% and lower when the SBS is more than 33.3%, indicating that salinity will affect the distribution of *T. japonicus* resources. Too high or too low salinity in seawater may affect the osmotic pressure adjustment and oxygen consumption of *T. japonicus*, thus affecting its growth. The normal growth of *T. japonicus* needs to be carried out within a certain salinity range.

Water depth affect the changes of factors such as light, pressure, and dissolved oxygen and can indirectly affect the habitat distribution of marine life and its bait [13]. Water depth is closely related to water mass movement of the fish community, processes related to fish life history (predation and competition), and bottom sediments [53]. Some studies have found that the depth of seawater directly affects the temporal and spatial changes of hydrological factors such as temperature, salinity, and transparency, thus directly affecting the distribution of organisms and the aggregation of fish [18]. Hu [54] found that water depth is one of the main factors affecting the diversity of fish communities in a *T. japonicus* reserve in spring and autumn. Zhang [55] found that water depth is one of the main environmental factors affecting the distribution of fish in the coastal waters of the Yangtze River Estuary, and the diversity of the fish community and the distribution of the fish may affect the feeding of *T. japonicus*, thus affecting the relative resources of *T. japonicus* to some extent. The above research results are similar to this study. In this study, the relative resources of *T. japonicus* showed an upward trend when the SWD was less than 72 m, and a slow downward trend after the SWD was higher than 72 m. This is consistent with *T. japonicus*'s habit of clustering near the bottom, which shows that there is the most suitable habitat environment for *T. japonicus* in the sea area with a water depth of less than 72 m. However, this study used past data and only considered five environmental factors, which has certain limitations. More environmental factors should be added for further research in the future.

4.3. Habitat Distribution Characteristics of *T. japonicus*

The habitat distribution of *T. japonicus* in the East China Sea and the central and southern parts of the Yellow Sea is characterized by high resource density in the southwest coastal and central waters and low resource density in the southeast and northern waters in spring, when *T. japonicus* is mainly distributed in the waters of 27.5–31° N and 122.5–125° E. In summer, the coastal waters in the northwest and southwest are characterized by high resource density, low resource density is found in the southeast, and *T. japonicus* is mainly distributed in the sea areas of 28–30° N, 122–124.5° E, 31.5–33.5° N, and 123–125° E. In autumn, the resource density in the southwest coast and central sea area is high, while the resource density in the north and southeast sea area is low, and *T. japonicus* is mainly distributed in the sea areas of 27.5–28.5° N, 121.5–123.5° E, 30–31° N, and 123.5–125° E. In winter, the resource density in the southwest is high, the resource density in other sea areas is low, and *T. japonicus* is mainly distributed in the sea areas of 27.5–29.5° N and 122–124.5° E. It can be found that with the increase of temperature, the distribution area of *T. japonicus* hotspots moves northward and outward, which is similar to the research conclusions of Yuan [48] and Zhu [51]. The distribution area of *T. japonicus* resource density hotspots spread to a certain extent with the seasonal changes and moved to the northern offshore, which is not only related to the rising water temperature in seasonal changes, but also related to the fact that the effective implementation of the summer fishing moratorium in the East China Sea and the central and southern parts of the Yellow Sea is beneficial to effectively replenish *T. japonicus* resources. According to the research of Yan [56], the summer fishing moratorium can protect the spawning groups and juveniles of major economic fish, reduce the fishing pressure, facilitate the cluster growth of *T. japonicus*, and distribute *T. japonicus* hotspots. Using the HSI index to predict the habitat distribution of *T. japonicus* in the East China Sea and south-central Yellow Sea in four seasons can make up for some missing data, and it also enables the effective assessment of stock [57].

5. Conclusions

By comparing three kinds of machine learning models, this study analyzed the habitat distribution characteristics of *T. japonicus* in the East China Sea and south-central Yellow Sea and their relationships with environmental factors. On the basis of the results and their interpretation, the following is noted: (1) The random-forest model had better fitting effect and prediction ability than the other two kinds of machine learning models; (2) Among the five environmental factors, SWD, SBT, and SSS had a great impact on the habitat distribution of belt fish. The relative resources of belt fish increased with the increase of the SBT and increased first and then decreased with the increase of the SWD and SSS;

(3) Based on the habitat index the habitat of *T. japonicus* in the East China Sea and the central and southern parts of the Yellow Sea was predicted, and the predicted results were similar to the actual survey results. The study of the habitat distribution characteristics of *T. japonicus* in the East China Sea and the south-central Yellow Sea and their relationships with environmental factors can provide some reference for the sustainable utilization and scientific management of *T. japonicus* resources. In future studies, comparative statistical methods and machine learning methods should be used to help explore models and methods that are more suitable for this species. In addition, this study only considered five environmental factors, namely SWD, SST, SBT, SSS, and SBS, and did not involve other factors, such as dissolved oxygen, chlorophyll, pH, flow rate, and mixed layer depth, which may affect the distribution of belt fish. More environmental factors and their effects on the habitat distribution of belt fish should be comprehensively analyzed. In order to further understand the relationship between the habitat distribution characteristics and environmental factors of *T. japonicus* in the East China Sea and south-central Yellow Sea and to provide reference for the protection and rational utilization of *T. japonicus* resources in the East China Sea and south-central Yellow Sea, further research is needed.

Author Contributions: Conceptualization, X.S. and W.Z.; methodology, W.Z.; writing—original draft preparation, X.S.; writing—review and editing, J.L., X.G., and Z.K.; supervision, W.Z. and Z.W.; project administration, W.Z. and Z.L. All authors have read and agreed to the published version of the manuscript.

Funding: This study was funded by the Key Technology and System Exploration of Quota Fishing, Ministry of Agriculture and Rural Affairs, Fishery Management Fund Project [Grant number: 36,2017], the National Key Research and Development Program of China (2019YFD0901505), the Zhejiang Provincial Key R&D Program project (2018C02026) and the Zhejiang Fishery Resources Survey Special Project [HYS-CZ-202314].

Institutional Review Board Statement: The data for this study were collected from a fixed bottom trawl survey in the East China Sea between 2008–2009 as part of a national science and technology support program. Since no actual fish were used, ethical approval and consent forms are not required, and there are no animal ethics concerns.

Informed Consent Statement: Informed consent was obtained from all subjects involved in the study.

Data Availability Statement: Data will be made available on request.

Acknowledgments: We thank the staff of the Zhejiang Marine Fisheries Research Institute for their help and support in our experiment.

Conflicts of Interest: The authors declare that they have no competing interests.

References

1. Zhang, Q.H.; Cheng, J.H.; Xu, H.X. *Fishery Resources and Their Sustainable Utilization in The East China Sea*; Fudan University Press: Shanghai, China, 2007; pp. 147–169.
2. Cheng, J.H.; Yan, L.P.; Lin, L.S. Analyses on the fishery ecological effect of summer close season in the East China Sea region. *J. Fish. Sci. China* **1999**, *4*, 81–85.
3. Luan, J.; Zhang, C.L.; Xu, B.D. Relationship between catch distribution of Portunid crab (Charybdis bimaculata) and environmental factors based on three species distribution models in Haizhou Bay. *J. Fish. China* **2018**, *42*, 889–901.
4. Jiang, Y.; Zhang, Y.L.; Pang, Z.W. Spatial distribution characteristics of Sepia esculenta in haizhou bay and adjacent waters and their relationship with environmental factors. *Acta Hydrobiol. Sin.* **2024**, *48*, 617–624.
5. Planque, B.; Loots, C.; Petitgas, P.; Lindstrøm, U.L.F.; Vaz, S. Understanding what controls the spatial distribution of fish populations using a multi-model approach. *Fish. Oceanogr.* **2011**, *20*, 1–17. [CrossRef]
6. Chen, X.J. *Fishery Resources and Fishery Oceanography*; China Ocean Press: Beijing, China, 2014; pp. 152–161.
7. Guo, Y.L.; Zhao, Z.F.; Qiao, H.J. Challenges and development trend of species distribution model. *Adv. Earth Sci.* **2020**, *35*, 1292–1305.
8. Zhu, W.B.; Zhu, H.C.; Zhang, Y.Z. Quantitative distribution of juvenile Engraulis japonicus and the relationship with environmental factors along the Zhejiang coast. *J. Fish. Sci. China* **2021**, *28*, 1175–1183.
9. Feng, B.; Chen, X.J.; Xu, L.X. Catch rate analysis of yellowfin tuna from longline fishery using generalized linear model in the Indian Ocean. *J. Fish. Sci. China* **2009**, *16*, 282–288.

10. Li, Z.G.; Wan, R.; Ye, Z.J. Use of random forests and support vector machines to improve annual egg production estimation. *Fish. Sci.* **2017**, *83*, 1–11. [CrossRef]
11. Haralabous, J.; Georgakarakos, S. Artificial neural networks as a tool for species identification of fish schools. *ICES J. Mar. Sci.* **1996**, *53*, 173–180. [CrossRef]
12. Yang, S.L.; Zhang, Y.; Zhang, H. Comparison and analysis of different model algorithms for CPUE standardization in fishery. *Trans. Chin. Soc. Agric. Eng.* **2015**, *31*, 259–264.
13. Cui, Y.H.; Liu, S.D.; Zhang, Y.L. Habitat characteristics of Octopus ocellatus and their relationship with environmental factors during spring in Haizhou Bay, China. *Chin. J. Appl. Ecol.* **2022**, *33*, 1686–1692.
14. Xu, M.Z.; Zhang, C.L.; Xue, Y. Relationship between species diversity and environmental factors in the fishery community of Shandong coastal waters. *J. Fish. China* **2022**, *46*, 1008–1017.
15. Chen, X.Z.; Fan, W.; Cui, X.S. Fishing ground forecasting of Thunnus alalung in Indian Ocean based on random forest. *Acta Oceanol. Sin.* **2013**, *35*, 158–164. (In Chinese)
16. Hou, J.; Zhou, W.F.; Fan, W. Research on fishing grounds forecasting models of albacore tuna based on ensemble learning in South Pacific. *South China Fish. Sci.* **2020**, *16*, 42–50.
17. Gao, F. *Fishing Ground Forecasting of Chub Mackerel in the East China Sea and Yellow Sea Using Boosted Regression Trees*; Shanghai Ocean University: Shanghai, China, 2016.
18. Song, L.M.; Ren, S.Y.; Zhang, M. Fishing ground forecasting of bigeye tuna (Thunnus obesus) in the tropical waters of Atlantic Ocean based on ensemble learning. *J. Fish. China* **2023**, *47*, 64–76.
19. General Administration of Quality Supervision; Inspection and Quarantine of the People's Republic of China; Standardization Administration of China. *Specification for Marine Survey-Part 6: Marine Biological Survey*; Standards Press of China: Beijing, China, 2008.
20. Breiman, L. Random forests. *Mach. Learn.* **2001**, *45*, 5–32. [CrossRef]
21. Fang, K.N.; Wu, J.B.; Zhu, J.P. A Review of Technologies on Random Forests. *Stat. Inf. Forum* **2011**, *26*, 32–38.
22. Dong, S.S.; Huang, Z.X. A brief theoretical overview of Random Forests. *J. Integr. Technol.* **2013**, *2*, 1–7.
23. Zhang, D. *Vehicle Logo Recognition Based on Convolutional Neural Network and K-Nearest Neighbor*; Xidian University: Xi'an, China, 2015.
24. Li, X.; Zhang, C.L. Analysis of LocalLDtree classification model based on K proximity algorithm. *Silicon Val.* **2013**, *6*, 33–146.
25. Ming, Y.S. Using clustering to improve the KNN-based classifiers for online anomaly network traffic identification. *J. Netw. Comput. Appl.* **2011**, *34*, 722–730.
26. Malunoud, M.; Chokri, B.A. Classification improvement of local feature vectors over the KNN algorithm. *Multimed. Tools Appl.* **2013**, *64*, 197–218.
27. Jagan, S.; Hanan, S.; Amitabh, V. A fast all nearest neighbor algorithm for applications involving large point clouds. *Comput. Graph.* **2007**, *31*, 157–174.
28. Friedman, J.H. Greedy function approximation: A gradient boosting machine. *Ann. Stat.* **2001**, *29*, 1189–1232. [CrossRef]
29. Gao, J.X.; Zhang, W.; Gao, M. Material calculation time prediction model based on gradient boosting decision trees. *Softw. Guide* **2024**, *23*, 15–20.
30. Zhu, Y.L.; Feng, X.Y.; Yan, Q.G. Spatial distribution and main controlling factors of soil organic carbon under cultivated land based on GBDT model in black soil region of Northeast China. *China Environ. Sci.* **2024**, *44*, 1407–1417. [CrossRef]
31. Kabacoff, R. R in Action: Data Analysis and Graphics with R. *Greenwich Manning Publ.* **2011**, *8*, 126–131.
32. Hyndman, R.J.; Koehler, A.B. Another look at measures of forecast accuracy. *Int. J. Forecast.* **2006**, *22*, 679–688. [CrossRef]
33. Li, J.W.; Chen, C.H.; Sun, Y. Total partial regression sum of squares method for variable selection in multiple linear regression models. *J. Math. Med.* **2007**, *20*, 126–127.
34. Xu, B.D.; Zhang, C.L.; Xue, Y. Optimization of sampling effort for a fishery-independent survey with multiple goals. *Environ. Monit. Assess.* **2015**, *187*, 252. [CrossRef]
35. Shim, J.S.; Kim, R.K.; Yoon, K.B. A basic research for the development of habitat suitability index model of Pelophylax chosenicus. *J. Korean Soc. Environ. Restor. Technol.* **2020**, *23*, 49–62.
36. Tian, S.Q.; Chen, X.J.; Chen, Y. Evaluating habitat suitability indices derived from CPUE and fishing effort data for Ommatrephes bratramii in the northwestern Pacific Ocean. *Fish. Res.* **2009**, *95*, 181–188. [CrossRef]
37. Gong, C.X.; Chen, X.J.; Gao, F. Review on habitat suitability index in fishery science. *J. Shanghai Ocean. Univ.* **2011**, *20*, 260–269.
38. Chen, F.; Li, N.; Fang, Z. Habitat distribution change pattern of Uroteuthis edulis during spring and summer in the coastal waters of Zhejiang Province. *J. Shanghai Ocean. Univ.* **2021**, *30*, 847–855.
39. Tanaka, K.; Chen, Y. Spatiotemporal variability of suitable habitat for American lobster (Homarus americanus) Long Island Sound. *J. Shellfish. Res.* **2015**, *34*, 531–543. [CrossRef]
40. Zhang, M.; Wang, X.H.; Cai, Y.C. Spatial aggregation and dispersion characteristics of Trichiurus haumela in the Beibu Gulf, northern South China Sea. *J. Fish. Sci. China* **2022**, *29*, 1647–1658.
41. Liu, X.Y. *Study on the Impacts of Climate Change on the Potential Suitable Habitat of Major Commercial Fish in Offshore China*; Zhejiang Ocean University: Zhoushan, China, 2022.
42. Liu, J.C.; Jia, M.X.; Feng, W.D. Spatial Temporal Distribution of Antarctic Krill (Euphausia superba) Resource and its Association with Environment Factors Revealed with RF and GAM Models. *Period. Ocean. Univ. China (Nat. Sci. Ed.)* **2021**, *51*, 20–29.
43. Zhang, K.; Zhang, J.; Li, J.J.; Liao, B.C. Model selection for fish growth patterns based on a Bayesian approach: A case study of five freshwater fish species. *Aquat. Living Resour.* **2020**, *33*, 17. [CrossRef]

44. Prchalova, M.; Kubecka, J.; Vašek, M. Distribution patterns of fishes in a canyon shaped reservoir. *J. Fish Biol.* **2008**, *73*, 54–78. [CrossRef]
45. Zou, Y.Y.; Xue, Y.; Ma, Q.Y. Spatial distribution of Larimichthys polyactis in Haizhou Bay Based on Habitat Suitability Index. *Period. Ocean. Univ. China (Nat. Sci. Ed.)* **2016**, *46*, 54–63.
46. Wang, Y.Z.; Jia, X.P.; Lin, Z.J. Responses of Trichiurus japonicus catches to fishing and climate variability in the East China Sea. *J. Fish. China* **2011**, *35*, 1881–1889.
47. Wang, Y.Z.; Qiu, Y.S. An analysis of interannual variations of hairtail catches in East China Sea. *South China Fish. Sci.* **2006**, *2*, 16–24.
48. Yuan, X.W.; Liu, Z.L.; Jin, Y. Inter-decadal variation of spatial aggregation of Trichiurus japonicus in East China Sea based on spatial autocorrelation analysis. *Chin. J. Appl. Ecol.* **2017**, *28*, 3409–3416. (In Chinese)
49. You, H.B.; Xu, R. Relationship between central fishing ground and water temperature and salinity in summer season. *Mar. Fish.* **1984**, *4*, 165–167.
50. Dai, L.B.; Chen, J.H.; Tian, S.Q. Prediction of fish species richness in the Yangtze River estuary using CART algorithm. *J. Fish. Sci. China* **2018**, *25*, 1082–1090. [CrossRef]
51. Zhu, D.K.; Yu, C.G. The relation on the environment of fishing ground with the occurrence of hairtail in winter off the middle part of Zhejiang. *J. Fish. Sci. China* **1987**, *11*, 195–203.
52. Wang, T.Z.; Han, Q.; Luo, N.J. Catches of several major demersal fish species catches inhabiting Zhejiangsea area and their relationships with main influencing factors. *Trans. Oceanol. Limnol.* **2021**, *43*, 77–85.
53. Azevedo, M.; Araújo, F.; Cruz-Filho, A.; Pessanha, A.; Silva, M.; Guedes, A. Demersal fishes in a tropical bay in southeastern Brazil: Partitioning the spatial, temporal and environmental components of ecological variation. *Estuar. Coast. Shelf Sci.* **2007**, *75*, 468–480. [CrossRef]
54. Hu, C.L.; Zhang, H.L.; Zhang, Y.Z. Fish community structure and its relationship with environment factors in the Nature Reserve of Trichiurus japonicus. *J. Fish. China* **2018**, *42*, 694–703.
55. Zhang, Y.Q. *Environmental Impact on the Fish Assemblage Structures Dissertation Submitted to in Adjacent Sea Area of the Yangtze River Estuary*; Graduate School of Oceanology, Chinese Academy of Sciences: Qingdao, China, 2012.
56. Yan, L.P.; Liu, Z.L.; Li, S.F. Effects of new summer close season of trawl fisheries on fishery ecology and resource enhancement in East China Sea. *Mar. Fish.* **2010**, *32*, 186–191.
57. Liao, B.; Ehsanul, K.; Kui, Z. Comparative performance of catch-based and surplus production models on evaluating largehead hairtail (Trichiurus lepturus) fishery in the East China Sea. *Reg. Stud. Mar. Sci.* **2021**, *48*, 102026. [CrossRef]

fishes

Article

Species Composition of Fish Larvae and Juveniles in the Nanji Islands, China

Xiaodong Wang [1,2], Dubin Zhao [2], Qiang Liu [2], Tianyu Lu [2], Junsheng Zhong [2], Wandong Chen [1], Shangwei Xie [1,*] and Shun Chen [3]

[1] Nanji Islands National Marine Nature Reserve Administration, Wenzhou 325400, China; xiaodonwang@shou.edu.cn (X.W.)
[2] Shanghai University Key Laboratory of Marine Animal Taxonomy and Evolution, Shanghai Ocean University, Shanghai 201306, China
[3] National Research Center for Marine Facility Aquaculture Engineering and Technology, Zhejiang Ocean University, Zhoushan 316022, China
* Correspondence: shangweixie1@163.com

Abstract: To describe the species composition of fish larvae and juveniles in the Nanji Islands, monthly collection was conducted at 12 stations around the Nanji Islands from March 2023 to February 2024 by using horizontal tow at the surface layer. The fish larvae and juveniles were collected by a larva net (1.3 m mouth diameter, 0.5 mm mesh aperture). A total of 6446 fish larvae and juveniles were collected, belonging to 59 species and 54 genera in 11 families. Most of them were preflexion larvae (93.31%). Gobiidae, with eight species, was the most diverse family. Based on the index of relative importance (IRI) result, the dominant species was *Sebastiscus marmoratus* (Cuvier, 1829), accounting for 84.52% of the total number of samples collected. The common species were *Stolephorus commersonii* Lacepède, 1803; *Omobranchus elegans* (Steindachner, 1876); *Nibea albiflora* (Richardson, 1846); *Parablennius yatabei* (Jordan & Snyder, 1900); *Lateolabrax maculatus* (McClelland, 1844); and *Odontamblyopus lacepediiand* (Temminck & Schlegel, 1845). Other species were all rare species. The highest species number was in September (18 species), and the highest density was in December. Only one fish larva was collected in April. Pielou's evenness index was highest in November, while Margalef's richness index and Shannon–Wiener diversity were highest in September. The result of cluster analysis showed that the 11 months (except April) could be divided into four groups. The species number, density, and three diversity indexes showed no significant correlation with temperature and salinity.

Keywords: fish larvae and juveniles; species composition; dominant species; species diversity; Nanji Islands National Nature Reserve

Key Contribution: This study was to investigate the early stages of fish composition in the Nanji Islands. The species number was lowest in spring and the density was highest in winter.

Citation: Wang, X.; Zhao, D.; Liu, Q.; Lu, T.; Zhong, J.; Chen, W.; Xie, S.; Chen, S. Species Composition of Fish Larvae and Juveniles in the Nanji Islands, China. *Fishes* **2024**, *9*, 421. https://doi.org/10.3390/fishes9110421

Academic Editor: Stylianos Somarakis

Received: 18 September 2024
Revised: 5 October 2024
Accepted: 11 October 2024
Published: 22 October 2024

1. Introduction

The fish larval and juvenile periods are important stages during the early developmental stage, which is a period of significant morphological, ecological, and physiological changes in the life history of fish, and the mortality rate is also higher than in other periods [1,2]. Fish larvae and juveniles are also one of the important links in the marine food chain. Their survival quantity is the basis for the replenishment of fish resources [3]. The study of fish larvae and juveniles provides basic information for assessing the abundance of fish resources and maintaining the sustainable utilization of fishery resources.

The Nanji Islands are located south-east of the sea in Zhejiang Province and are a very representative protected area in China. The Nanji area comprises 52 islands (area > 500 m^2), numerous vibrant reefs, and surrounding waters spanning a total of 201.06 km^2, and it is approximately 55 km away from the nearest point on the mainland to the west. The hydrological environment is mainly influenced by the interaction between the Taiwan Warm Current and the Jiangzhe Coastal Currents. The special topography and complex current system have also formed upwelling currents. The water quality in Nanji is fertile, and nutrient-rich and bait organisms are abundant and suitable for a variety of economic fish, shrimps, crabs, shellfish, and algae. It is commonly known as "the kingdom of shellfish and algae" [4,5]. At present, research on the Nanji Islands mainly focuses on zooplankton [6,7], benthos [8,9], shellfish [10,11], and algae [12,13]. In contrast, not so many studies have been conducted on the biological resources of fish in the Nanji Islands [14–17], and studies on the resources of fish in this area have not yet been reported.

In this study, the species composition of fish larvae and juveniles in the Nanji Islands was examined, and the relationships between fish larvae and juveniles and environmental factors were analyzed so as to provide the main basis for the management of fish resources in the Nanji Islands.

2. Materials and Methods

At 12 station sites (L1–L12, Figure 1) around Nanji Islands, fish larvae and juveniles were collected monthly by a larva net (1.3 m mouth diameter, 1 mm mesh aperture) from March 2023 to February 2024.

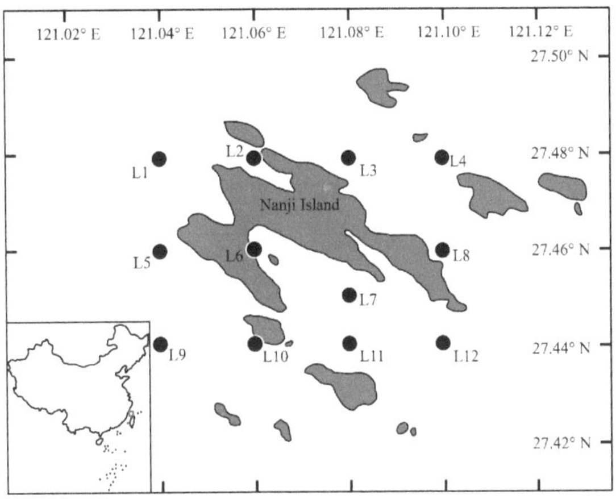

Figure 1. Sampling stations in the Nanji Islands. Solid circles indicate the sampling stations.

All samples were fixed in a 5% seawater formalin solution and then preserved in 75% ethanol. Fish were counted and identified to the lowest possible taxon with the information currently available [18–21]. The temperature (°C) and salinity were measured on site with a digital salinometer during the sampling period.

The number of individuals (n) was converted to density (D: inds. 1000 m^{-3}) using the following equation:

$$D = \frac{n \times 1000}{A \times L \times C}$$

where A is the area (m^2) of the larva net, L is the number of revolutions of the flowmeter (No. 438 100, Hydro-Bios) attached to the larva net during the tow, and C is the calibration factor in meters per revolution of the flowmeter (C = 0.3 m revolution^{-1}).

The index of relative importance (IRI) was calculated for each species by

$$IRI = N\% \times F\% \times 10000$$

where $N\%$ is the percentage of the number of each species among the total number of fish species and $F\%$ is the species frequency per species sampled in percentages. Accordingly, each species was defined as dominant ($IRI \geq 100$), common ($100 > IRI \geq 10$), or rare ($IRI < 10$) [22].

Diversity studies were analyzed using Margalef's richness index (D'), Shannon–Wiener's diversity index (H'), and Pielou's evenness index (J).

$$D' = \frac{S-1}{\ln N}$$

$$H' = -\sum P_i \ln P_i$$

$$J = \frac{H'}{\ln S}$$

where S is the number of fish larvae and juveniles; N is the number of species; and P_i is the proportion of individuals in taxa i.

The density data were transformed by $\log_{10}(x + 1)$ to reduce the effects of extremum and then analyzed using non-metric multidimensional scaling ordination (n-MDS) with Bray–Curtis's similarity index. Based on this data analysis, non-linear multidimensional scaling (NMDS) for the standard two-dimensional sequence diagram uses cluster points (CLUSTER) as a branch map. The analysis of similarities (ANOSIM) was used to test differences between communities by month, and the percentage similarity analysis (SIMPER) was used to find out the similarity and the typical species within a similar community structure. Primer 6.0 was used to perform this analysis.

To analyze the relationships between environmental and biotic variables, Pearson correlation coefficients were carried out using SPSS 19.0.

3. Results

3.1. Environmental Conditions

The water temperature showed clear seasonal variations. It first increased from later winter to summer, then decreased during autumn and winter. The peak water temperature was in September 2023. The temperature was lower during March, December, and February. Salinity showed variability. It was around 25 ppt from March to August. In September, the salinity was relatively low due to heavy rains. In autumn and winter, it returned to ca. 30 ppt (Figure 2).

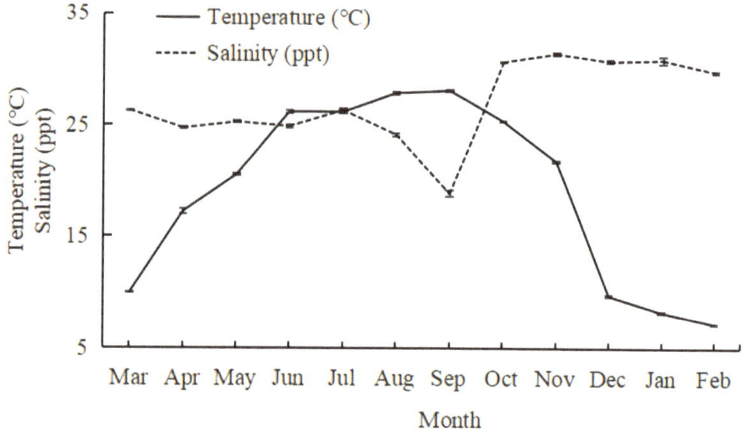

Figure 2. Water temperature and salinity from March 2023 to February 2024.

3.2. Species Composition

A total of 6446 larval and juvenile fishes were collected in the present study, belonging to 59 species and 54 genera in 11 families (Table 1). Among them, 93.31% were preflexion larvae, followed by postflexion larvae (2.91%) and flexion larvae (2.87%). There were very few yolk-sac larvae, juvenile fish, and young fish. Gobiidae, with eight species, was the most diverse family. Blenniidae had six species; followed by four species of Sciaenidae and Engraulidae; three species of Myctophidae and Cynoglossidae; and two species of Mugilidae, Belonidae, Scombridae, Terapontidae, Sillaginidae, and Sparidae. The rest of the families had only one species each (Table 1). There were 34 warm-temperate species, 15 warm-water species, and 2 cold-temperate species. According to the ecological type, there were 22 demersal species, 9 reef-associated species, 9 pelagic species, and 9 benthopelagic species. There were 33 commercial species, accounting for 93.27% of the total number of individuals.

Table 1. Species composition of fish larvae and juveniles in the Nanji Islands from March 2023 to February 2024.

Species	Abundance %	Developmental Stage	IRI	Temperature Adaption	Ecological Type	Month
Engraulidae						
Coilia sp. *	0.06	Pre	0.09	UN	UN	2
Engraulis japonicus *	0.85	Pre-Y	7.70	WW	P	5, 6, 9, 1, 2
Stolephorus commersonii *	2.19	Yolk-J	45.57	WW	P	7–2
Thryssa mystax	0.02	Y	0.01	WT	P	2
Chanidae						
Chanos chanos *	0.08	Post	0.22	WT	BP	9
Salangidae						
Salanx ariakensis	0.09	J, Y	0.32	WT	D	2
Synodontidae						
Harpadon nehereus *	0.03	Pre	0.02	WT	BP	11
Myctophidae						
Benthosema pterotum	0.03	Post	0.04	WT	BP	12
Diaphus sp.1	0.03	Pre	0.04	UN	UN	3
Diaphus sp.2	0.03	Pre	0.04	UN	UN	3
Bregmacerotidae						
Bregmaceros pseudolanceolatus	0.14	J	0.68	WT	P	12, 1
Gobiidae						
Acanthogobius flavimanus	0.02	Post	0.01	CT	D	1
Amblychaeturichthys hexanema	0.09	Pre, Post-J	0.19	WT	D	1, 2
Ctenotrypauchen chinensis	0.02	J	0.01	WW	D	12
Eviota sp.	0.02	Post	0.01	WW	UN	6
Odontamblyopus lacepedii	1.13	Yolk-Post	14.94	WT	BP	6–10, 12–1
Paratrypauchen microcephalus	0.09	Pre, Y	0.13	WT	D	7, 9
Tridentiger barbatus	0.02	Post	0.01	WT	D	7
Gobiidae sp.	0.02	Pre	0.01	UN	UN	9
Ambassidae						
Ambassis gymnocephalus	1.15	Pre-Post	8.77	WT	D	7
Pomacentridae						
Pomacentridae sp.	0.02	Pre	0.01	WW	R	6
Mugilidae						
Mugil cephalus *	0.03	J, Y	0.04	WW	BP	2
Planiliza haematocheilus *	0.03	Post	0.04	WT	P	7, 10
Atherinidae						
Hypoatherina valenciennei	0.05	Pre-F	0.06	WT	P	6, 7
Hemiramphidae						
Hemiramphus far *	0.02	J	0.01	WT	R	10
Belonidae						
Strongylura anastomella *	0.02	Post	0.01	WT	P	9
Tylosurus crocodilus *	0.05	J, Y	0.10	WT	R	6, 7

Table 1. *Cont.*

Species	Abundance %	Developmental Stage	IRI	Temperature Adaption	Ecological Type	Month
Blenniidae						
Entomacrodus stellifer	0.16	Pre	0.65	WW	R	5, 7–9
Istiblennius sp.	0.09	Pre	0.26	UN	UN	8, 9
Laiphognathus longispinis	0.02	J	0.01	WW	D	6
Omobranchus elegans	2.37	Pre-Post	36.26	WT	D	6–8, 10
Parablennius yatabei	1.04	Pre-Post	21.65	WT	D	6, 7, 9–11
Petroscirtes sp.	0.03	Post	0.04	UN	UN	11
Carangidae						
Decapterus maruadsi *	0.12	Pre-F	0.34	WW	R	6
Pleuronectidae						
Pleuronectidae sp.	0.02	Pre	0.01	UN	UN	9
Cynoglossidae						
Cynoglossus gracilis *	0.02	Pre	0.01	WT	D	8
Cynoglossus lighti *	0.09	Pre, Post	0.19	WT	D	7, 9
Cynoglossus robustus *	0.02	Post	0.01	WT	D	10
Scombridae						
Euthynnus affinis *	0.02	Pre	0.01	WW	P	6
Rastrelliger kanagurta *	0.06	Pre	0.09	WW	P	5
Gerreidae						
Gerres japonicus *	0.03	F-Post	0.04	WT	D	7
Terapontidae						
Pelates quadrilineatus *	0.09	J	0.13	WT	R	5
Terapon jarbua *	0.03	J	0.02	WT	D	9
Haemulidae						
Plectorhinchus cinctus *	0.02	Pre	0.01	WT	R	9
Scatophagidae						
Scatophagus argus *	0.03	Pre	0.04	WT	D	8
Scorpaenidae						
Sebastiscus marmoratus *	84.52	Pre-F	2875.94	WT	D	3, 11–2
Hexagrammidae						
Hexagrammos sp. *	0.19	Yolk-Post	1.03	CT	UN	3, 12–2
Moronidae						
Lateolabrax maculatus *	1.15	Pre-Post	15.94	WT	R	3, 1–2
Sciaenidae						
Chrysochir aureus *	0.02	J	0.01	WW	BP	9
Larimichthys crocea *	0.11	Pre, Post	0.15	WT	BP	10, 1
Nibea albiflora *	1.72	Pre-Post	33.48	WT	BP	7–10
Pennahia argentata *	0.05	Pre, Post	0.06	WT	BP	7, 9
Sillaginidae						
Sillago parvisquamis *	0.61	Pre-Post	2.94	WT	D	7
Sillago sihama *	0.92	Pre-Post	4.45	WT	R	6, 7
Lethrinidae						
Lethrinidae sp.	0.08	Pre	0.16	UN	UN	8
Sparidae						
Acanthopagrus latus *	0.06	Pre, Post	0.17	WW	D	9, 11, 1
Acanthopagrus schlegelii *	0.02	Post	0.01	WW	D	4
Monacanthidae						
Stephanolepis cirrhifer *	0.02	Y	0.01	WW	D	5
Tetraodontidae						
Takifugu xanthopterus	0.02	Y	0.01	WT	D	12

Notes: Yolk: yolk-sac larvae; Pre: preflexion larvae; F: flexion larvae; Post: postflexion larvae; J: juvenile; Y: young; CT: cold-temperate species; WT: warm-temperate species; WW: warm-water species; P: pelagic; BP: benthopelagic; D: demersal; R: reef-associated; UN: unknown; *: commercial species.

3.3. Temporal Variations

The density was relatively low from March to November, the highest density was in December, and then it gradually decreased until the following February. The highest species number was in September (18 species), followed by July (17 species). Very few specimens were collected in April and November (Figure 3). The Margalef's richness index ranged from 0 to 3.72, and the mean value of each month was 1.75 ± 0.23. The Pielou's evenness index ranged from 0 to 0.84 (mean 0.46 ± 0.08), and the Shannon–Wiener diversity index ranged from 0 to 2.08 (mean 0.95 ± 0.20). The Margalef's richness index and Shannon–Wiener diversity index were the highest in September, while the Pielou's evenness index was highest in November. All diversity indexes were lowest in April (Figure 4).

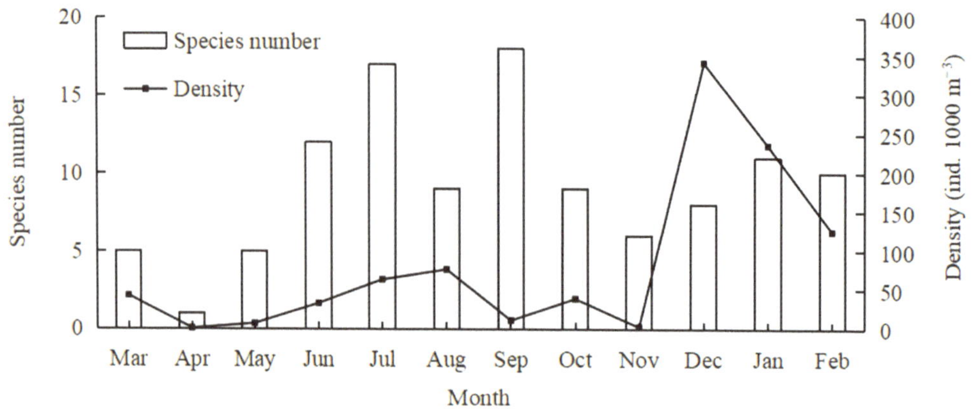

Figure 3. Temporal variations in species number and density.

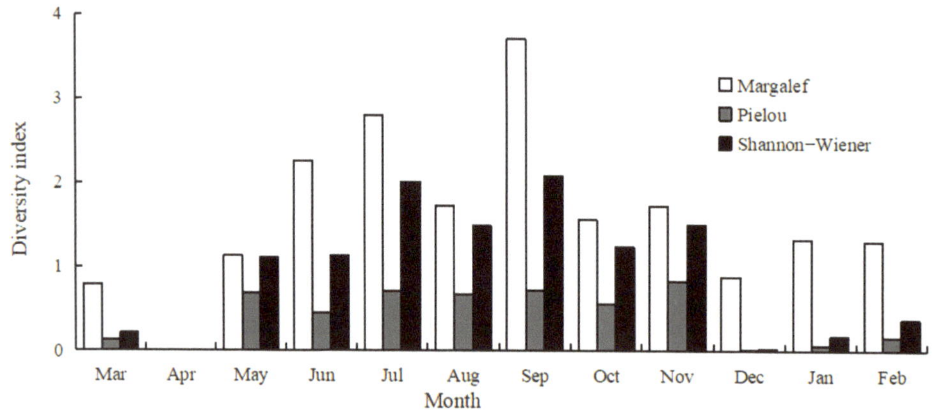

Figure 4. Temporal variations in three diversity indexes.

Based on the IRI result, the dominant species was *Sebastiscus marmoratus* (Cuvier, 1829), accounting for 84.52% of the total number of individuals collected from March 2023 to February 2024. The common species were *Stolephorus commersonii* Lacepède, 1803; *Omobranchus elegans* (Temminck & Schlegel, 1845); *Nibea albiflora* (Richardson, 1846); *Parablennius yatabei* (Jordan & Snyder, 1900); *Lateolabrax maculatus* (McClelland, 1844); and *Odontamblyopus lacepediiand* (Temminck & Schlegel, 1845). Other species were all rare species (Table 1). *S. marmoratus* was the dominant species in five months, followed by *O. lacepediiand* (four months), *N. albiflora* and *P. yatabei* (three months), and *Engraulis japonicus* Temminck & Schlegel, 1846, *O. elegans*, *S. commersonii*, and *L. maculatus* (two months). The remaining species was the dominant species in only one month (Table 2).

Table 2. The dominant species of fish larvae and juveniles in the Nanji Islands (*IRI* > 100).

Species	Month											
	Mar	Apr	May	Jun	Jul	Aug	Sep	Oct	Nov	Dec	Jan	Feb
Sebastiscus marmoratus	8823								370	9958	9712	9243
Acanthopagrus schlegelii		833										
Engraulis japonicus			1571	1346								
Pelates quadrilineatus			286									
Rastrelliger kanagurta			190									
Omobranchus elegans				4788	1574							
Decapterus maruadsi				205								
Ambassis gymnocephalus					1820							
Sillago sihama					951							
Sillago parvisquamis					533							
Odontamblyopus lacepedii					350	178	412	5099				
Stolephorus commersonii					131	1812						
Nibea albiflora						3495	962	1864				
Lethrinidae sp.						121						
Parablennius yatabei							3591	937	2593			
Chanos chanos							172					
Istiblenniussp.							129					
Petroscirtes sp.									185			
Lateolabrax maculatus											112	387

Because only one larva was collected in April, this month's data were excluded from the monthly clustering to prevent rare data from affecting the overall analysis. From the result of monthly cluster analysis, the 11 months could be divided into four groups by the similarity 20. May–June was clustered as group I, July–October was group II, November was group III, and December–March was group IV (Figure 5). The n-MDS showed that the stress was 0.07. ANOSIM analysis values, R = 0.989, and P = 0.01 indicated that the differences between the four groups were significant. Results from SIMPER analysis revealed that group I had an average similarity of 28.93%, with *Engraulis japonicus* contributing 100%. Group II had an average similarity of 35.10%, with *Nibea albiflora* contributing 27.68%, *Stolephorus commersonii* contributing 24.07%, *Odontamblyopus lacepedii* contributing 22.42%, and *Parablennius yatabei* contributing 11.05%. Group III has only one month. Group IV had an average similarity of 64.35%, with *Sebastiscus marmoratus* contributing 88.26%.

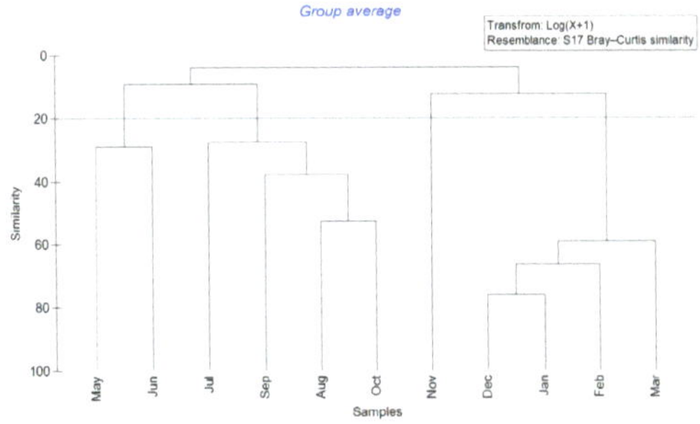

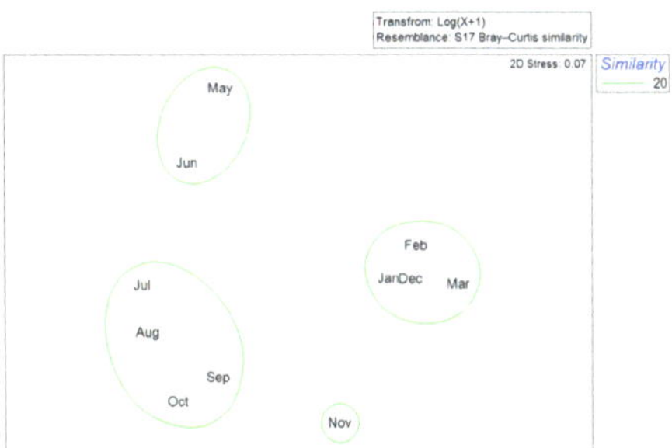

Figure 5. Cluster analysis of larval and juvenile fish community.

3.4. The Relationships between Environmental and Biotic Variables

The species number, density, Margalef's richness index, Shannon–Wiener's diversity index, and Pielou's evenness index showed no significant correlations with temperature and salinity ($p > 0.05$, Table 3).

Table 3. Matrix showing the Pearson correlation coefficients for environmental and biotic variables.

	Species Number	Density	Margalef	Pielou	Shannon–Wiener
Temperature	0.127	−0.557	0.199	0.221	0.307
Salinity	−0.125	0.069	0.061	0.216	0.11

4. Discussion

In the present study, species composition of fish larvae and juveniles in the Nanji Islands showed that the density was low in spring and summer and high in autumn and winter, and species number was high in summer, followed by autumn and winter, and was lowest in spring (Figure 3). The three diversity indexes showed similar seasonal variation (Figure 4). However, no significant correlation was found between the community of fish larvae and environmental factors (Table 3). Usually, the spring season is the most abundant season for ichthyofauna [23–26], but this is the opposite of the results of the present study and the survey in Jeju Island, Korea [27]. Some studies showed that the species number and density were the highest in the estuary in spring [25,28]. In the Oujiang estuary, an estuary close to the Nanji Islands on the mainland, the fish eggs were more abundant inside than outside of the Oujiang estuary, while fish larvae showed the opposite trend [29]. He et al. [30] and Zhang et al. [31] found that the species number of adult fish was also the lowest in spring in the Nanji Islands.

Most of the collected fish larvae and juveniles were warm-temperate and warm-water species, and the warm-temperate species account for 93.51% of the total number of collected fish larvae and juveniles (Table 1). The Nanji Islands are located in the warm-temperate zone, with a subtropical monsoon climate. Under the influence of the Taiwan Warm Current, a large number of warm-temperate and warm-water fish species inhabit the area due to the continuous intrusion of high-temperature oceanic water [32]. Cold-temperate fish, represented by *Hexagrammos* sp., appeared in winter. This is affected by the north-east monsoon along the coast of Zhejiang and the southward flow of the Jiangzhe Coastal Current in winter [5,33].

The dominant species was different in different seasons. In spring, *Engrualis japonicus* was the dominant species. The local names of this species larvae and juveniles are Dingxiangyu and Xiangyan, respectively. Both of them are important commercial fish in the Nanji Islands [5]. The spawning season of this species was a very long period, from April to November [34]. When the surface water temperature reaches 13 °C in the Zhoushan Islands in April, the adult *E. japonicus* begins to enter the spawning ground. When the water temperature is more than 20 °C in June, *E. japonicus* starts spawning. The spawning ground surface water temperature ranges from 13 to 24 °C, and salinity ranges from 23.13 to 31.88 ppt [35], which is similar to the result in the present study.

In summer and autumn, under the influence of the south-west monsoon, the upwelling was strengthened, and the nutrient-rich bottom water was continuously transferred to the upper layer [5,33], which was attractive to the adult fish for spawning in the Nanji Islands [30,31] and might increase the species number of fish larvae and juveniles. The dominant species were *Omobranchus elegans*, *Odontamblyopus lacepedii*, *Stolephorus commersonii*, *Nibea albiflora*, and *Parablennius yatabei*. Among them, *S. commersonii* and *N. albiflora* are commercial species. *O. elegans* occurred in the Nanji Islands from May to September, with peak abundance in June. It is reported that the spawning season of this fish is from June to August, with water temperature 23.6–32.8 °C and salinity 9–28.5 ppt in Jiulong River [36]. *O. lacepedii* was the dominant species from July to October. This species spawns twice a year: its spring spawning season is from February to April and its autumn spawning season is from July to September [37]. The fish in the Nanji Islands should belong to autumn spawners. *S. commersonii* was dominant in July and August, and it is a euryhaline fish with a long spawning season, from February to November [38,39]. Li et al. reported that *S. commersonii* density was also higher in the Oujian estuary in August, with water temperature 21–29 °C and salinity 25–30 ppt [29]. *N. albiflora* was the dominant species from August to October, while this was *P. yatabei* from September to November. Both of the species' dominance occurred two months earlier than those in Korea [40] and Japan [41].

In winter, *Sebastiscus marmoratus* and *Lateolabrax maculatus* were the dominant species. *Sebastiscus marmoratus* was the most dominant species and appeared in the Nanji Islands from November to March. It is an ovoviviparous, reef-associated and commercial species, living in a limited area without long-distance migratory habits, and the spawning season is from winter to the following spring [42]. *L. maculatus* is a migratory and commercial fish and was the dominant species in January and February. The spawning season of this species is from December to February [43]. The postflexion larvae enter the surf zone of an adjacent island or estuary with growth [44]. A future study should conduct a survey in both areas and try to find the connection of specific species in the water and surf zone.

5. Conclusions

In this study, a total of 6446 fish larvae and juveniles, belonging to 59 species, were collected in 12 stations of the Nanji Islands, of which over 84% were *Sebastiscus marmoratus* preflexion larvae. The species number, density, and three diversity indexes showed seasonal variation, but there was no significant correlation between the community of fish larvae and environmental factors.

Author Contributions: Conceptualization, S.C.; methodology, J.Z.; software, T.L.; validation, Q.L.; formal analysis, D.Z. and T.L.; investigation, X.W., D.Z., Q.L., T.L. and W.C.; resources, W.C. and S.X.; data curation, Q.L. and T.L.; writing—original draft preparation, X.W.; writing—review and editing, X.W., J.Z. and S.X.; visualization, Q.L.; supervision, J.Z. and S.C.; project administration, W.C.; funding acquisition, J.Z. All authors have read and agreed to the published version of the manuscript.

Funding: This study was supported by Nanji Islands National Marine Nature Reserve Administration.

Institutional Review Board Statement: This study complies with the Specifications for oceanographic survey—Part 6: Marine biological survey and laws of China. Larval and juvenile fish samples were collected with a larva net, as this sampling method results in the mortality of the sampled fishes; they were not reared or maintained in the laboratory. According to the above guidelines, ethical approval was not needed for this study.

Informed Consent Statement: Not applicable.

Data Availability Statement: The data are available from the corresponding author upon reasonable request.

Acknowledgments: We extend our sincere gratitude to Zhiqing Chen and the staff from Nanji Islands National Marine Nature Reserve Administration for their generous assistance with sample collection. We also want to thank Roland Passmore for improving the English of this manuscript.

Conflicts of Interest: The authors declare no conflicts of interest.

References

1. Yin, M.C. Advances and studies on early life history of fish. *J. Fish. China* **1991**, *15*, 348–358. (In Chinese with English abstract)
2. D'Alessandro, E.K.; Sponaugle, S.; Cowen, R.K. Selective mortality during the larval and juvenile stages of snappers (Lutjanidae) and great barracuda *Sphyraena barracuda*. *Mar. Ecol. Prog. Ser.* **2013**, *474*, 227–242. [CrossRef]
3. Houde, E.D. Fish early life dynamics and recruitment variability. *Am. Fish. Soc. Symp.* **1987**, *2*, 17–29.
4. Pan, Y.; Ye, S.F.; Liu, X.; Wu, Y.Q. Studies on marine ecosystem services and valuation: A case of Nanji Archipelagos Natural Marine Reserve. *Mar. Envi. Sci.* **2009**, *28*, 176–180. (In Chinese with English abstract)
5. Cai, H.C. *An Overview of Nanji Islands National Marine Nature Reserve*; China Ocean Press: Beijing, China, 2021. (In Chinese)
6. Ji, H.H.; Ye, S.F.; Liu, X.; Hong, J.C. The species composition and diversity of zooplankton in Nanji Islands National Nature Reserve. *Biodivers. Sci.* **2006**, *14*, 206–215. (In Chinese with English abstract) [CrossRef]
7. Zhang, X.H.; Zhou, Y.; Long, H.; Yang, Y.L.; Huang, J.Q.; Yu, J.; Zhang, C. Investigation on zooplankton in Nanji Marine Protected Area (MPA). *Chin. J. Zool.* **2006**, *41*, 83–86. (In Chinese with English abstract)
8. Tang, Y.B.; Liao, Y.B.; Shou, L.; Zeng, J.N.; Gao, A.G.; Chen, Q.Z. Intertidal zone of the Nanji Islands is a niche for dominant species of the macrobenthic community. *Acta Ecol. Sin.* **2016**, *36*, 489–498. (In Chinese with English abstract)

9. Zhang, H.W.; Pen, X.; Liu, J.F.; Xue, F. Community structure of macrobenthos in the intertidal zones of Nanji Islands in winter and summer. *J. Shanghai Ocean Univ.* **2018**, *27*, 133–140. (In Chinese with English abstract)
10. Chen, G.H.; Zhou, X.D.; Zhao, H.B.; Wang, Y.N. Age structure and growth of the limpet *Cellana grata* (Gould, 1859) of the Nanji Islands, China. *J. Shellfish Res.* **2023**, *42*, 281–288.
11. Li, Z.; Huang, J.; Gu, Z.Q.; Ye, Y.Y. Genetic diversity and genetic structure of *Macridiscus multifarius* Based on ITS Gene Sequence. *J. Zhejiang Ocean Univ. (Nat. Sci.)* **2023**, *42*, 103–108. (In Chinese with English abstract)
12. Jiang, T.; Xu, Y.X.; Li, Y.; Jiang, T.J.; Wu, F.; Zhang, F. Seasonal dynamics of *Alexandrium tamarense* and occurrence of paralytic shellfish poisoning toxins in bivalves in Nanji Islands, East China Sea. *Mar. Freshw. Res.* **2013**, *65*, 350–358. [CrossRef]
13. He, E.Y.; Wang, D.; Lu, Y.D.; Li, X.; Ji, X.L. Analysis of environmental factors for typical *Prorocentrum donghaiense* blooms. *Ecol. Sci.* **2023**, *42*, 63–73. (In Chinese with English abstract)
14. Pan, L.W.; Wang, H.J.; Zhang, Z.; Zhang, Y.P.; Zou, Q. Discovery of the amphioxus in Zhejiang Province, China. *Chin. J. Zool.* **2022**, *57*, 742–750. (In Chinese with English abstract)
15. Dai, D.X.; Yu, C.G.; Liu, H.; Yan, W.C.; Sun, B.B.; Jian, K.K.; Xin, Y.; Zhang, W.J. Niche and interspecific association of major nekton in the sea area to the east of the Nanji Islands. *Chin. J. Appl. Ecol.* **2019**, *30*, 4249–4258. (In Chinese with English abstract)
16. Sarr, C.; Yu, C.G.; Ndiaye, O.; Diadhiou, H.; Ndour, N. Community structure of fish in Nanji Islands National Nature Reserve and its relationship with environmental variation. *J. Water Resour. Prot.* **2023**, *15*, 377–392. [CrossRef]
17. Sarr, C.; Yu, C.G.; Ndiaye, O.; Diadhiou, H.; Ndour, N. Fish assemblage and abundance distribution in Nanji Islands Marine Nature Reserve in relation to season change. *J. Agric. Chem. Environ.* **2024**, *13*, 13–32. [CrossRef]
18. Leis, J.M.; Carson-ewart, B.M. *The Larvae of Indo-Pacific Coastal Fishes*; Brill: Leiden, The Netherlands, 2004.
19. Okiyama, M. *An Atlas of Early Stage Fishes in Japan*, 2nd ed.; Donghai University Publishing Association: Kanagawa, Japan, 2014. (In Japanese)
20. Wan, R.J.; Zhang, R.Z. *Fish Eggs, Larvae and Juveniles in the Offshore Waters of China and Their Adjacent Waters*; Shanghai Scientific & Technical Publishers: Shanghai, China, 2016. (In Chinese)
21. Hou, G.; Zhang, H. *Illustrations of Fish Larvae and Juvenile in the South China Sea*; China Ocean University Press: Qingdao, China, 2021. (In Chinese)
22. Xia, W.T.; Miao, Z.B.; Wang, S.; Chen, K.; Liu, Y.L.; Xie, S.G. Influence of tidal and diurnal rhythms on fish assemblages in the surf zone of sandy beaches. *Fish. Oceanogr.* **2023**, *32*, 448–460. [CrossRef]
23. Yu, C.M.; Jiang, R.J.; Yin, R.; Li, P.F.; Zhu, C.H. Species composition of ichthyoplankton and its relationship with environmental factors in Dachenyang Spawning Ground Reserve. *J. Zhejiang Ocean Univ. (Nat. Sci.)* **2021**, *40*, 477–485. (In Chinese with English abstract)
24. Wan, R.J.; Sun, S. The category composition and abundance of ichthyoplankton in the ecosystem of the Yellow Sea and the East China Sea. *Acta Zool. Sin.* **2006**, *52*, 28–44. (In Chinese with English abstract)
25. Yan, J.; Xu, Z.J.; Li, T.J.; Wang, H.J.; Zhang, B.Y.; Qian, H. Spatiotemporal dynamics of larvae and juveniles in Zhoushan coastal waters and their relationship with environmental factors. *Oceanol. Limnol. Sin.* **2023**, *54*, 799–810. (In Chinese with English abstract)
26. Liu, L.W.; Sui, Y.Z.; Jiang, R.J.; Xu, K.D.; Zhu, W.B.; Zhou, Y.D. Species composition and abundance distribution of ichthyoplankton in Zhoushan Fishing Ground in spring. *J. Zhejiang Ocean Univ. (Nat. Sci)* **2021**, *40*, 1–8. (In Chinese with English abstract)
27. Lee, S.J.; Go, Y.B.; Kim, B.J. Seasonal variation of species composition and distribution of fish eggs and larvae in the western part of Jeju Island, Korea. *Korean J. Ichthyol.* **2006**, *18*, 129–140. (In Korean with English abstract)
28. Yan, X.; Zhong, J.S.; Wang, M.X.; Li, A.D.; Jiang, X.L.; Wang, C.Q.; Qiu, Y.F. Study on the species composition and diversity of fish larvae and juveniles in the north of Hangzhou Bay. *J. Shanghai Ocean Univ.* **2014**, *32*, 765–773. (In Chinese with English abstract)
29. Li, S.Q.; Ye, S.; Wang, Y.H.; Tang, W.; Qin, S.; Fan, Q.S.; Qiu, W.Z.; Ai, L.W.; Zhen, C.F.; Liu, W.C. Community structure and correlation of dominant species in earlyfish resources in the Oujiang Estuary. *Acta Hydrobiol. Sin.* **2024**, *48*, 898–909.
30. He, X.B.; Zhang, F.J.; Lin, L.; Chen, W.D.; Cai, H.C.; Yu, C.G. Species composition and quantitative distribution of fishes in island-reef water of Nanji islands. *Oceanol. Limnol. Sin.* **2013**, *44*, 453–459. (In Chinese with English abstract)
31. Zhang, W.J.; Yu, C.G.; Jian, K.K.; Yan, W.C.; Zhen, J. Study on the characteristics of fish community structure in the east of Nanji Islands Marine Nature Reserve. *J. Zhejiang Ocean Univ. (Nat. Sci.)* **2019**, *38*, 487–494. (In Chinese with English abstract)
32. Zhao, F.; Jiang, R.J.; Yin, R.; Liu, M.Z.; Long, X.Y.; San, X.L.; Zhang, H.Q.; Hao, H.B. Species composition and relationship with environmental factors of fish eggs and larvae in Qixing Islands spawning ground reserve. *J. Fish. Sci. China* **2023**, *30*, 1311–1326. (In Chinese with English abstract)
33. Yu, C.G.; Cai, H.C.; Liu, L.S.; Lin, L.S. *Nanji Island Marine Nature Reserve Shallow Sea Ecological Environment and Fishery Resources*; Science Press: Beijing, China, 2021. (In Chinese)
34. Funamoto, T.; Aoki, I.; Wada, Y. Reproductive characteristics of Japanese anchovy, *Engraulis japonicus*, in two bays of Japan. *Fish. Res.* **2004**, *70*, 71–81. [CrossRef]
35. Zhou, Y.D.; Jin, H.W.; Zhang, H.L.; Jiang, R.J.; Pan, G.L. Species composition and abundance distributions of ichthyoplankton along the south coast of Zhejiang Province in spring and summer. *Oceanol. Limnol. Sin.* **2013**, *44*, 590–598. (In Chinese with English abstract)
36. Jiang, S.F.; Yan, W.B. Larvae of *Omobranchus elegans* (Steindachner) found in Luo Yuan Bay and Jiulong River, Estuary Fujian. *J. Xiamen Univ. (Nat. Sci.)* **1989**, *28*, 672–676. (In Chinese with English abstract)

37. Zhuang, P.; Zhang, T.; Li, S.F.; Ni, Y.; Wang, Y.H.; Deng, S.M. *Fishes of the Yangtze Estuary*, 2nd ed.; China Agriculture Press: Beijing, China, 2018.
38. Wu, G.Z. The ecological characteristics of distribution of eggs, larvae and juveniles of the *Engraulis japonicus* and *Anchoviella commersonii* in the Changjiang River Estuary. *Oceanol. Limnol. Sin.* **1989**, *20*, 217–229. (In Chinese with English abstract)
39. Tian, F.G.; Zheng, Y.J.; Xiao, Y.Z.; Fang, H.D. Spawning season and variation of the eggs of *Stolephorus commersonii* in Guanghaiwan Bay. *J. Appl. Oceanogr.* **2017**, *36*, 395–402. (In Chinese with English abstract)
40. Kim, H.J.; Jeong, J.M.; Park, J.H.; Huh, S.H.; Baeck, G.W. Species composition of larval fishes in the eastern Jinhae Bay, Korea. *J. Korean Soc. Fish. Ocean Technol.* **2018**, *54*, 81–88. (In Korean with English abstract) [CrossRef]
41. Kume, G.; Yagishita, N.; Furumitsu, K.; Nakata, H.; Suzuki, T.; Handa, M.; Yamaguchi, A. The role of molecular methods to compare distribution and feeding habits in larvae and juveniles of two co-occurring sciaenid species *Nibea albiflora* and *Pennahia argentata*. *Estuar. Coast. Shelf S.* **2015**, *167*, 516–525. [CrossRef]
42. Jin, X.B. *Fauna Sinica Ostichthyes: Scorpaeniformes*; Science Press: Beijing, China, 2006. (In Chinese)
43. Ni, Y.; Wu, H.L. *Fishes of Jiangsu Province*; China Agriculture Press: Beijin, China, 2006. (In Chinese)
44. Mao, C.Z.; Chen, Y.G.; Zhong, J.S.; Hua, W.H.; Jiao, X.M.; Zhang, X.M.; Zhang, X.Y.; Yuan, G.W. Differences and association of larval and juvenile fish community in surf zone of south branch of Yangtze River and the adjacent waters. *Resour. Environ. Yangtze Basin* **2018**, *27*, 125–134. (In Chinese with English abstract)

fishes

Article

Use of Environmental DNA to Evaluate the Spatial Distribution of False Kelpfish (*Sebastiscus marmoratus*) in Nearshore Areas of Gouqi Island

Rijin Jiang [1,2,3], Huibo Hao [1,2], Rui Yin [1,2,3], Peng Zhao [1,2], Feng Chen [1,2,3,*], Yongdong Zhou [1,2,3] and Xuejun Chai [1,2]

1 Zhejiang Marine Fisheries Research Institute, Zhoushan 316021, China; jiangridge@163.com (R.J.); hhb86629@163.com (H.H.); yr18368091951@163.com (R.Y.); zhaopeng8786@163.com (P.Z.); zyd511@126.com (Y.Z.); chaixj6530@sohu.com (X.C.)

2 Marine and Fishery Institute, Zhejiang Ocean University, Zhoushan 316021, China

3 Key Laboratory of Sustainable Utilization Research for Fishery Resources of Zhejiang Province, Zhoushan 316021, China

* Correspondence: fchen@zjou.edu.cn

Abstract: This study aims to explore the spatial distribution of false kelpfish (*Sebastiscus marmoratus*) in the mussel farming area, artificial reef areas of Gouqi Island (Shengsi, China), and natural areas using eDNA detection methods. Surface and bottom water samples were collected at 12 stations in November 2022 and April 2023, totaling 52 samples. We used species-specific primers and probes for quantitative PCR (qPCR) detection of *Sebastiscus marmoratus* eDNA. The eDNA concentrations differed seasonally ($p < 0.05$) and did not differ ($p > 0.05$) among the three sampling areas and two water layers. The greatest eDNA concentrations occurred in the surface layer during the spring. Higher concentrations of *Sebastiscus marmoratus* eDNA were also found in the mussel aquaculture area. Temperature exhibited a significant positive correlation with *Sebastiscus marmoratus* eDNA concentration ($p < 0.05$). Additionally, we developed linear equations predicting the relationship between environmental factors and environmental factors, providing a reference for future fishery resource surveys.

Keywords: QPCR; linear mixed effects model; multiple linear regression

Key Contribution: This study analyzed the spatial distribution differences of *Sebastiscus marmoratus* across different seasons and water layers in mussel farming areas, artificial reef areas, and natural waters, as well as the relationship between the environmental DNA concentration of *Sebastiscus marmoratus* and environmental factors. The research provides crucial insights for developing conservation strategies for Sebastiscus marmoratus and lays the groundwork for further studies.

check for updates

Citation: Jiang, R.; Hao, H.; Yin, R.; Zhao, P.; Chen, F.; Zhou, Y.; Chai, X. Use of Environmental DNA to Evaluate the Spatial Distribution of False Kelpfish (*Sebastiscus marmoratus*) in Nearshore Areas of Gouqi Island. *Fishes* **2024**, *9*, 418. https://doi.org/10.3390/fishes9100418

Academic Editor: Rodolphe Elie Gozlan

Received: 20 August 2024
Revised: 1 October 2024
Accepted: 16 October 2024
Published: 19 October 2024

1. Introduction

The drastic reduction in Earth's biodiversity has led to irreparable losses in both ecological functionality [1] and the human environment, amounting to an increasingly prominent ecological crisis in the 21st century. Marine biodiversity is a critical component of this global diversity, and due to excessive fishing and environmental pollution in coastal areas, marine ecosystems have seen negative impacts due to human exploitation of the marine environment.

Knowledge about the distribution and abundance of key species is crucial for understanding the functioning of marine ecosystems and implementing effective fisheries resource management [2]. Traditionally, the acquisition of aquatic organism distribution data relies on a series of standardized field survey activities aimed at accurately recording the actual quantity and locations of target species to derive scientific results. However, traditional survey methods have shown several limitations in practice. First, they often require a lot of human and material resources, leading to time-consuming and expensive surveys. Second, due to the differences in sampling environments and the ecological habits of different study subjects, it is sometimes necessary to develop and apply targeted sampling strategies and methods, which can increase the complexity and difficulty of the investigation [3]. Moreover, for fish inhabiting specific habitats, such as reefs, traditional methods are often ineffective and inaccurate for assessing fish distribution due to factors such as habitat complexity and difficulty of approach. In recent years, researchers have used acoustic methods combined with traditional trawling to investigate the biomass of specific fish species before and after the deployment of artificial fish reefs in marine reserves, which has mitigated some disadvantages of traditional survey methods but still has the drawback of complexity [4]. Therefore, it is necessary to explore and develop more advanced and efficient survey methods to meet the needs of aquatic organism distribution monitoring under different environmental conditions; eDNA is a novel ecological monitoring technology where DNA fragments from tissues, such as hair, skin, and scales shed by organisms during their life processes persist in the environment for a period of time. Analysis via high-throughput sequencing or PCR can provide insights into the species identity and quantity of these organisms. Previous studies have shown that the real-time quantitative polymerase chain reaction (qPCR) quantification of eDNA concentration is positively correlated with the biomass of target species, which can include fish [5–10]. Hence, eDNA concentration can reflect the abundance or biomass of target organisms. The use of eDNA technology for biological resource monitoring can cover a wide range of areas in a relatively short time period without disturbing the target species or their habitats [11,12]. Additionally, compared to traditional monitoring, eDNA allows for easier sampling. This method is currently being applied to endangered species [13,14], the monitoring of invasive species [15,16], and general biodiversity monitoring [17].

The false kelpfish (*Sebastiscus marmoratus*) is a reef-dwelling fish species found along the coasts of Japan, Korea, the Philippines, and China's Bohai, Yellow, and East China Seas, as well as to the east of the Leizhou Peninsula [18]. The false kelpfish (*Sebastiscus marmoratus*) is a primary catch in recreational fisheries, with peak production during fishing moratoriums (May to August in the East China Sea) reaching up to 55 tons at Shengshan Island and Gouqi Island, accounting for 66% of the annual yield. Ecologically, the false kelpfish (*Sebastiscus marmoratus*) also plays a crucial role in energy transfer within the marine ecosystem of Gouqi Island. Due to its preference for habitats near reefs, it is highly favored by anglers. This fish species is rarely caught by general trawling (Single Trawler or Double Trawler) and, instead, is captured by fishing, crab pots, and gill nets. Its capture with more labor-intensive sampling methods in part contributes to its high market price.

Existing survey methods and predictions suggest that *Sebastiscus marmoratus* is widely distributed near reefs, but the specific distribution is not known. Further exploration of the resource distribution of *S. marmoratus* is beneficial for understanding its ecological habits and implementing scientific fisheries management. There is already an ongoing research basis for monitoring *S. marmoratus* eDNA, with experiments conducted by Tianxiang et al. [19] showing a strong positive correlation between *S. marmoratus* aquaculture density and the duration and eDNA concentration of indoor water tanks. This indicates that eDNA concentration can be used to understand and predict the spatial distribution of *S. marmoratus* to some extent. This study aims to analyze the spatial distribution of *S. marmoratus* and its seasonal variation using eDNA methods, providing technical support for population surveys, including among populations that are difficult to monitor.

2. Materials and Methods

2.1. Sample Collection

The sampling area (Figure 1) is primarily located around Gouqi Island and is divided into three main zones: the mussel farming area, the artificial reef area, and the natural area. Gouqi Island Marine Reserve has open nearshore waters rich in nutrients as well as abundant reefs and biological resources. The surrounding waters of Gouqi Island are the main area for mussel (*Mytilus galloprovincialis*) cultivation in Zhejiang Province, China with the aquaculture area nearly doubling from 2003 to 2016. By 2019, the mussel farming area had reached 10 km^2, with an output exceeding 180,000 metric tons (SI units), ranking third in the country. The mussel farming area is characterized by the presence of mussel cultivation structures and dense clusters of mussel strings. From 2010 to 2015, a total area of 0.5 km^2 of artificial fish reefs (a square reef made of cement) were deployed in the marine reserve [20]. The mussel farming area is characterized by the presence of mussel cultivation structures and dense clusters of mussel strings. Research has shown that the similarity of fish community structure and richness between artificial fish reef areas and natural sea areas is relatively high, suggesting that these artificial reefs may be playing a positive role in fish aggregation and enhancement [21]. So, the artificial reef area features man-made structures designed to enhance marine habitats, often used to attract fish and other marine life. The natural area represents an unaltered, natural marine environment without human-made cultivation or enhancement structures.

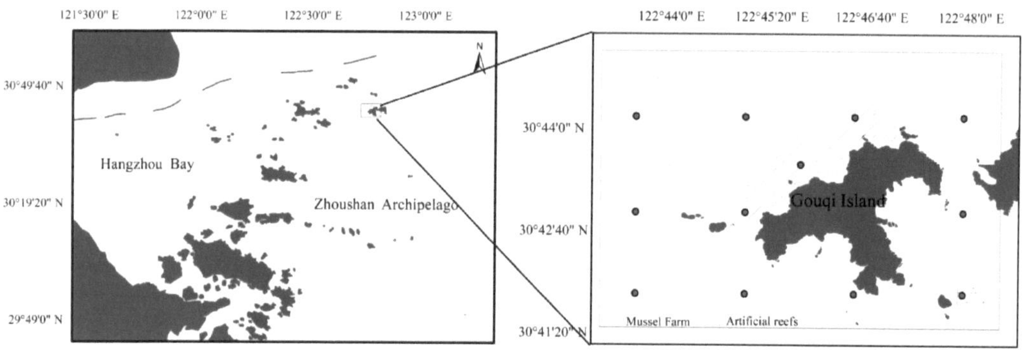

Figure 1. Sampling station map (the circles on the right represent the locations of the 12 sampling points, the diagonal shadow represents the mussel farming area, the dotted shadow represents the artificial reef area, and the blank background points are the natural areas).

A total of 52 environmental samples were collected in November 2022 and April 2023 in this marine area, from 12 sampling points: natural areas (7), mussel farming areas (4), and artificial reef areas (1). The distance between units is between 2 km and 2.8 km and there is no clustering of similar type sites per regional site. The mussel farming area has its sampling points concentrated in the main regions of the farming zone, with mussel strings densely distributed between the sampling points. In contrast, the natural area, which is a natural sea area, does not have cluster sampling points but, instead, is distributed around the nearshore areas of Gouqi Island. As the artificial reef area is relatively small, sampling was conducted twice at the same location. Sampling times were chosen during low tide each month, and sampling occurred only on sunny days. Surface and bottom seawater samples were collected using a water sampler and SBE 911 Plus CTD system (Sea Bird Scientific, Washington, DC, USA). Surface seawater was obtained at a depth of 1 m and bottom seawater samples were collected at depths between 7 and 30 m. Each seawater sample was collected in 2 L volumes and stored in single-use sterile bags. To measure nutrient content, 1 L of water was collected at each season and each sampling point in the surface water layer. The entire sampling process was completed within 5 h. After sampling, the samples were stored at −20 °C, and filtration was completed within 24 h. Subsequently, 2 L of seawater from each sample was filtered through 47 mm diameter glass fiber filters (0.45 µm pore size) using a vacuum pump in a nearby laboratory. To minimize cross-contamination, filter funnels were rinsed with tap water, immersed in sodium hypochlorite solution for 10 min, rinsed with tap water again, and finally rinsed with distilled water before filtering each sample. Distilled water and tap water were used as negative controls. The entire filtration process was completed within 24 h after sampling. All filtrates, frozen and stored at −196 °C in liquid nitrogen, were transported to the laboratory and stored at −80 °C until further eDNA extraction.

2.2. DNA Extraction

Due to the presence of a large number of coarse particles and sediment in the water sample that were not effectively removed before filtration, they accumulated on the filter membrane, prompting the use of a soil-specific kit for DNA extraction. The eDNA was extracted from filters using the SoilDNAKIT (D5625Omega) kit following the manufacturer's extraction protocol. The extracted DNA was then stored at −20 °C until required for subsequent PCR analysis.

2.3. QPCR

Quantitative PCR (qPCR) to detect *S. marmoratus* eDNA was performed using the LightCycler 480 II fluorescence quantitative PCR instrument (Roche, Rotkreuz, Switzerland). The primers and probes used in this experiment were designed based on the COI gene sequence of the false kelpfish (*Sebastiscus marmoratus*) and the specificity of the primers was verified. The forward primer sequence was H-F1-5′-CTGTTCTAATTACCGCTGTCCTT-3′, the reverse primer sequence was H-R1-5′-CGGCTGGGTCAAAGAATGT-3′, the primer length was 111 bp, and the TaqMan™ probe sequence was HPro-6524: TCTCCCTACCAGTT CTTGCTGCTGGCAT-5′FAM-3′BHQ1, designed to specifically bind to the target region. Sequencing of three positive PCR products was successfully conducted, and the obtained sequences were subjected to BLAST search in the NCBI database. They matched perfectly with the COI sequence of *Sebastiscus marmoratus*, confirming that the amplified sequences in this experiment are from *Sebastiscus marmoratus*. The primers, probe, and sequencing were synthesized by Sangon Biotech (Shanghai, China).

Each 20 μL PCR reaction mixture consisted of 2 μL of template DNA extract, 900 nM each of forward and reverse primers, 125 nM of probes, and 1× PCR master mix (FastStart Essential DNA Probes master, Roche Diagnostics, Basel, Switzerland). The amplification conditions were as follows: initial denaturation at 94 °C for 3 min, followed by denaturation at 94 °C for 15 s, annealing at 57 °C for 15 s, and extension at 72 °C for 30 s. After completing these steps, the 96/384-well plate containing the prepared samples was placed in the LightCycler 480 II instrument (Roche, Basel, Switzerland) for amplification. Three replicates of PCR were performed for each water sample (i.e., each DNA extract). Additionally, triplicates of four concentrations of standards were amplified, including artificial DNA fragments with 3×10^1, 3×10^2, 3×10^3, and 3×10^4 copies inserted into the 129 bp target region. The DNA concentration was quantified based on the calibration curve obtained from the standard results. To check for cross-contamination during the PCR process, three non-template controls were included in each 96-well reaction plate.

2.4. Acquisition of Environmental Factor Data

In this study, the environmental factors included temperature, salinity, sampling station depth, chlorophyll, and nutrient concentrations, including NH_3^-, NO_2^-, NO_3^-, and PO_4^{3-}. These environmental factors were collected in each season, and only surface water layers were collected except for the depth of sampling. Temperature, salinity, station depth, and chlorophyll were obtained using a CTD device during field sampling. The concentrations of NH_3^-, NO_2^-, NO_3^-, and PO_4^{3-} in seawater were determined using a continuous flow injection method with a continuous flow analyzer (SEAL Analytical AS4 V001, SEAL Analytical Instruments (Shanghai) Co., Ltd., Shanghai, China).

2.5. Data Analysis Methods

To understand the relationship between the eDNA concentration of *Sebastiscus marmoratus* and the sampling area, water layer, and season, a linear mixed effects model was constructed. In this model, eDNA concentration served as the response variable, while sampling area, water layer, and season were treated as fixed effects, and sampling points were considered random effects. This analysis was performed using the R software version 4.1.3 and the lmer function of the LME4 package [22]. The model residuals did not meet the parametric assumptions and the fish eDNA concentrations were log-transformed (log(x + 1)) before analysis. The formula for the model is as follows:

$$y = \beta_0 + a \times \beta_1 + b \times \beta_2 + c \times \beta_3 + (1|site)$$

Among them,

y is eDNA concentration;
B_0 is the intercept;
the sampling seasonal variable is a;
the sampling layer variable is b;
the sampling area variable is c;
(1 | Site) indicates the sampling point as a random effect.

Kriging interpolation was applied to grid the false kelpfish (*Sebastiscus marmoratus*) eDNA concentration data at each sampling site, and a mask was used to exclude grid data from the study area. It can be used to show qualitative trends in *Sebastiscus marmoratus* eDNA concentration across different seasons, layers, and areas, and this visualization was completed using ArcGIS 10.8 software.

Correlation analyses between environmental variables were calculated using the "Pheatmap" package [23] and the cor function in R software version 4.1.3 [24].

In the study, multiple linear regression (MLR) models were used to describe the relationships between various environmental factors, allowing for the evaluation of the strength of these relationships and the overall explanatory power of the model. This analysis involves the use of sampling points that were repeatedly sampled, and as such, it is pseudoreplicated. Linear mixed effects model analysis would be ideal; however, we used multiple linear regression analysis to ensure our resulting predictive equations could be used and understood by a greater number of fishery professionals. The R software version 4.1.3 was employed to perform the MLR analysis on data obtained from eDNA concentrations and environmental factors. The specific method is to first construct all models containing 1 to 8 variables, then remove models that include environmental factors with moderate to strong multicollinearity ($|r| > 0.5$), and finally select the best model based on R^2 and RMSE values. The MLR model was constructed using the stats package in R software version 4.1.3 and the lm function.

3. Results

3.1. Plasmid Construction

The amplicon length of the quantified plasmid is 111 bp, and the size of the plasmid is 2803 bp. The concentration of the plasmid standard is 178.5 ng/μL. In the qPCR results, the Ct values of the plasmid standard range from 14.3 to 32. The coefficient of determination (R^2) of the curve is 99.99%. The amplification efficiency is 109.7%. Within 40 cycles, only samples with false kelpfish (*Sebastiscus marmoratus*) DNA as the template were detected, with a detection limit of 31.9, while no other species or negative controls were detected. The results indicate that within the standard range of diluted plasmids, there is a consistent and predictable relationship between the fluorescent signal and the concentration of the target DNA. This indicates stable PCR amplification efficiency, making it effective for quantitative analysis of samples, and the established standard curve accurately reflects the amplification of COI.

3.2. The Concentration of False Kelpfish (*Sebastiscus marmoratus*) eDNA

The results of the linear mixed effects model show that the cross-validation mean square error is 0.89. The intercept of the model, the estimates of the fixed effects, the standard deviation, and the t-value are shown in Table 1. The eDNA concentrations were significantly affected by sampling season ($|t| \geq 2.58$) but less by sampling area and water layer ($|t| < 1.96$). Figure 2 shows the average Edna concentration in different seasons.

Table 1. Linear mixed-effects model effect parameter.

Effect	Estimate	Standard Error	T Value
(Intercept)	6.21	0.80	7.74
type	0.53	0.31	1.73
layer	−0.49	0.28	−1.75
season	−1.35	0.28	−4.85

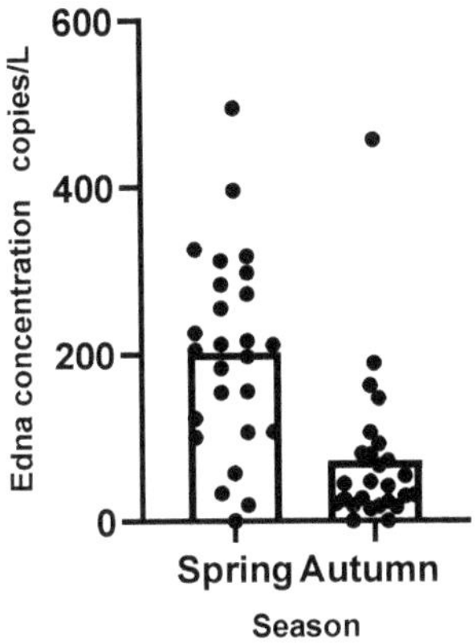

Figure 2. Mean eDNA concentrations in spring and fall.

Overall, out of 52 eDNA samples, only 3 samples did not detect S. marmoratus DNA, located, respectively, in the northeast outside the mussel aquaculture area in spring and autumn (2 samples) and southwest of Gouqi Island in autumn (1 sample). High eDNA concentrations (copies/L > 300) occurred at five stations in spring and one in autumn, with the highest frequency of occurrence in the mussel aquaculture area (five stations).

The results of the linear mixed effects model show that the cross-validation mean square error is 0.89. The intercept of the model and the estimates of the fixed effects, the standard deviation, and the t-value are shown in Table 1. The eDNA concentrations were significantly affected by sampling season ($|t| \geq 2.58$) but less by sampling area and water layer ($|t| < 1.96$). Figure 2 shows the average eDNA concentration in different seasons. The average eDNA concentration in spring was 202.35 ± 118.75 copies/L, while in autumn it was 71.50 ± 92.54 copies/L.

The visualized heat map shows the spatial and temporal distribution of eDNA concentration in *S. marmoratus* (Figure 3). The eDNA concentration of *S. marmoratus* was higher on the east side than on the west side of Gouqi Island in both spring and autumn (Figure 3). In spring, surface eDNA was mainly distributed within the aquaculture area and on both sides of Gouqi Island, with the highest concentration in the center of the aquaculture area (Figure 3); bottom layer eDNA was mainly distributed within the aquaculture area and on the east side of Gouqi Island. In autumn, surface eDNA was mainly distributed in the eastern part of Gouqi Island, with noticeably lower concentrations in the west; bottom layer eDNA was mainly distributed in the northern part of the aquaculture area and the southern part of Gouqi Island, moving toward deeper offshore areas compared to spring.

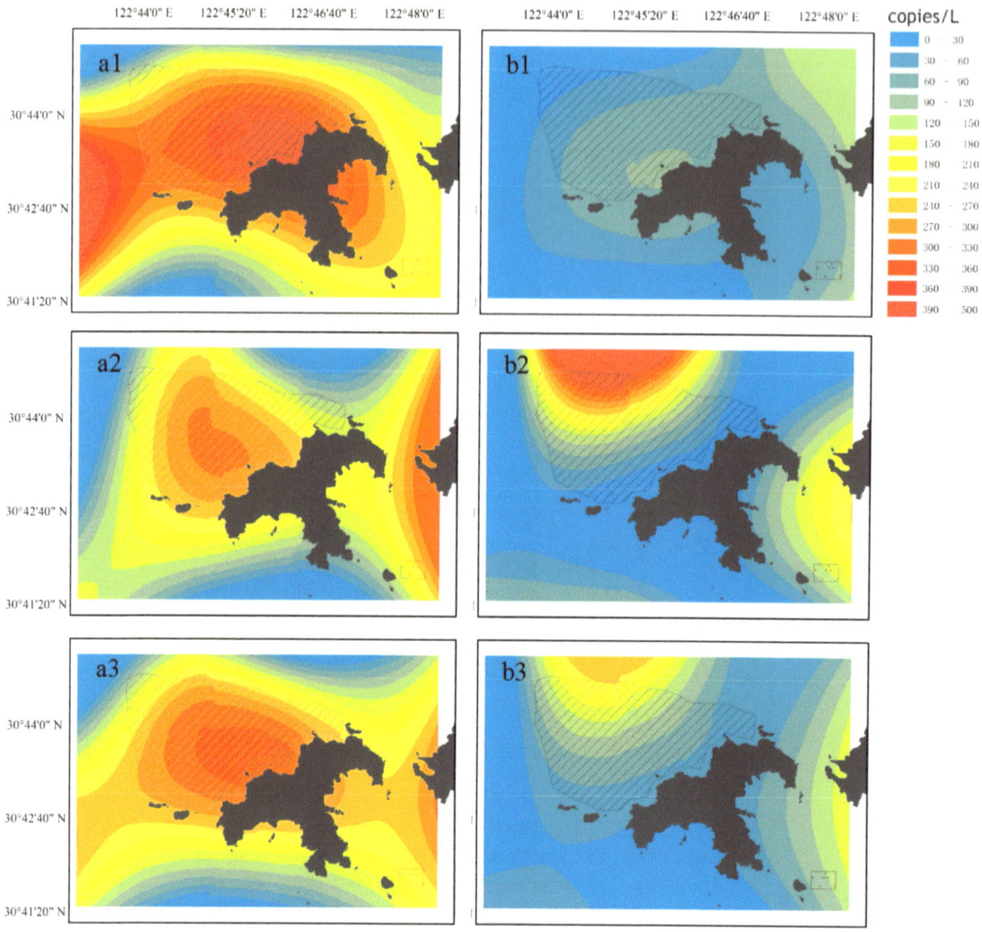

Figure 3. The eDNA concentration in the sampling area of *Sebastiscus marmoratus*. (**a1**) Surface in spring; (**a2**) bottom in spring; (**a3**) average in spring; (**b1**) surface in autumn; (**b2**) bottom in autumn; (**b3**) average in autumn.

3.3. The Relationship Between False Kelpfish (S. marmoratus) eDNA Concentration and Environmental Factors

The correlation analysis of eight environmental variables at sampling sites revealed that in spring, the presence and concentration of *S. marmoratus* eDNA were significantly correlated with temperature, salinity, depth, and nitrite concentration at the sampling points ($p < 0.05$). However, in autumn, there was no significant correlation between the presence or concentration of *S. marmoratus* eDNA and any of the environmental factors.

The correlations among independent variables presented in Figure 4 indicate that moderate to strong multicollinearity ($|r| > 0.5$) occurs between the following: (1) temperature and salinity; (2) NH_3^- and station depth; (3) salinity and NO_2^-; (4) salinity and NO_3^-; (5) NO_2^- and NO_3^-; and (6) salinity and PO_4^{3-}. Models containing combinations of these environmental factors were first excluded, and then the best models were selected based on R^2 and RMSE values (Table 2). Additionally, since this study included only 52 samples, only models with fewer than five variables were retained.

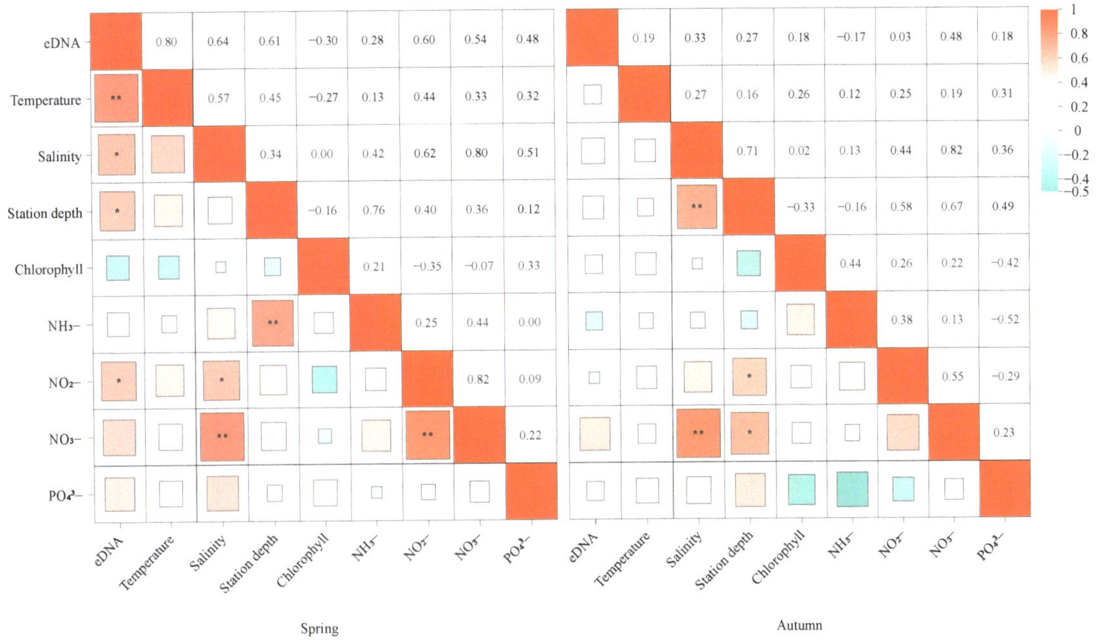

Figure 4. Correlation between *S. marmoratus* eDNA concentration and various environmental factors of *Sebastiscus marmoratus*. The significance level is marked with * (p-value < 0.1), ** (0.05 < p-value < 0.1).

Table 2. The filtered MLR model.

MLR Model	R^2	RMSE
Y = 460.16 − 19.10 × **Tem**	0.39	83.00
Y = 39.60 − 19.18 × **Tem** + 6.25 × **Dep**	0.52	73.36
Y = 314.88 − 18.52 × **Tem** + 5.39 × Dep + 2314.32 × PO_4^{3-}	0.55	71.56
Y = 324.86 − 19.19 × **Tem** + 4.32 × Dep + 2760.17 × NO_2^- + 2928.13 × PO_4^{3-}	0.55	70.85
Y = 476.70 − 28.26 × **Tem** − 6.25 × Chl + 279.08 × NH_3^- + 341.35 × NO_3^- + 4142.67 × PO_4^{3-}	0.58	68.64

Notes: eDNA concentration, temperature, salinity, sampling station depth, chlorophyll abbreviated as Y, Tem, Sal, Dep, and Chl. The independent variables with p-values < 0.05 were presented in bold.

4. Discussion

4.1. Distribution Characteristics of *S. marmoratus*

Monitoring eDNA concentrations can help predict the potential distribution of target species across the landscape, reducing workload and research costs and enabling more effective ecological monitoring [10,25]. In this study, eDNA of false kelpfish (*S. marmoratus*) was detected at all 12 sampling stations, suggesting its widespread distribution in the larger coastal waters near Gouqi Island. Previous research by Zeng Xu et al. [26] on the habitat of false kelpfish (*Sebastiscus marmoratus*) around Gouqi Island found substantial abundance near Gouqi Island's coast. Although their survey was limited to within 200 m offshore, their results also demonstrated the species' dominant status within the local community structure, consistent with findings from this study. The results also confirm the reliability of environmental DNA technology.

The survey results showed that *S. marmoratus* concentrations were higher in the spring season, which may be closely related to seasonal ecological changes. Spring is typically a season of vigorous reproduction for algae and plankton in marine ecosystems, providing richer food resources and thus leading higher abundance of *S. marmoratus* in spring. The spawning period of *S. marmoratus* is from December to April each year, and the survey time of this study was in April, so it is probable that the spawning period of *S. marmoratus* was coming to an end, and the biomass had reached a high level with fish eggs. Therefore, the high eDNA concentration in spring may be due to the increased biomass resulting from the spawning behavior of *S. marmoratus*. This conclusion indicates that eDNA surveys can infer fish spawning behavior by monitoring eDNA concentrations. Research by Hayer and Michael et al. [27,28] also confirms this conclusion. Additionally, in autumn, there is a trend for the concentration of *S. marmoratus* eDNA to move toward deeper offshore areas, similar to the findings of Changwen et al. [29], possibly due to a relative decrease in bait availability near Gouqi Island in autumn, intensifying food competition and leading *S. marmoratus* to move outward in search of more food sources. (The swimming ability of *S. marmoratus* juveniles is poor, and they are more susceptible to the influence of ocean currents, so the distribution of eDNA in spring may be larger than the actual range).

In addition to being influenced by their own habits, the distribution of marine fishery resources is also affected by abiotic factors such as physical and chemical factors. Therefore, environmental factors play a key role in fishery resource surveys [30]. According to the MLR model equation obtained, temperature is significantly correlated with *S. marmoratus* eDNA concentration. Water temperature is one of the main environmental factors affecting fish ecological habits, which not only affects the growth, development, reproduction, and metabolism of fish, but also influences the migratory distribution of fish, the timing of fishing periods, the duration of fishing seasons, the position of central fishing grounds, and the aggregation behavior of fish [31]. Water temperature can also indirectly affect the migratory distribution of fish by influencing the behavior, distribution, and abundance of bait organisms [32]. As the main carrier of phytoplankton photosynthesis, chlorophyll is related to oxygen content in seawater [33]. The content of chlorophyll can reflect the change in plankton abundance, and then reflect the feeding condition of *S. marmoratus* to some extent; however, plankton abundance may be affected by many other external factors, such as temperature or other low-trophic level fish predation [34,35]. NH_3 and NO_3 are important sources of nitrogen for aquatic plants. Many Aquatic animals obtain their nitrogen indirectly by feeding on plants or algae. NH_3 and NO_3 play a key role in the material cycle and energy flow of the aquatic food chain. In ecosystems, the availability of PO_4 is often an important factor limiting the growth of aquatic organisms, especially in water bodies with low phosphorus levels, where PO_4 supplementation may significantly promote growth. Levels of NH_3, NO_3, and PO_4 should not be too high, as they are also key factors in algal bloom [36].

4.2. Application of eDNA Surveys in Marine Fish: Potential, Limitations, and Prospects

This study reveals the spatial distribution and characteristics of *S. marmoratus* in the marine reserve of Gouqi Island, including its seasonal variability. The species is widely distributed near Gouqi Island's coast. Biomass is lower in autumn compared to spring, with a tendency for outward migration. These results can help fishery managers or commercial fisherpeople choose the right fishing time in spring and the right fishing location. In addition, a five-variable combination Edna concentration prediction model was established, which will help fisheries managers predict *S. marmoratus* abundance. For example, measurements of temperature, chlorophyll, NH_3, NO_3, and PO_4 will help identify high areas of *S. marmoratus* eDNA and potentially can help catch more *S. marmoratus* for research purposes or consumption. Also, MLR models suggest that even a simple measurement of temperature and water depth might be sufficient and that false kelpfish appear associated with cooler water temperatures at deeper water depths. Overall, this study helps measure aspects of *S. marmoratus* behavior and abundance that were difficult to achieve with traditional surveys, such as the distribution of *Sebastiscus marmoratus* in different water layers. Conventional surveys using trawl nets near the coast are limited by habitats and net height, making it difficult to monitor various water layers and most of the marine area, while also causing disturbances to the ecosystem. Therefore, in some cases, eDNA surveys have higher efficiency compared to traditional methods [37–39], such as biodiversity surveys, especially as the cost of genetic research has decreased in recent decades, making genetic-based resource surveys more popular [28]. The advantages of eDNA are evident and will be applied in more research.

However, there are still some issues in current research, particularly the inability to obtain actual survey data for *Sebastiscus marmoratus* due to the limitations of traditional survey methods. Therefore, it is not possible to accurately assess the correlation between eDNA concentration and the actual biomass of *Sebastiscus marmoratus*. Additionally, with eDNA technology, we cannot confirm whether the detected species are juveniles or adults, nor can we determine their survival status. Therefore, there is an urgent need for further innovation in eDNA technology to distinguish the sources of eDNA and reveal the basic dynamics of eDNA in the marine environment [40]. Moreover, standardized methods for field surveys and analysis need to be established. Considering the shortcomings in the duration and influencing factors of eDNA signals, eDNA research should be considered as a supplement rather than an independent tool, especially in cases involving marine flow environments [5,41]. Therefore, more research is needed to further improve our understanding of the parameters and processes that may affect eDNA distribution in the ocean.

In recent years, despite the decline in false kelpfish (*Sebastiscus marmoratus*) resources, certain achievements have been made through restocking and marine reserve construction. Considering the high fecundity of *Sebastiscus marmoratus*, their production has been maintained to a certain extent. The designation of marine reserves has been proven to play an important role in ecosystem maintenance. To better develop marine reserves, it is necessary to conduct more baseline assessments of ecosystems, such as community structure in mussel farming areas and biomass of various organisms. In these aspects, eDNA technology can play a crucial role.

5. Conclusions

This study utilized eDNA methods to investigate the distribution of *S. marmoratus* around the marine ranch near Gorgonian Island. The distribution of *S. marmoratus* exhibited marked seasonal variations. *S. marmoratus* showed significantly higher abundance in spring compared to autumn due to spawning occurring in the winter and spring. The study of environmental factors revealed a significant correlation between *S. marmoratus* distribution and temperature. eDNA has proven to be an effective tool for biological monitoring in marine ranches, highlighting the important roles of facilities such as artificial reefs and mussel farms. eDNA holds considerable potential for marine biological monitoring and could be increasingly applied to life history assessments of marine species.

Author Contributions: R.J., H.H. and F.C.: methodology, formal analysis, and writing—original draft preparation. R.Y., P.Z., Y.Z. and X.C.: software and data curation. F.C., Y.Z. and R.J.: writing—review and editing, visualization, supervision, and project administration. All authors have read and agreed to the published version of the manuscript.

Funding: This study was supported by the National Key R & D Program of China (2023YFD2401901).

Institutional Review Board Statement: Not applicable.

Informed Consent Statement: Not applicable.

Data Availability Statement: The datasets that support the findings of this study are available from the corresponding author upon reasonable request.

Acknowledgments: The authors express their gratitude to the staff of the Zhejiang Marine Fisheries Research Institute and Key Laboratory of Sustainable Utilization Research for Fishery Resources of Zhejiang Province for help with conducting the experiments.

Conflicts of Interest: The authors declare no conflicts of interest.

References

1. Valdez, J.W.; Callaghan, C.T.; Junker, J.; Purvis, A.; Hill, S.L.L.; Pereira, H.M. The undetectability of global biodiversity trends using local species richness. *Ecography* **2023**, *2023*, e06604. [CrossRef]
2. Rullens, V.; Stephenson, F.; Lohrer, A.M.; Townsend, M.; Pilditch, C.A. Combined species occurrence and density predictions to improve marine spatial management. *Ocean Coast. Manag.* **2021**, *209*, 105697. [CrossRef]
3. Itakura, H.; Wakiya, R.; Yamamoto, S.; Kaifu, K.; Sato, T.; Minamoto, T. Environmental DNA analysis reveals the spatialdistribution, abundance, and biomass of Japaneseeels at the river-basin scale. *Aquat. Conserv. Mar. Freshw. Ecosyst.* **2019**, *29*, 331–520. [CrossRef]
4. Yuan, X.; Jiang, Y.; Zhang, H.; Jin, Y.; Ling, J. Quantitative assessment of fish assemblages on artificial reefs using acoustic and conventional netting methods, in Xiangshan Bay, Zhejiang Province, China. *J. Appl. Ichthyol.* **2021**, *37*, 389–399. [CrossRef]
5. Yamamoto, S.; Minami, K.; Fukaya, K.; Takahashi, K.; Sawada, H.; Murakami, H.; Tsuji, S.; Hashizume, H.; Kubonaga, S.; Horiuchi, T.; et al. Environmental DNA as a 'snapshot' of fish distribution: A case study of Japanese jack mackerel in Maizuru Bay, Sea of Japan. *PLoS ONE* **2016**, *11*, e0149786.
6. Doi, H.; Inui, R.; Akamatsu, Y.; Kanno, K.; Yamanaka, H.; Takahara, T.; Minamoto, T. Environmental DNA analysis for estimating the abundance and biomass of stream fish. *Freshw. Biol.* **2016**, *62*, 30–39. [CrossRef]
7. Dougherty, M.M.; Larson, E.R.; Renshaw, M.A.; Gantz, C.A.; Egan, S.P.; Erickson, D.M.; Lodge, D.M. Environmental DNA (eDNA) detects the invasive rusty crayfish Orconectes rusticus at low abundances. *J. Appl. Ecol.* **2016**, *53*, 722–732. [CrossRef]
8. Minamoto, T.; Fukuda, M.; Katsuhara, K.R.; Fujiwara, A. Environmental DNA reflects spatial and temporal jellyfish distribution. *PLoS ONE* **2017**, *12*, e0173073. [CrossRef]
9. Pilliod, D.S.; Goldberg, C.S.; Arkle, R.S.; Waits, L.P. Estimating occupancy and abundance of stream amphibians using environmental DNA from filtered water samples. *Can. J. Fish. Aquat. Sci.* **2013**, *70*, 1123–1130. [CrossRef]
10. Buxton, A.S.; Groombridge, J.J.; Zakaria, N.B.; Griffiths, R.A. Seasonal variation in environmental DNA in relation to population size and environmental factors. *Sci. Rep* **2017**, *7*, 46294. [CrossRef]
11. Darling, J.A.; Mahon, A.R. From molecules to management: Adopting DNA-based methods for monitoring biological invasions in aquatic environments. *Environ. Res.* **2011**, *111*, 978–988. [CrossRef] [PubMed]
12. Jerde, C.L.; Mahon, A.R.; Chadderton, W.L.; Lodge, D.M. "Sight-unseen" detection of rare aquatic species using environmental DNA. *Conserv Lett* **2017**, *4*, 150–157. [CrossRef]
13. Feist, S.M.; Jones, R.L.; Copley, J.L.; Pearson, L.S.; Berry, G.A.; Qualls, C.P. Development and validation of an environmental DNA method for detection of alligator snapping turtle (*Macrochelys temminckii*). *Chelonian Conserv. Biol.* **2018**, *17*, 271–279. [CrossRef]
14. Xu, N.; Xiong, M.H.; Shao, K.; Que, Y.F.; Li, J.Y. Preliminary study on environmental DNA metabarcoding for detecting biodiversity in the middle and lower reaches of the Yangtze River. *Res. Environ. Sci.* **2020**, *33*, 1187–1196.
15. Xu, M.Z.; Yang, Y.; Zhang, J.H.; Fu, X.D. Application Challenges of Environmental DNA Technology in Monitoring the Invasion of Bivalves. *Acta Ecol. Sin.* **2023**, *43*, 4423–4433.
16. Wang, X.Y.; Zhang, H.B.; Lu, G.Q.; Gao, T.X. Detection of an invasive species through an environmental DNA approach: The example of the red drum *Sciaenops ocellatus* in the East China Sea. *Sci. Total Environ.* **2022**, *815*, 152865. [CrossRef]
17. Gao, T.; Chen, Z.; Wang, X. A New Method for Surveying Fish Diversity in Coastal Areas: Environmental DNA Analysis Technology. *J. Zhejiang Ocean Univ. (Nat. Sci. Ed.)* **2018**, *37*, 1–7.
18. Zhao, S.L.; Xu, H.X.; Zhong, J.S. *Fauna of Marine Fishes in Zhejiang*; Zhejiang Science and Technology Press: Hangzhou, China, 2016.
19. Gao, T.-X.; Chen, Z.; Wang, X.-Y.; Zhang, H.-B.; Shi, H.-L. Discussion on the Correlation between Density of *Sebastes schlegelii* Cultivation and its eDNA. *J. Hydrobiol.* **2022**, *46*, 1007–1015.
20. Pei, W.H.; Hu, C.Y.; Zhang, X.M.; Chen, Y.Y.; Liu, Y.D. Study on the Community Structure of Macrobenthos in the Mussel Breeding Area of Gouqi Island. *Oceanol. Et Limnol. Sin.* **2023**, *54*, 526–536.
21. Li, F.; Xu, M.; Liu, Q.; Wang, Z.F.; Xu, W.J. Ecological restoration zoning for a marine protected area: A case study of Haizhouwan National Marine Park, China. *Sci. Total Environ.* **2014**, *98*, 158–166. [CrossRef]
22. Bates, D.; Maechler, M.; Bolker, B.; Walker, S. Fitting linear mixed-effects models using lme4. *J. Stat. Softw.* **2015**, *67*, 1–48. [CrossRef]
23. Kolde, R. Pheatmap: Pretty Heatmaps. 2019. Available online: https://cran.r-project.org/web/packages/pheatmap/pheatmap.pdf (accessed on 10 April 2024).
24. Wickham, H. *ggplot2: Elegant Graphics for Data Analysis*; Springer: New York, NY, USA, 2016.
25. Erickson, R.A.; Rees, C.B.; Coulter, A.A.; Merkes, C.M.; McCalla, S.G.; Touzinsky, K.F.; Walleser, L.; Goforth, R.R.; Amberg, J.J. Detecting the Movement and Spawning Activity of Bigheaded Carps with Environmental DNA. *Mol. Ecol. Resour.* **2016**, *16*, 957–965. [CrossRef] [PubMed]
26. Zeng, X.; Zhang, S.; Wang, Z.; Lin, J.; Wang, K. Habitat Suitability Assessment for *Sebasticus marmoratus* in the Saddle Islands. *Acta Ecol. Sin.* **2016**, *36*, 3765–3774.
27. Hayer, C.-A.; Bayless, M.F.; George, A.E.; Thompson, N.; Richter, C.A.; Chapman, D. Use of Environmental DNA to Detect Grass Carp Spawning Events. *Fishes* **2020**, *5*, 27. [CrossRef]
28. Tillotson, M.D.; Kelly, R.P.; Duda, J.J.; Hoy, M.; Kralj, J.; Quinn, T.P. Concentrations of environmental DNA (eDNA) reflect spawning salmon abundance at fine spatial and temporal scales. *Biol. Conserv.* **2018**, *220*, 1–11. [CrossRef]

29. Wu, C.W. Biological Research on *Sebastiscus marmoratus* in the Nearshore Waters of Zhoushan, Zhejiang Province. *J. Zhejiang Ocean. Univ. (Nat. Sci. Ed.)* **1999**, *3*, 185–190+226.
30. Mukherjee, S.; Chaudhuri, A.; Kundu, N.; Mitra, S.; Homechaudhuri, S. Comprehensive analysis of fish assemblages in relation to seasonal environmental variables in an estuarine river of Indian sundarbans. *Estuaries Coasts* **2012**, *36*, 192–202. [CrossRef]
31. Yin, M.L. *Fish Ecology*; China Agricultural Press: Beijing, China, 1995.
32. Li, X.D. The Relationship between Seawater Temperature and Fishing Grounds. *Acta Oceanol. Sin.* **1982**, *1*, 103–113. (In Chinese)
33. Belkin, I.M.; O'Reilly, J.E. An algorithm for oceanic front detection in chlorophyll and SST satellite imagery. *J. Mar. Syst.* **2009**, *78*, 319–326. [CrossRef]
34. Andersson, A.; Haecky, P.; Hagström, A. Effect of temperature and light on the growth of micro-nano-and pico-plankton: Impact on algal succession. *Mar. Biol.* **1994**, *120*, 511–520. [CrossRef]
35. Mazumder, A.; Taylor, W.D.; McQueen, D.J.; Lean, D.R. Effects of fish and plankton on lake temperature and mixing depth. *Science* **1990**, *247*, 312–315. [CrossRef] [PubMed]
36. Qin, B.Q. Approaches to Mechanisms and Control of Eutrophication of Shallow Lakes i n the Middle and Lower Reaches of the Yangze River. *J. Lake Sci.* **2002**, *14*, 193–202.
37. Ji, Y.; Ashton, L.; Pedley, S.M.; Edwards, D.P.; Tang, Y.; Nakamura, A.; Kitching, R.; Dolman, P.M.; Woodcock, P.; Edwards, F.A.; et al. Reliable, verifiable and efficient monitoring of biodiversity via meta barcoding. *Ecol. Lett.* **2013**, *16*, 1245–1257. [CrossRef] [PubMed]
38. Valentini, A.; Taberlet, P.; Miaud, C.; Civade, R.; Herder, J.; Thomsen, P.F.; Bellemain, E.; Besnard, A.; Coissac, E.; Boyer, F.; et al. Next-generation monitoring of aquatic biodiversity using environmental DNA metabarcoding. *Mol. Ecol.* **2016**, *25*, 929–942. [CrossRef] [PubMed]
39. Kelly, R.P.; O'donnell, J.L.; Lowell, N.C.; Shelton, A.O.; Samhouri, J.F.; Hennessey, S.M.; Feist, B.E.; Williams, G.D. Genetic signatures of ecological diversity along an urbanization gradient. *PeerJ* **2016**, *4*, e2444. [CrossRef]
40. Barnes, M.A.; Turner, C.R. The ecology of environmental DNA and implications for conservation genetics. *Conserv Genet* **2016**, *17*, 1–17. [CrossRef]
41. Murakami, H.; Yoon, S.; Kasai, A.; Minamoto, T.; Yamamoto, S.; Sakata, M.K.; Horiuchi, T.; Sawada, H.; Kondoh, M.; Yamashita, Y.; et al. Correction to: Dispersion and degradation of environmental DNA from caged fish in a marine environment. *Fish Sci.* **2019**, *85*, 1109. [CrossRef]

fishes

MDPI

Article

Seasonal Distribution of Key Small-Sized Fish in the South Inshore of Zhejiang, China

Minghao Xu [1], Xiaodi Gao [2], Weicheng Liu [3,4,*] and Jiaqi Wang [5]

[1] Faculty of Arts and Social Science, The University of Sydney, Sydney, NSW 2050, Australia; xuminghao0520@outlook.com
[2] East China Sea Fisheries Research Institute, Chinese Academy of Fishery Sciences, Shanghai 200090, China
[3] Zhejiang Mariculture Research Institute, Wenzhou 325005, China
[4] Zhejiang Key Laboratory of Coastal Biological Germplasm Resources Conservation and Utilization, Wenzhou 325005, China
[5] College of Marine Living Resource Sciences and Management, Shanghai Ocean University, Shanghai 201306, China
* Correspondence: lwch80@126.com

Abstract: Small-sized fish are a vital food source for large predatory commercial fish and play a key role in marine food webs, bridging lower and higher trophic levels. They are indispensable in maintaining the energy flow and material cycling within aquatic ecosystems. This study utilized bottom-trawl survey data from 2017 to 2020 along the south inshore of Zhejiang, China, complemented by concurrent environmental data, to examine the influence of environmental factors on the resource density and seasonal distribution patterns of four dominant small-sized fish species. The research findings indicated that SSH (sea surface height) and Chl (chlorophyll-a concentration) emerged as the key environmental factors influencing resource densities, with all four species exhibiting similar preferences toward these variables. However, significant disparities were observed in their preferences for SST (sea surface temperature), SSS (sea surface salinity), and DO (dissolved oxygen). The various species' resource density and distribution patterns underwent significant seasonal variations. Additionally, the seasons and regions with the highest resource densities consistently aligned, occurring predominantly in autumn within the northern waters of the study area. This research further elucidated the environmental predilections and seasonal spatial distribution traits of small-sized fish in the south inshore of Zhejiang, an important feeding ground for economic fish species in the East China Sea. This provides scientific backing for forecasting alterations in coastal fishery resources under environmental and climate change scenarios and supports ecosystem-based fisheries management strategies.

Keywords: small-sized fish; habitat preference; seasonal variation; the south inshore of Zhejiang

Key Contribution: The small-sized fish resources found in the southern coastal region of Zhejiang, China, exhibited notable seasonal fluctuations in their distribution patterns and converged in the northern waters, attaining their peak biomass during autumn. Primary productivity stood as the pivotal factor influencing both the proliferation and distribution of these vital fish resources.

Citation: Xu, M.; Gao, X.; Liu, W.; Wang, J. Seasonal Distribution of Key Small-Sized Fish in the South Inshore of Zhejiang, China. *Fishes* **2024**, *9*, 412. https://doi.org/10.3390/fishes9100412

Academic Editor: Susana Franca

Received: 5 September 2024
Revised: 8 October 2024
Accepted: 12 October 2024
Published: 13 October 2024

1. Introduction

Small-sized fish or forage fish play a pivotal role in the marine food chain, serving as essential links in the effective transfer of matter and energy from lower to higher trophic levels within marine ecosystems. This transfer process is crucial for maintaining the structure and function of marine ecosystems [1–3]. By examining the habitat distribution of these species, we can gain profound insights into the interactions among different organisms in the ocean ecosystem. Moreover, the dynamic changes in their population distribution are critical for devising effective area-based marine management strategies.

Additionally, considering the variations induced by climate and other environmental factors, understanding species' environmental preferences and spatiotemporal distribution patterns can help predict the potential impact of future changes on ecosystems [4].

That traditional economic fish resources in China's coastal waters are in decline due to environmental pollution, overfishing, climate change, and interspecific interactions appears to be a consensus among ecologists, management professionals, and those working in the fishing industry [5]. In response, researchers have initiated systematic and protracted surveys of fishery resources. With a wealth of data gathered and analyzed, these experts have delved into discussions and examinations of the anthropogenic and environmental factors contributing to resource fluctuations and the remedial actions required, from an ecosystem, community, and population level. A majority of studies suggest that the structure of coastal fish communities continues to evolve in an unfavorable direction, characterized primarily by the miniaturization of fish species and shifts in the trophic structure of fish communities [6–11]. The status of key economic species remains grim. Conversely, small fish species, characterized by their early maturation, rapid growth, and high reproductive capacity, are increasingly becoming dominant in these ecosystems [12,13].

Located in the warm temperate zone, the south inshore waters of Zhejiang are under the perennial influence of coastal currents on the western nearshore side, while the eastern side is affected by the Taiwan Warm Current [14]. The varying strength of the Taiwan Warm Current significantly impacts this maritime region. Numerous studies have examined the fish community and population levels within this region [15]. Existing researches confirm a continuing decline in the resources of the area's economic species; small-sized fish have emerged as the dominant species here [12,13]. However, the majority of attention has been on the resource status of principal economic species, their environmental preferences, and spatiotemporal distribution characteristics, with few researchers delving into studying the small-sized fish in these waters.

Thus, this research capitalized on trawl survey data collected from the south inshore waters of Zhejiang, China, integrating environmental variables, to devise distribution models for four pivotal small-sized fish species: *Amblychaeturichthys hexanema* (AMH), *Apogon lineatus* (also known as *Jaydia lineata*, APL), *Benthosema pterotum* (BEP), and *Bregmaceros mcclellandi* (BRM). These species were selected due to their prevalence as small baitfish and their pivotal roles within the intricate food web [12,13]. This study delved deeply into the environmental preferences of these species in the region, scrutinizing the spatiotemporal patterns of their resource distribution and interspecific variations, ultimately identifying crucial bait areas. Such invaluable insights will not only fortify scientific forecasting capabilities but also provide a solid foundation for the development of adaptive marine management strategies.

2. Materials and Methods

2.1. Data Collection

The data for this study were collected through bottom-trawling surveys conducted in four distinct seasons, spring (May), summer (August), autumn (November), and winter (January or February), spanning the years 2017 to 2020, along the south inshore of Zhejiang, China (Figure 1). The selection of sampling stations adhered to a predetermined fixed pattern. A total of 478 stations were investigated during the surveying periods. At each sampling location, trawling operations were executed at a consistent tow speed ranging between 2 and 4 knots for 30 min. The dimensions of the trawl mouth were standardized to a width of 40 m and a height of 7.5 m. Species were meticulously identified, and the catch was quantified. To ensure uniformity in data representation, the resource density of each species was standardized to g/3 nmi (grams per three nautical miles). All sampling procedures and biological measurements complied with the standards prescribed by the Specifications for Oceanographic Survey, ensuring reliability and consistency in the data collection process [16].

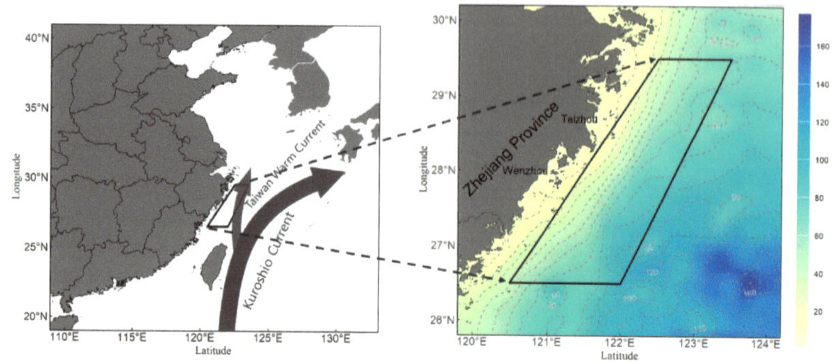

Figure 1. The trawl survey area for fishery resources in the south inshore of Zhejiang, China. The fill color within the right plot signifies the bathymetry, measured in meters (m), and the dashed line within it represents the bathymetric contour.

The environmental data utilized in this study were obtained from the Copernicus Marine Environment Monitoring Service (CMEMS, https://marine.copernicus.eu/, accessed on 8 July 2024). Salinity, temperature, and dissolved oxygen are universally acknowledged as significant factors influencing fish physiology, serving as primary catalysts for fish migration. The concentration of chlorophyll-a, directly or indirectly, mirrors the level of primary productivity in aquatic environments. These variables are commonly employed in species distribution models as explanatory factors [17–19]. The data encompass sea surface temperature (SST, °C), sea surface salinity (SSS), sea surface height (SSH, m), chlorophyll-a concentration (Chl, $mg \cdot m^{-3}$), and dissolved oxygen (DO, $mmol \cdot m^{-3}$), all at a spatial resolution of $0.25^{\circ} \times 0.25^{\circ}$ and a temporal resolution of monthly intervals.

2.2. Species Distribution Model

A boosted regression tree (BRT) melds two algorithms from the realms of statistics and machine learning: a classification and regression tree (CART) and boosting. This fusion propels the construction of numerous simple decision tree models to elevate predictive prowess. An advantage of utilizing BRTs lies in their capacity to efficiently manage the correlation and collinearity effects among environmental variables, thereby eliminating the necessity for preliminary evaluations of predictor variables [20]. Renowned for its superior forecasting abilities, BRT has garnered substantial endorsement and application within the ecological research literature [21]. In numerous studies, BRT has outperformed regression-based models, such as generalized linear models (GLMs) and generalized additive models (GAMs), in analyzing complex species–habitat relationships [20,22,23]. In this study, we utilized the resource density of key small-sized fish species as response variables and environmental factors as predictor variables to develop a BRT model.

To achieve an optimal BRT model, it is essential to fine-tune the learning rate (lr), tree complexity (tc), and bag fraction. The learning rate, also known as the shrinkage parameter, plays a crucial role in preventing overfitting. A practical guideline for setting the learning rate is to ensure that it enables the BRT model to produce at least 1000 trees. Tree complexity denotes the number of nodes in each decision tree, whereas the bag fraction represents the proportion of data utilized for model construction at each step, typically ranging between 0.5 and 0.75 [20]. In this research, learning rates were set to 0.001, 0.005, and 0.01; tree complexities were configured to 2, 4, 6, 8, and 12; and the bag fraction was fixed at 0.75. A ten-fold cross-validation approach was employed to evaluate model robustness and to identify the parameter combination that results in the lowest average estimation bias while generating more than 1000 decision trees, thereby establishing the optimal model.

The potentially significant difference in the resource density of each species among seasons and areas was determined by a Wilcoxon test ($\alpha = 0.05$).

All analyses were carried out in the R-4.3.1 software [24].

3. Results

3.1. Model Performance

The optimal parameters for each stock's model are detailed in Table 1. The findings revealed that the ideal number of trees varied from 1050 to 2950, and the deviance explained by models ranged from 25.6% for BRM to 59.5% for APL. These results underscored that the models adeptly captured the association between catch rates and environmental variables for the majority of populations within the examined spatiotemporal context (Table 1).

Table 1. Optimal parameter tuning for BRT models and the variance explained by these models.

Species	Code	Number of Trees	Tree Complexity (tc)	Learning Rate (lr)	Bagging Fraction	Explained Variance%
Amblychaeturichthys hexanema	AMH	1250	12	0.001	0.75	40.0
Apogon lineatus (or *Jaydia lineata*)	APL	2550	12	0.001	0.75	59.5
Benthosema pterotum	BEP	2150	12	0.001	0.75	52.3
Bregmaceros mcclellandi	BRM	1500	6	0.001	0.75	25.7

The model results indicated that the impact of environmental factors on the resource distribution of different fish varies significantly. The relative contribution rate of environmental factors to the model's explanatory power revealed that SSH was the most significant influencer of catch rates, with an average relative contribution rate of 29.1%, followed by Chl at 23.4%, SST at 17.4%, SSS at 16.5%, and DO at 13.7% (Figure 2).

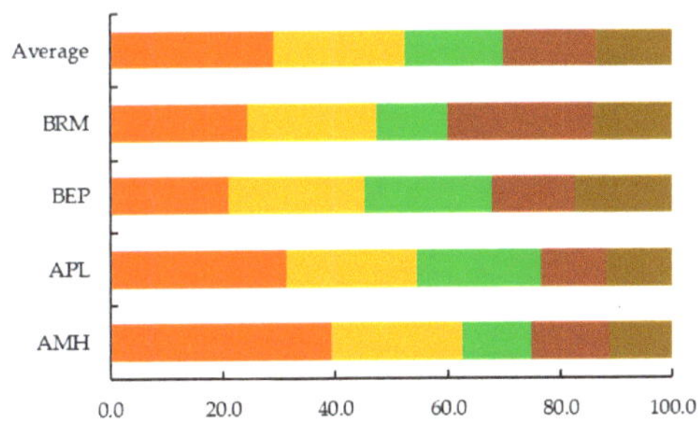

	AMH	APL	BEP	BRM	Average
SSH	39.5	31.5	21.0	24.4	29.1
Chl	23.1	22.9	24.3	23.1	23.4
SST	12.2	22.1	22.7	12.5	17.4
SSS	13.9	11.5	14.6	25.9	16.5
DO	11.3	12.0	17.5	14.1	13.7

Figure 2. The relative contribution rate of environmental factors to the model's explanatory power. The *y*-axis represents species, with AMH referring to *Amblychaeturichthys hexanema*, APL to *Apogon lineatus* (or *Jaydia lineata*), BEP to *Benthosema pterotum*, and BRM standing for *Bregmaceros mcclellandi*. SST, SSS, SSH, Chl, and DO are abbreviations for specific environmental factors, which represent sea surface temperature, sea surface salinity, sea surface height, chlorophyll-a concentration, and oxygen concentration, respectively.

3.2. Environmental Preference

The relationship between covariates and response variables is illustrated in Figure 3. The distribution of AMH was predominantly influenced by SSH and Chl, followed by SSS; SST, and DO. AMH tended to concentrate more in waters with elevated SSH (>0.15 m), higher concentrations of Chl (>1.2 mg·m^{-3}), and comparatively lower SSS, while resource density diminished when SST exceeded 28 °C. In the case of APL, the primary environmental factors shaping its resource distribution were SSH, Chl, and SST, succeeded by

DO and SSS. The environmental characteristics of regions with high APL density were as follows: SSH was elevated (>0.15 m), Chl concentration hovered around 0.5 mg·m^{-3} or 1.2 mg·m^{-3}, SST fell within the range of 16–19 °C, while DO (<240 mmol·m^{-3}) and SSS (<28) remained relatively low. As for BEP, Chl, SST, and SSH stood as the paramount environmental determinants of its resource distribution, with each factor making a nearly equal contribution to the model's explanatory power, succeeded by DO and SSS. Areas rich in BEP resources were marked by Chl > 0.5 mg·m^{-3}, SST spanning 19~25 °C, SSH fluctuating between 0 and 0.2 m, and relatively enhanced DO (>250 mmol·m^{-3}) and SSS (>32). Regarding BRM, the principal environmental influencers of its resource distribution were SSS, SSH, and Chl, followed by DO and SST. Waters boasting increased BRM resource density were distinguished by elevated SSS (>32), SSH oscillating between 0.15 and 0.20 m, heightened Chl (>1.2 mg·m^{-3}), and relatively reduced DO (<240 mmol·m^{-3}) and SST (<25 °C) (Figure 3).

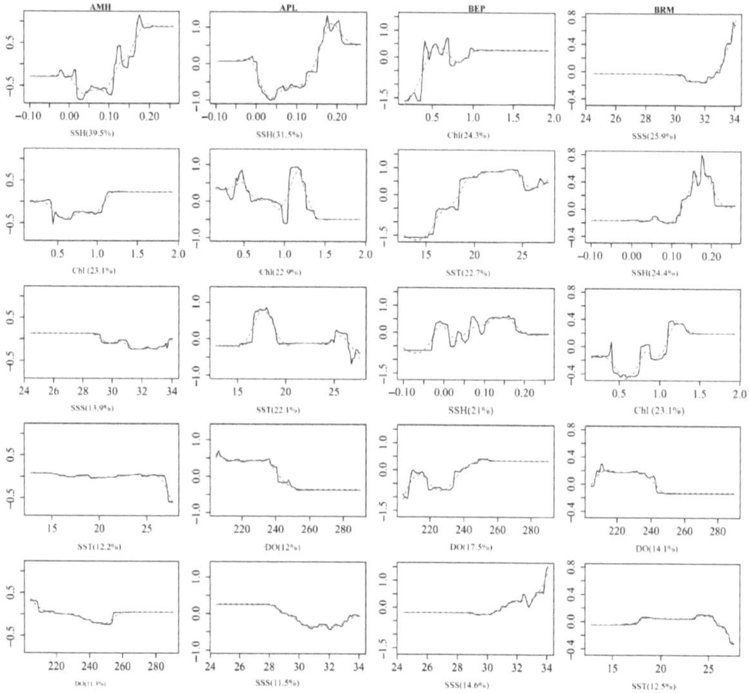

Figure 3. Environmental factor effects on the relative resource density of the four small-sized fish in the south inshore of Zhejiang, China. The solid black lines depict the index's response curve in relation to environmental changes, whereas the dashed red lines serve as a smoothed rendition of the former.

3.3. Spatiotemporal Distribution

The spatial and temporal distributions of the resource density for the four small-sized fish species are presented in Figure 4. For AMH, the peak resource density was observed in autumn, followed by winter, summer, and spring. During autumn, the regions with relatively high resource density for AMH were located in the northern waters (zones a and b), in summer and winter mainly in zone a, and in spring in zone c. In the case of APL, the highest resource density was in autumn, followed by spring, winter, and summer. For autumn, the areas with relatively high resource density for APL were situated in zone a, in spring and winter in zone b, and in summer, the distribution was more uniform without significant high-density regions. Regarding BEP, the season with the highest resource density was autumn, followed by summer, spring, and winter. In autumn, the areas with relatively high resource density for BEP were found in zone b and a small portion of zone a, in summer in zone c, and in spring and winter in zone d. As for BRM, the resource density reaches its zenith in autumn, followed by winter, spring, and summer. In autumn, the regions with relatively high resource density for BRM were in zones b and d, with a minor portion in zone c, in winter and spring primarily in zone d, and in summer the distribution was notably uniform. In summary, the four small-sized fish examined in this study were predominantly distributed in the northern part of the south inshore of Zhejiang, China, exhibiting the highest resource density in autumn (Figures 4 and 5).

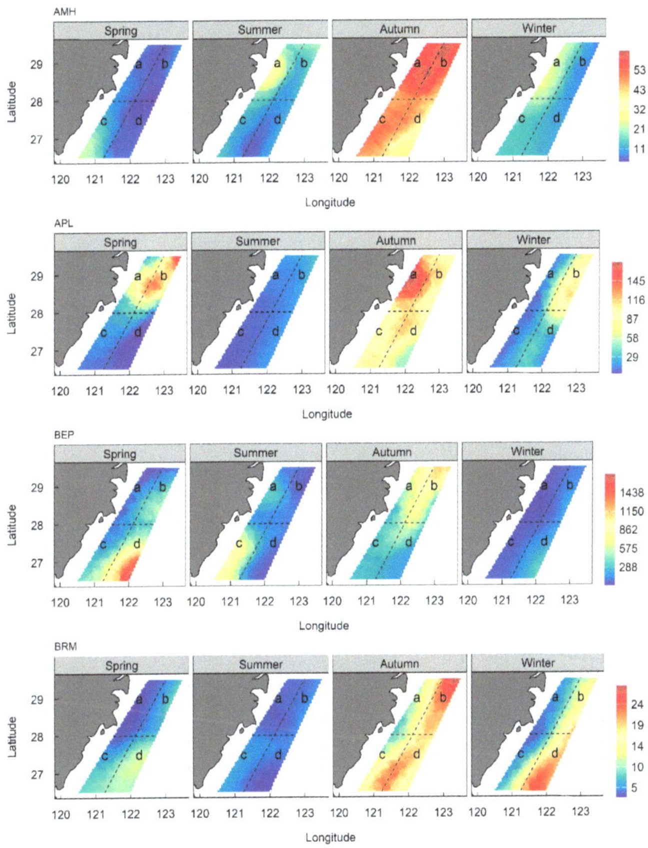

Figure 4. The seasonal distribution of the four small-sized fish in the south inshore of Zhejiang, China. The hues depicted in the plot signify the average relative resource density spanning from 2017 to 2020, quantified in units of grams per three nautical miles (g/3 nmi). The dashed lines in the figure represent the regional boundary. The north–south boundary is delineated by the 28° north latitude line, while the east–west boundary is defined by the line connecting the points (26.5° N, 121.25° E) and (29.5° N, 122.75° E). In the figure, a, b, c, and d, respectively, represent the northwest, northeast, southwest, and southeast waters of the study area.

(A)

(B)

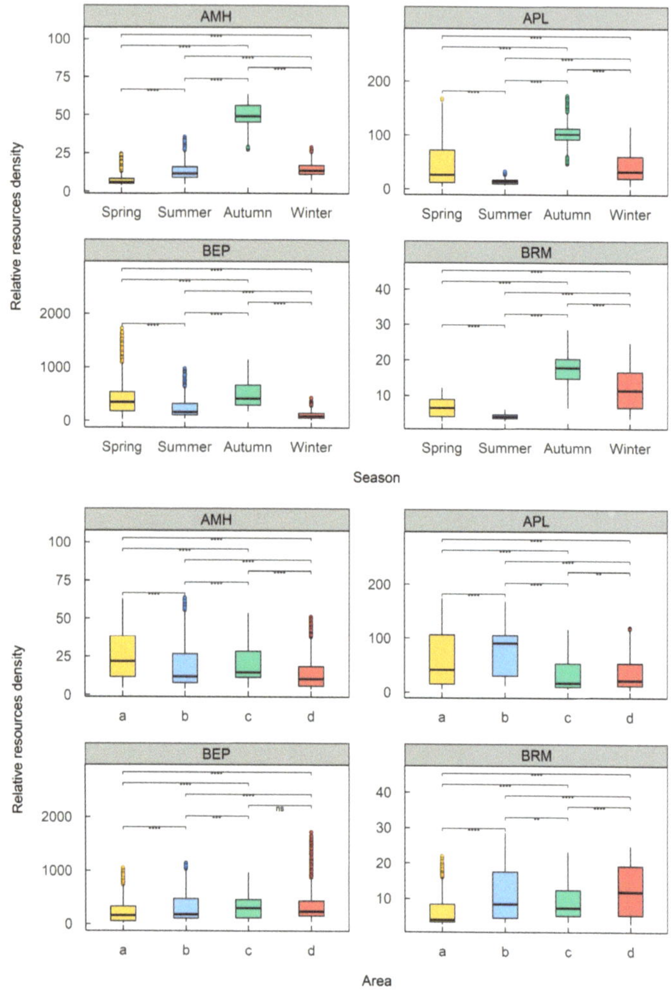

Figure 5. Seasonal (**A**) and regional (**B**) differences in resource density of the four small-sized fish in the south inshore of Zhejiang, China (ns: $p > 0.05$, **: $p <= 0.01$, ***: $p <= 0.001$, ****: $p <= 0.0001$).

4. Discussion

While numerous studies have underscored substantial changes in the fish community structure of the research area, characterized by the rising predominance of small-sized fish, our comprehension of these fish species' habitat preferences remains profoundly limited. This study harnessed trawl survey data from the south inshore of Zhejiang, in conjunction with crucial environmental factors, to scrutinize the environmental preferences and spatiotemporal variation aspects of the key small-sized fish inhabiting these important foraging waters. The findings suggested that, despite variations in the environmental preferences among different species, SSH and Chl overwhelmingly influenced the resources and distribution of these fish species. The northern portion of the research area boasted more abundant resources compared to its southern counterpart, with resource densities peaking in autumn.

4.1. Dominant Environmental Variables

Primary productivity is a key factor influencing the resource density and individual size of fish larvae or small-sized fish that feed on plankton [25–27]. In the Northwest Pacific, changes in temperature and the intensity of upwelling can delay phytoplankton reproduction and subsequently affect zooplankton size [28]. These changes directly impact the recruitment of small pelagic fish, thereby influencing fish population sizes [29]. Since the mid-2000s, significant shifts in environmental conditions in the Mediterranean Lion Bay have affected the Chl, SST, upwelling, and frontal activity, which may be important contributors to the decline in resource density and size of small pelagic fish in that region [27]. In this study, SSH and Chl were identified as key environmental factors, revealing a high similarity in their utilization among the four small-sized fish species. Areas with high resource density exhibited relatively high levels of both SSH and Chl. SSH is generally associated with heat flux, wind, and eddy currents, which influence the transport of marine materials and indirectly indicate levels of primary productivity [25]. Conversely, Chl directly reflects phytoplankton biomass and the level of primary productivity in the ocean, subsequently affecting the food sources for zooplankton and herbivorous fish [25,26]. This highlights that the richness of primary productivity was also a critical determinant of small forage fish resource density in the south inshore waters of Zhejiang.

Salinity, temperature, and dissolved oxygen are well known to significantly influence fish physiology and serve as key drivers of fish migration [17]. In our analysis, the four small-sized fish species exhibited notable differences in their utilization of SST, SSS, and DO. For instance, AMH and APL preferred lower SSS environments, while the high resource densities of BEP and BRM were found in areas with higher SSS. With the exception of BRP, the resource densities of the other three species decreased to varying extents at elevated SST (>25 °C). These factors may be the primary reasons behind the observed differences in the temporal and spatial distribution of these small forage fish species.

4.2. Differences in Spatial and Temporal Distribution

4.2.1. *Amblychaeturichthys hexanema* (AMH)

AMH is widely distributed in the coastal waters of the northwest Pacific. In the broader East China Sea region, AMH exhibits relatively low dominance and is not considered a dominant small-sized fish species [30]. However, in the waters around Dachen, AMH has been observed to dominate across all seasons except summer, indicating its consistent presence in that area [31]. Additionally, it showed a relatively high dominance in the south inshore of Zhejiang, with a trend of increasing prevalence year by year [13]. This study also revealed that its resource density centroid was primarily located in nearshore waters. These findings further affirmed that AMH is a coastal or nearshore fish species, suggesting that the south inshore of Zhejiang may be one of its key habitats.

Beyond SSH and Chl, SSS and SST are significant factors influencing AMH distribution. Research has indicated that when salinity exceeded 30, there was a marked decline in the relative resource density of AMH in Haizhou Bay [19], which aligned with the results of this study. Excessively high salinity may disrupt AMH's osmoregulation and metabolic functions, affecting the normal life processes of the fish and leading to a decrease in resource density. The optimal water temperature for AMH in Haizhou Bay ranged between 19 and 20 °C; when temperatures exceeded 19.8 °C, a significant decline in resource density occurred [32]. In contrast, the suitable SST range for AMH in the studied area was broader. The relatively stable SST conditions in the south inshore of Zhejiang may be a key factor contributing to the widespread distribution of this species in that region.

4.2.2. *Apogon lineatus* (or *Jaydia lineata*, APL)

Unlike AMH, APL is one of the main small fish species in the East China Sea, alongside BEP and *Champsodon capensis* [30], predominantly found in the southeastern waters of the East China Sea [33]. APL has a limited seasonal movement range, engaging in only short-distance migrations, with summer and autumn being their primary spawning season. It is a seasonally dominant species in the nearshore waters of eastern Zhoushan [34], central Zhejiang [35], and south Zhejiang [13]. This study found that during winter, APL was primarily distributed in the northeastern waters of the research area, with a trend shifting the resource centroid northward in spring. In summer, resource density in this region was at its lowest, likely because APL primarily spawns in the Yangtze River estuary to the north of the study area [19]. Following spawning, adults migrate toward the nutrient-rich nearshore waters of south Zhejiang in search of food.

In addition to the interspecific differences in environmental preferences highlighted by this study, significant variations in the environmental preferences were also observed among different waters and different life stages of APL. For instance, in Haizhou Bay during autumn, the optimal SSS for slim-flanked bream was between 31.4 and 32.0, with water temperatures ranging from 22.8 to 27.0 °C [36]. In contrast, APL in the studied area seemed to be better adapted to relatively cooler (16–19 °C) and lower-salinity (<28) waters. On the other hand, the eggs of APL in the East China Sea exhibited varying temperature preferences throughout the months; the SST in June was between 18.89 and 19.00 °C, while in July, they were primarily found in waters with an SST ranging from 23.89 to 29.60 °C [19]. This SST preference was higher than that observed for adult APL in this study. The salinity preferences for the eggs and adults of APL were found to be similar.

4.2.3. *Benthosema pterotum* (BEP)

Previous studies have found that the highest resource density of BEP occurs in autumn, primarily in the offshore waters of the northern East China Sea (30.5°–32.5° N, 124°–126.5° E), with some scattered distribution in the nearshore areas of south Zhejiang (27°–28° N, 122°–123° E) [37]. It tended to short-distance migration in response to seasonal changes; the center of distribution for the northern East China Sea population shifts short distances from southwest to northeast and from northeast to southwest between summer and autumn [37]. In this study, BEP also displayed the characteristic of having the highest resource density in autumn, with the center of this resource density located in the northern part of the study area, aligning with the observed short-distance movement of the northern population during the transition from summer to autumn. The suitable SST for the nearshore population of south Zhejiang was found to be between 24 and 25 °C [37], along with a salinity range of 31–33. During June and July, the optimal temperature and salinity for BEP larvae in the East China Sea were identified as 18–26 °C and 30–34, respectively [18], which correlates with the favorable SST and SSS characteristics for adult BEP identified in this research.

In contrast to the findings of Li et al. (2006) [37], this study discovered that BEP was widely distributed in the south inshore of Zhejiang. This discrepancy may result from differences in the fishing efficiency of the survey gear used for this species or may be attributed to environmental changes influencing BEP distribution. This suggests that, over the long term, there may not have been significant changes in the environmental adaptability of BEP. However, as nearshore environmental conditions have changed, its spatial and temporal distribution characteristics throughout the East China Sea may have undergone considerable alteration.

4.2.4. *Bregmaceros mcclellandi* (BRM)

BRM is primarily distributed in tropical and subtropical waters, mainly found in the southeastern regions of the East China Sea in China, where it is a seasonal dominant species during autumn and winter. It is rarely encountered in coastal waters [34,35], but it does appear as a seasonal dominant species in the offshore areas of central [38] and southern [13] Zhejiang, extending to the northern Taiwan Strait [39]. Consistent with previous studies, this research indicated that the resource density distribution centroids of the three species, excluding BRM, shifted between nearshore areas (zones a and d) and offshore waters (zones b and c) with changing seasons (Figure 4). In contrast, BRM is consistently found in more distant offshore waters.

4.3. Area-Based Marine Management Consideration

Area-based management tools (ABMTs) are regarded as pivotal in achieving marine conservation objectives. A crucial aspect of this strategy involves pinpointing key areas [40]. The south inshore of Zhejiang, China, holds immense significance as a breeding ground, offering a wealth of nourishment for fish and acting as a vital locale for the maturation and expansion of numerous species [15]. Safeguarding these areas ensures a consistent supply of fishery resources. Our study has enriched and clarified the spatial distribution and seasonal shifts of essential small-sized fish resources within this territory. The findings revealed that the northern segment of Zhejiang's southern inshore harbored the densest concentration of bait resources, with seasonal fluctuations underscoring the need for temporally adaptable ABMTs aimed at preserving small-sized fish resources to bolster management efficacy.

Moreover, small-sized fish, in contrast to large commercial species, are more acutely influenced by environmental alterations [29], ultimately reverberating throughout the ecosystem via the food chain. Consequently, tracking the dynamic variations in small-sized fish resources within key areas can foster insight into the mechanisms driving fluctuations in commercial fish stocks and serve as a vital predictor of the impact of climatic and environmental changes on marine living resources [41]. It further constitutes a fundamental cornerstone for devising appropriate response strategies.

5. Conclusions

This study enhances our understanding of the environmental preferences and spatial and temporal distributions of small-sized fish in the south inshore of Zhejiang, China. Overall, primary productivity level was the core factor influencing the resource density and distribution of small forage fish species in this region. Although different species exhibited varying trends in resource density with increasing primary productivity, there was an overall upward trend, with the highest resource densities found in areas abundant in primary productivity. Additionally, differences in adaptability to SST, SSS, and DO may be the primary factors causing variations in the temporal and spatial distribution of these fish species. Notably, the coastal waters near Taizhou in autumn (the northern part of the study area) were the most abundant in small-sized fish resources. In the intricate interplay of numerous influencing factors, preserving the ecological service functions of nearshore ecosystems and fostering the sustainable exploitation of fishery resources have emerged as pivotal concerns in contemporary marine management. The invaluable insights garnered from this study, about the environmental preferences and spatiotemporal distribution patterns of crucial small fish species, offer crucial guidance for pinpointing key nearshore regions and facilitating informed conservation decisions pertaining to fishery resources.

Author Contributions: Conceptualization, M.X., X.G. and W.L.; methodology, M.X.; formal analysis, M.X. and J.W.; data curation, W.L.; writing—original draft preparation, M.X.; writing—review and editing, X.G., W.L. and J.W.; funding acquisition, W.L. All authors have read and agreed to the published version of the manuscript.

Funding: This research was funded by the Wenzhou Estuary Fishery Resources Conservation Pilot Special Investigation and Study (JS2022071) and the Evaluation of Fishery Resources and Environmental Effects of Important Fishery Waters in South Zhejiang Province and Study on Enhancement Capacity (2024YS005).

Institutional Review Board Statement: Not applicable.

Informed Consent Statement: Not applicable.

Data Availability Statement: The data presented in this study are available on reasonable request from the corresponding author.

Acknowledgments: We would like to thank the teachers and students from the laboratory of Shanghai Ocean University for their work and help in sample collection and data analysis and valuable comments on the revision of the paper.

Conflicts of Interest: The authors declare no conflicts of interest.

References

1. Konar, M.; Qiu, S.; Tougher, B.; Vause, J.; Tlusty, M.; Fitzsimmons, K.; Barrows, R.; Cao, L. Illustrating the Hidden Economic, Social and Ecological Values of Global Forage Fish Resources. *Resour. Conserv. Recycl.* **2019**, *151*, 104456. [CrossRef]
2. Nissar, S.; Bakhtiyar, Y.; Arafat, M.Y.; Andrabi, S.; Bhat, A.A.; Yousuf, T. A Review of the Ecosystem Services Provided by the Marine Forage Fish. *Hydrobiologia* **2023**, *850*, 2871–2902. [CrossRef]
3. Pikitch, E.K.; Rountos, K.J.; Essington, T.E.; Santora, C.; Pauly, D.; Watson, R.; Sumaila, U.R.; Boersma, P.D.; Boyd, I.L.; Conover, D.O. The Global Contribution of Forage Fish to Marine Fisheries and Ecosystems. *Fish Fish.* **2014**, *15*, 43–64. [CrossRef]
4. Hilborn, R.; Amoroso, R.O.; Bogazzi, E.; Jensen, O.P.; Parma, A.M.; Szuwalski, C.; Walters, C.J. When Does Fishing Forage Species Affect Their Predators? *Fish. Res.* **2017**, *191*, 211–221. [CrossRef]
5. Halpern, B.S.; Walbridge, S.; Selkoe, K.A.; Kappel, C.V.; Micheli, F.; D'Agrosa, C.; Bruno, J.F.; Casey, K.S.; Ebert, C.; Fox, H.E.; et al. A Global Map of Human Impact on Marine Ecosystems. *Science* **2008**, *319*, 948–952. [CrossRef]
6. Cheng, J.; Ding, F.; Li, S.F.; Yan, L.; Li, J.; Liang, Z. Changes of Fish Community Structure in the Coastal Zone of the Northern Part of East China Sea in Summer. *J. Nat. Resour.* **2006**, *21*, 775–781.
7. Jin, X. Long-Term Changes in Fish Community Structure in the Bohai Sea, China. *Estuar. Coast. Shelf Sci.* **2004**, *59*, 163–171. [CrossRef]
8. Jiang, Y.Z.; Cheng, J.H.; Li, S.F. Temporal Changes in the Fish Community Resulting from a Summer Fishing Moratorium in the Northern East China Sea. *Mar. Ecol. Prog. Ser.* **2009**, *387*, 265–273. [CrossRef]
9. Zhang, K.; Guo, J.; Xu, Y.; Jiang, Y.; Fan, J.; Xu, S.; Chen, Z. Long-Term Variations in Fish Community Structure under Multiple Stressors in a Semi-Closed Marine Ecosystem in the South China Sea. *Sci. Total Environ.* **2020**, *745*, 140892. [CrossRef]
10. Zhai, L.; Pauly, D. Construction and Interpretation of Particle Size Distribution Spectra from 19 Ecopath Models of Chinese Coastal Ecosystems. *Front. Mar. Sci.* **2020**, *7*, 298. [CrossRef]
11. Li, J.L.; Cao, K.; Ding, F.; Yang, W.B.; Shen, G.M.; Li, Y. Changes in Trophic-Level Structure of the Main Fish Species Caught by China and Their Relationship with Fishing Method. *J. Fish. Sci. China* **2017**, *7*, 109–119. [CrossRef]
12. Wang, Y.; Kindong, R.; Gao, C.; Wang, J. Identification of Keystone Species in Ecological Communities in the East China Sea. *Fishes* **2023**, *8*, 224. [CrossRef]
13. Wang, J.; Gao, C.; Tian, S.; Han, D.; Ma, J.; Dai, L.; Ye, S. Shifts in Composition and Co-Occurrence Patterns of the Fish Community in the South Inshore of Zhejiang, China. *Glob. Ecol. Conserv.* **2023**, *44*, e02502. [CrossRef]
14. Su, J.L. Circulation Dynamics of the China Seas North 18°N. *Glob. Coast. Ocean. Reg. Stud. Synth.* **1998**, *11*, 483–506.
15. Zheng, Y.; Chen, X.; Cheng, J.; Wang, Y.; Shen, X.; Chen, W.; Li, C. *East China Sea Shelf Environment and Living Resources*; Shanghai Scientific & Technical Publishers: Shanghai, China, 2003.
16. *GB/T 12763.6-2007*; Specifications for Oceanographic Survey—Part 6: Marine Biological Survey. China National Standardization Management Committee: Beijing, China, 2007.
17. FRY, F.E.J. The Effect of Environmental Factors on the Physiology of Fish. In *Fish Physiology*; Elsevier: Amsterdam, The Netherlands, 1971; Volume 6, pp. 1–98, ISBN 1546-5098.
18. Zhang, H.; Yuan, X.; Ling, J.; Jiang, Y. Larval Ecology of Gobiid Fishes in a Subtropical Embayment: Environmental Preferences and Spatiotemporal Habitat Partitioning. *Front. Mar. Sci.* **2022**, *9*, 929919. [CrossRef]
19. Xu, M.; Liu, Z.; Wang, Y.; Jin, Y.; Yuan, X.; Zhang, H.; Song, X.; Otaki, T.; Yang, L.; Cheng, J. Larval Spatiotemporal Distribution of Six Fish Species: Implications for Sustainable Fisheries Management in the East China Sea. *Sustainability* **2022**, *14*, 14826. [CrossRef]
20. Elith, J.; Leathwick, J.R.; Hastie, T. A Working Guide to Boosted Regression Trees. *J. Anim. Ecol.* **2008**, *77*, 802–813. [CrossRef]
21. Yu, H.; Cooper, A.R.; Infante, D.M. Improving Species Distribution Model Predictive Accuracy Using Species Abundance: Application with Boosted Regression Trees. *Ecol. Model.* **2020**, *432*, 109202. [CrossRef]
22. Shabani, F.; Kumar, L.; Ahmadi, M. A Comparison of Absolute Performance of Different Correlative and Mechanistic Species Distribution Models in an Independent Area. *Ecol. Evol.* **2016**, *6*, 5973–5986. [CrossRef]
23. Norberg, A.; Abrego, N.; Blanchet, F.G.; Adler, F.R.; Anderson, B.J.; Anttila, J.; Araújo, M.B.; Dallas, T.; Dunson, D.; Elith, J. A Comprehensive Evaluation of Predictive Performance of 33 Species Distribution Models at Species and Community Levels. *Ecol. Monogr.* **2019**, *89*, e01370. [CrossRef]
24. Team, R.C.; Team, M.R.C.; Suggests, M.; Matrix, S. *Package Stats*; The R-4.3.1 Stats Package: Vienna, Austria, 2018.
25. Ayers, J.M.; Lozier, M.S. Physical Controls on the Seasonal Migration of the North Pacific Transition Zone Chlorophyll Front. *J. Geophys. Res. Oceans* **2010**, *115*, C05001. [CrossRef]
26. Rykaczewski, R.R. Changes in Mesozooplankton Size Structure along a Trophic Gradient in the California Current Ecosystem and Implications for Small Pelagic Fish. *Mar. Ecol. Prog. Ser.* **2019**, *617*, 165–182. [CrossRef]
27. Feuilloley, G.; Fromentin, J.-M.; Stemmann, L.; Demarcq, H.; Estournel, C.; Saraux, C. Concomitant Changes in the Environment and Small Pelagic Fish Community of the Gulf of Lions. *Prog. Oceanogr.* **2020**, *186*, 102375. [CrossRef]
28. Kudela, R.M.; Cochlan, W.P.; Peterson, T.D.; Trick, C.G. Impacts on Phytoplankton Biomass and Productivity in the Pacific Northwest during the Warm Ocean Conditions of 2005. *Geophys. Res. Lett.* **2006**, *33*. [CrossRef]

29. Cushing, D.H. Plankton Production and Year-Class Strength in Fish Populations: An Update of the Match/Mismatch Hypothesis. In *Advances in Marine Biology*; Elsevier: Amsterdam, The Netherlands, 1990; Volume 26, pp. 249–293, ISBN 0065-2881.
30. Lin, L.; Zheng, Y.; Liu, Y.; Zhang, H. The Ecological Study of Small Sized Fish in the East China Sea I-the Species Composition and Seasonal Variation of Small Sized Fish. *Mar. Sci.* **2006**, *30*, 58–63. (In Chinese)
31. Zou, Q.; Wang, Z.; Zhang, S.; Cheng, X.; Chen, Y.; Zhou, Y. Characteristics of Fish Community Structure in the Sea Area of Dachen Island. *Chin. J. Ecol.* **2024**, *43*, 795–803. (In Chinese)
32. Shen, D.; Zhang, Y.; Cui, Y.; Yu, H.; Zhang, C.; Xu, B.; Zhang, C.; Ji, Y.; Xue, Y. Study on the Influencing Factors of Fish Spatial Distribution Using Three Bayesian Models: A Case Study of *Amblychaeturichthys hexanema* in Haizhou Bay. *Haiyang Xuebao* **2023**, *45*, 88–100. (In Chinese) [CrossRef]
33. Jiang, S.; Chen, J. Research Progress on Small-Scale Marine Fishes. *Mar. Fish.* **2006**, *28*, 336–341. (In Chinese)
34. Liu, K.; Yu, N.; Yu, C.; Zheng, J.; Xu, Y.; Yan, W.; Han, L.; Liu, H.; Sun, B.; Dai, D. The Spatial Niche and Differentiation of Major Fish Species in the Waters East of the Zhoushan Islands in Spring And Autumn. *J. Fish. Sci. China* **2021**, *28*, 100–111. (In Chinese) [CrossRef]
35. He, J.-N.; Xu, Y.-J.; Zhang, H.L. Study on the Relationship between Fish Diversity, Quantity and Environmental Factors in the Middle Area of Zhejiang Offshore. *J. Zhejiang Ocean. Univ. (Nat. Sci.)* **2017**, *36*, 283–288. (In Chinese)
36. Tang, L.; Zhang, Y.; Xu, B.; Zhang, C.; Ren, Y.; Xue, Y. Hbitat Suitability Analysis of *Apogon lineatus* during Autumn in Haizhou Bay. *Peridical Ocean. Univ. China* **2021**, *51*, 154–160. (In Chinese)
37. Li, J.; Hu, F.; Li, S.; Cheng, J. Quantity Distribution of Benthosema Pterotum and in Relationship with Surface Layer Water Temperature and Salinity in the East China Sea Region. *Mar. Fish.* **2006**, *28*, 105–110. (In Chinese)
38. Lu, Z.H.; Miao, Z.Q.; Lin, N. The Structure and Diversity Fish Communities in Spring in the Middle Sea Area of Zhejiang Province and Adjacent Region. *J. Zhejiang Ocean. Univ. (Nat. Sci.)* **2009**, *28*, 51–56. (In Chinese)
39. Liu, Z.L.; Yang, L.L.; Yan, L.P.; Yuan, X.W.; Chen, J. Fish Assemblages and Environmental Interpretation in the Northern Taiwan Strait and Its Adjacent Waters in Summer. *J. Fish. Sci. China* **2016**, *23*, 1399–1416. (In Chinese)
40. Hilborn, R.; Agostini, V.N.; Chaloupka, M.; Garcia, S.M.; Gerber, L.R.; Gilman, E.; Hanich, Q.; Himes-Cornell, A.; Hobday, A.J.; Itano, D.; et al. Area-Based Management of Blue Water Fisheries: Current Knowledge and Research Needs. *Fish Fish.* **2021**, *23*, 492–518. [CrossRef]
41. Peck, M.A.; Alheit, J.; Bertrand, A.; Catalán, I.A.; Garrido, S.; Moyano, M.; Rykaczewski, R.R.; Takasuka, A.; van Der Lingen, C.D. Small Pelagic Fish in the New Millennium: A Bottom-up View of Global Research Effort. *Prog. Oceanogr.* **2021**, *191*, 102494. [CrossRef]

fishes

Article

Construction and Comparison of Machine-Learning Forecast Models of Albacore *Thunnus alalunga* Fishing Grounds in the South Pacific Ocean

Jianxiong Li [1,2], Feng Chen [1,2,3,*], Qian Dai [2,3], Wenbin Zhu [2,3], Dewei Li [2,3], Wei Yu [4] and Weifeng Zhou [5]

1 Marine and Fishery Institute, Zhejiang Ocean University, Zhoushan 316021, China; lijianxiong@zjou.edu.cn
2 Zhejiang Marine Fisheries Research Institute, Zhoushan 316021, China; daiqian@zjou.edu.cn (Q.D.); wenbinzhu@zjou.edu.cn (W.Z.); lidewei@zjou.edu.cn (D.L.)
3 Key Laboratory of Sustainable Utilization of Technology Research for Fishery Resource of Zhejiang Province, Zhoushan 316021, China
4 College of Marine Biological Resources and Management, Shanghai Ocean University, Shanghai 201306, China; wyu@shou.edu.cn
5 East China Sea Fisheries Research Institute, Chinese Academy of Fishery Sciences, Shanghai 200090, China; zhouwf@ecsf.ac.cn
* Correspondence: fchen@zjou.edu.cn

Abstract: The traditional methods for predicting the distribution of albacore (*Thunnus alalunga*) fishing grounds have low performance and accuracy. Uneven sampling can result in unreasonable evaluation indicators. To address these issues, three methods, equi-frequency, K-means clustering algorithm, and 1-R split, were applied to discretize the catch per unit effort (CPUE) of albacore in the South Pacific from 2016 to 2021 and partition the fishing grounds into abundance levels. Eight machine learning models were used to predict the fishing grounds. In addition to the traditional evaluation index based on confusion matrix, top-k index was also used to evaluate the accuracy of fishery abundance predictions. The results showed that (1) When sampling is unbalanced, the reported accuracy does not fully represent the actual performance of the model in predicting the abundance of albacore in the fishing ground. *F*1 value can be used as the index of the model effect and stability. (2) In binary classification, the quartile stacking algorithm has the best stacking performance, with *F*1 0.89. (3) The top-1 prediction accuracy of three-category fishery forecasting is the highest at 0.74, and the top-1 prediction accuracy of five-category fishery forecasting is the highest at 0.54. (4) The top-k accuracy of classification of fisheries with multiple abundance using K-means is significantly better than that of equal frequency discretization ($p < 0.001$). The top-k evaluation index was used to predict the fishing grounds of albacore across multiple abundance levels for the first time in this study, which is significant for pioneering a new method for this application and which provides a demonstration of the development of artificial intelligence techniques for fisheries in the future.

Keywords: albacore; fishing ground prediction; unsupervised learning; machine learning; discretization; south pacific

Key Contribution: This paper compared and analyzed the influence of different discrete methods on the prediction of fishing grounds by several machine learning methods. Additionally, a top-k index was first used to describe the fault tolerance rate of different discrete methods for albacore fishing ground forecasts. These methods' applications provide a more abundant prediction method for tuna fishery forecasts based on artificial intelligence in the future.

Citation: Li, J.; Chen, F.; Dai, Q.; Zhu, W.; Li, D.; Yu, W.; Zhou, W. Construction and Comparison of Machine-Learning Forecast Models of Albacore *Thunnus alalunga* Fishing Grounds in the South Pacific Ocean. *Fishes* **2024**, *9*, 375. https://doi.org/10.3390/fishes9100375

Academic Editor: Susana Franca

Received: 10 August 2024
Revised: 13 September 2024
Accepted: 17 September 2024
Published: 25 September 2024

1. Introduction

The albacore (*Thunnus alalunga*) is a pelagic fish that inhabits open ocean waters in tropical and temperate regions worldwide, including the Atlantic, Pacific, and Indian

Oceans [1]. As a top predator in the marine food chain, albacore play an important role in marine ecology and are also an important economic target in fisheries [2]. As a highly migratory species, the movements and habits of albacore also provide valuable insights into the complex dynamics of open marine ecosystems [3]. However, the impacts of climate change, overfishing, and bycatch on albacore populations highlight the need for increased measures to protect this valuable species and its widespread habitat [4]. Therefore, understanding how to establish a stable and accurate model for predicting the spatiotemporal distribution of albacore through remote sensing technology has become an urgent problem.

The distribution of albacore is influenced by various environmental factors, and previous studies typically included various marine environmental variables as indicators of the spatiotemporal distribution of tuna [5]. Vertical seawater temperature (ST) directly affects the spatial and vertical distribution of albacore [6]. Chlorophyll-a concentration (CHL) has an indirect effect on the distribution of albacore [7] and sea surface height anomaly (SSHA) may also affect their distribution [8]. The salinity of seawater is considered to play an important role in modeling the distribution of albacore tuna [9]. In addition, sea surface wind speed (WS) may cause changes in mixed layer depth (MLD), which in turn could result in changes in the vertical distribution of albacore. Eddy kinetic energy (EKE) causes changes in other marine environments, indirectly affecting the population of albacore [10]. Large scale climate events such as El Niño and La Niña can have an impact on the migration and distribution of albacore [4,11], as seen in the observed correlation between the Southern Oscillation Index and the fishing of albacore tuna. In addition, research on the life cycle of albacore tuna has shown that the Pacific intergenerational oscillation index affects the sensitivity of mid- to upper-sea-level fish [12].

In the era of big data, artificial intelligence algorithms are rapidly developing and maturing. In recent years, researchers have begun using machine learning algorithms, and especially supervised learning algorithms, to construct prediction models for fishing grounds [13,14]. Early artificial neural networks were used to predict albacore fishing grounds in the South Pacific [15], but were limited to algorithms to explore the relationship between environmental factors and catch per unit effort (CPUE). Moreover, due to the easy overfitting of neural networks, the low accuracy of actual deployment models requires further improvement in their robustness and accuracy. Ensemble learning algorithms, such as random forest [16] in the Indian Ocean and boosted regression trees [17] in the Tonga EZZ have been applied to the prediction of tuna fisheries. STK (stacking) models are unique in their prediction accuracy around fisheries, and their performance for fishing ground forecasting was better than other single models and integrated learning [18].

In previous studies on predicting tuna fishing grounds, the traditional classification method of high and low fishing grounds was the third percentile method [16]. This method takes the first point of the third percentile of CPUE as the dividing line and divides the fishing grounds into low-yield fishing grounds and high-yield fishing grounds through CPUE [19]. However, this method cannot avoid the problem of imbalanced sampling [20]. An imbalance in the proportion of majority and minority classes can cause a model to bias predictions in favor of the majority class, resulting in high accuracy but often limiting recall and accuracy [21]. Although imbalanced sampling can sometimes be addressed with data augmentation methods such as oversampling, undersampling, and reinforcement learning based on generative adversarial networks, the lack of clear boundaries in the multi-dimensional feature space of the abundance level of fishing grounds prevents the accuracy of the model from being improved solely through simple feature engineering [22].

This study hypothesizes that a new method of dividing the abundance levels of fishing grounds will improve the predictive performance of the model (Figure 1). Three different methods of dividing fishing grounds, equal frequency division, cluster division, and 1-R division, were used to classify fishing ground abundance prediction as a binary or multi-class classification problem. Eight machine learning methods, including advanced ensemble learning, were applied to the dataset, and the top-k evaluation index was used as a novel metric for evaluating the predictions of fishing ground abundance levels. The relationship between the accuracy of the model's prediction of albacore fishing ground abundance and the division method was explored, and the concept of model fault tolerance was added to determine how unsupervised learning can guide the classification of fishing ground levels to improve the power of the model's prediction capabilities.

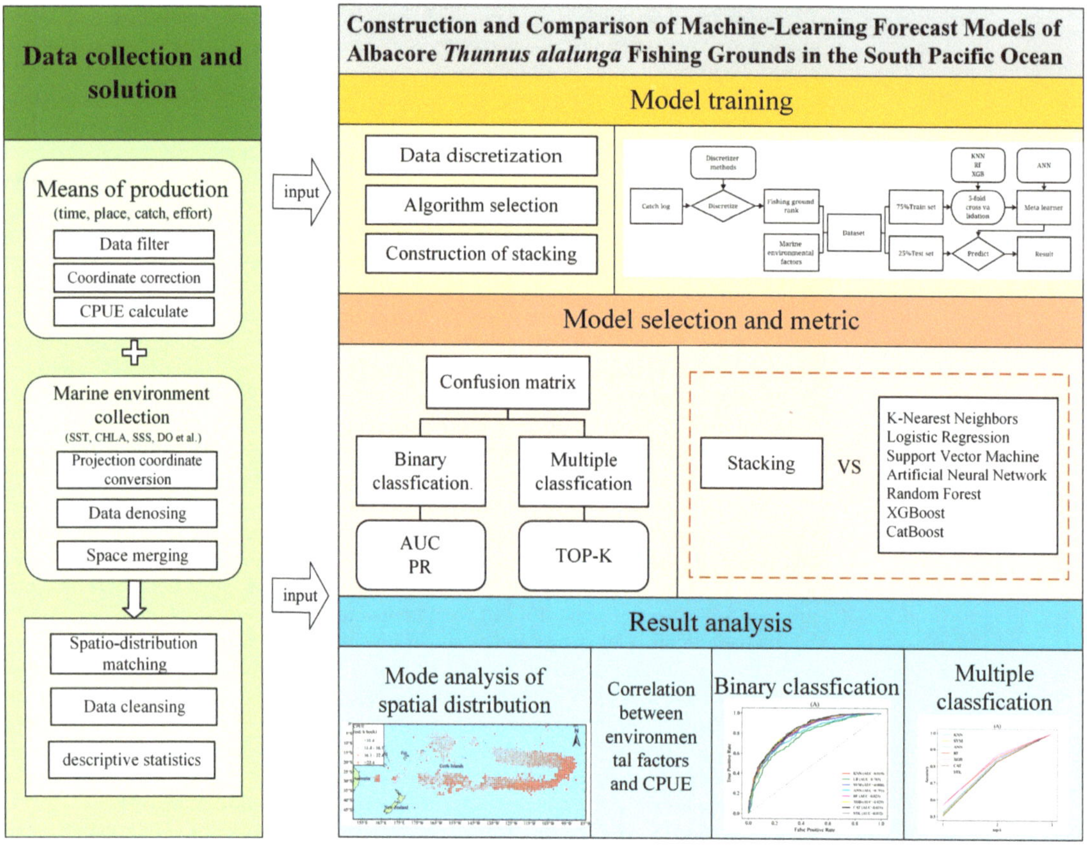

Figure 1. Flowcharts of research on comparing the models of albacore fishing grounds in the South Pacific Ocean.

2. Materials and Methods

2.1. Data Sources

This study used tuna production data from a total of 34 longline vessels provided by Pingtairong Marine Fisheries Group Co., Ltd. (Zhoushan, China) and Jiangsu Yuanyou Marine Fisheries Co., Ltd. (Qidong, China) (Figure 2). The operating ocean area was the South Pacific (0° S–40° S, 160° E–90° W). A total of 27,126 actual fishing records were matched to operational fishing records to a grid of 1° × 1° by month, yielding a total of 4621 samples.

The following 17 marine environmental factors were collected to build the fishing ground prediction models: sea surface temperature (SST); vertical sea surface temperature at a depth of 50–200 m (abbreviated as ST50, ST100, ST200); sea surface salinity (SSS); vertical salinity at a depth of 50–200 m (abbreviated as SS50, SS100, SS200); chlorophyll-a concentration (CHL); sea surface dissolved oxygen concentration (DO); sea surface height anomalies (SSHA); mixed layer depth (MLD); eddy kinetic energy (EKE); Southern

Oscillation Index (SOI); photosynthetically active radiation (PAR); wind speed (WS); and Pacific Intergenerational Oscillation Index (PDO). SST, SSS, SOI, and PAR were downloaded from database of National Oceanic and Atmospheric Administration (https://coastwatch.pfeg.noaa.gov (accessed on 20 August 2023)). ST50, ST100, ST200, SS50, SS100, SS200, CHL, DO, SSHA, MLD, EKE, PDO, and WS were downloaded from the Copernicus marine environmental monitoring website (https://data.marine.copernicus.eu (accessed on 20 August 2023)). The temporal resolution of all environmental factors was monthly. The spatial resolution of SST and PAR is 0.04° × 0.04°, and the spatial resolution of all other environmental data was 0.25° × 0.25°. All environmental spatial data were downscaled in Python 3.8 to match the 1° × 1° resolution of the longline fishing data.

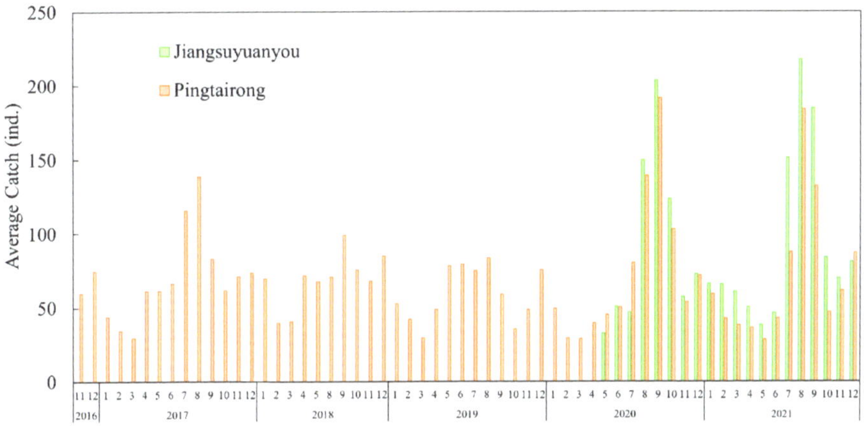

Figure 2. Monthly average catch of Pingtairong Marine Fisheries Group Co., Ltd. and Jiangsu Yuanyou Marine Fisheries Co., Ltd.

2.2. Data Processing

The catch per unit effort (CPUE) was calculated for each 1° × 1° grid cell monthly using the formula [23]:

$$CPUE_{y,m,i,\ j} = \frac{Catch_{y,m,i,j}}{Hook_{y,m,i,j}} \times 1000 \quad (1)$$

where, $Catch_{y,m,i,j}$ refers to the count of individuals (ind.) of albacore caught in the unit grid of longitude *i* and latitude *j* in year *y* and month *m*. $Hook_{y,m,i,j}$ is the number of hooks in the unit grid of longitude *i* and latitude *j* in year *y* and month *m*.

Three different partitioning methods were used to discretize CPUE: equal frequency partitioning, K-means-clustering partitioning [24], and 1-R discretization [25]. Equal-frequency partitioning can achieve segmentation boundaries of the third quartile, fourth quartile, and median, and can divide fishing grounds into binary low- and high-abundance areas. The K-means-clustering discretization method divides a continuous variable into several discrete intervals based on the inherent distance and similarity of the data. The 1-R discretization method is a decision tree-based discretization method that divides a continuous variable into several discrete values according to the construction method of the decision tree. Low- and high-abundance fishing grounds were thus represented as "0"and "1" respectively.

2.3. Feature Engineering

This study used Pearson correlation tests to analyze the relationships between environmental factors, screen out highly correlated explanatory variables, and identify correlations between the spatiotemporal and environmental factors and CPUE.

The calculation method was as follows:

$$\rho_{X,Y} = \frac{E[(X - \mu_X)(Y - \mu_Y)]}{\sigma_X \sigma_Y} \quad (2)$$

X and Y refer to two sets of variables, respectively, μ and σ refer to expected and standard deviation, respectively.

In order to offset the impact between different scales, data were normalized using the following formula:

$$X' = \frac{X - X_{min}}{X_{max} - X_{min}} \quad (3)$$

where, X', X, X_{min}, and X_{max} refer to the normalized values, actual values, minimum values, and maximum values of the variables, respectively.

This study used a large number of environmental factors as variables to predict albacore fishing ground abundances. Principal Component Analyses (PCAs) were used to analyze the included environmental variables to eliminate data noise and display differences in samples from fishing areas across multiple abundances.

2.4. Modeling Methods

Eight machine learning methods (Table 1) were tested to build prediction models of albacore fishing grounds. The 5-fold cross-validation was used for the selection of model hyperparameters, while main parameters were set in Table 1 for the algorithms.

Table 1. Introduction and parameters of machine learning models in the research.

Model	Introduction	Parameters
K-Nearest Neighbors (KNN) (reproduced from Springer, 2020) [26]	The KNN classification algorithm is an algorithm that only considers the K training set data closest to the new data.	K = 9, weight: distance K is the amount of nearest neighbors.
Logistic Regression (LR) (reproduced from American Heart Association, 2008) [27]	LR is a classic classification algorithm widely used in fields such as life sciences, social sciences, and finance.	penalty coefficient = 1.2, max_iterate = 100.
Support Vector Machine (SVM) (reproduced from Springer, 1998) [28]	SVM is a powerful classification algorithm which finds an optimal line or surface (also known as a hyperplane), divides the data into two categories, and maximizes the spacing between the two categories.	penalty coefficient = 1
Artificial Neural Network (ANN) (reproduced from *Fish. Res*, 2008) [29]	ANN is a model widely used in fields such as image, audio, and natural language processing.	HiddenLayerSizes = 20, learning rate: adaptive
Random Forest (RF) (reproduced from *Mach. Learn*, 2001) [30]	Random Forest is an ensemble learning algorithm based on decision trees.	n_estimators = 300
Extreme Gradient Boosting (XGB) (reproduced from CRAN, 2015) [31]	XGBoost is an efficient implementation of gradient lifting trees that combines various techniques such as tree pruning and feature subsampling.	n_estimators = 200, learning rate = 0.05
CatBoost (CAT) (reproduced from *Ad. Neural. Infor. Process. Syst.*, 2018) [32]	CatBoost is a machine learning algorithm based on gradient lifting methods that supports tasks such as classification, regression, and sorting.	n_estimators = 200, learning rate = 0.05
Stacking (STK) (reproduced from *Mach.Learn*, 2004) [33]	Stacking is an integrated learning technique that combines different models to achieve higher accuracy in the final prediction.	Estimators = {KNN, RF, XGB}, meta_learner = ANN, cv = 5

The STK model combines multiple base learners to form a learner with a stronger generalization ability. The base learner is suitable for strong and stable models such as KNN, RF and XGB. The meta-learner is suitable for models with simple structure and strong generalization ability. ANN has a simple structure and can effectively process the nonlinear part of the data with strong generalization ability. KNN, RF, and XGB were applied as base learners, ANN as meta learners, and the system uses 5-fold cross validation for training. The algorithm flowchart is shown in Figure 3.

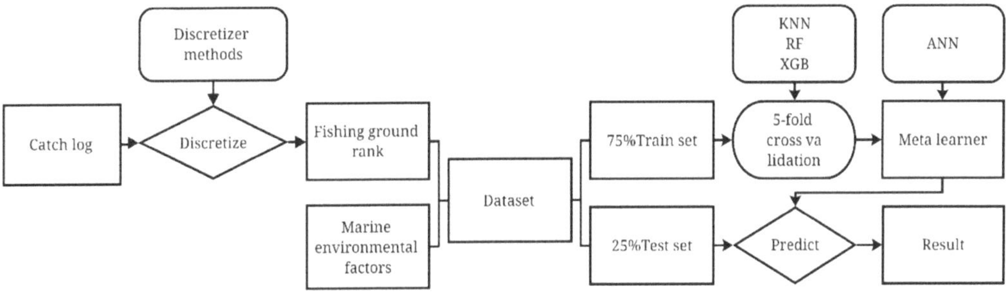

Figure 3. Flowchart of Stacking algorithm.

2.5. *Evaluating Indicators*

The confusion matrix (Table 2) summarizes how a classification model performs in the comparison of the predicted and actual values in a dataset [21].

Table 2. Confusion matrix.

	Actual Positive Class	Actual Negative Class
Predict positive class	TP (True positives)	FP (False positives)
Predict negative class	FN (False negatives)	TN (True negatives)

Accuracy (A), Precision (P), recall (R), and harmonic coefficient ($F1$) scores are performance indicators used to evaluate the effectiveness of classification models. The calculation method was as follows:

$$A = \frac{TP + TN}{TP + TN + FP + FN} \tag{4}$$

$$P = \frac{TP}{TP + FP} \tag{5}$$

$$R = \frac{TP}{TP + TN} \tag{6}$$

$$F1 = \frac{2 \times P \times R}{P + R} \tag{7}$$

The RoC curve (receiver operation characteristic curve) is a graphical representation of the performance of binary classification models showing when threshold changes are detected [34]. The area under the RoC curve (AUC) is a single scalar that summarizes the overall performance of the model, without considering the selected classification threshold. AUC = 1.0 indicates a perfect classifier, while AUC = 0.5 indicates that the performance of the classifier is not better than random chance [35].

Top-k accuracy is one of the key indicators for evaluating the classification performance of a model in multi-classification problems [36]. It refers to the probability of the model correctly classifying in the first K-value predictions, where K is a pre-specified positive integer. Top-k accuracy refers to the proportion of the first K results with the highest probability in the prediction results that contain correct labels. Therefore, as the K-value approaches the total number of labels, the model's fault tolerance increases. When K is 1, the accuracy is equal to the model's prediction accuracy for the test set. Usually, a higher top-k accuracy implies better classification performance of the model. Compared to traditional accuracy, top-k accuracy is more suitable for classification problems, especially in evaluating model performance when there are a large number of categories [37].

2.6. Statistical Analysis

Python 3.9 was primarily used for statistical analysis in this study. In Python, the modules NumPy and pandas were used for statistical analysis and calculation of large datasets and SciPy was used for Pearson correlation analysis. Feature engineering and machine learning methods came from the scikit-learn, CatBoost, and XGBoost modules. Matplotlib, seaborn, and basemap modules were used for data visualization and geographic information processing.

3. Results

3.1. Distribution of Albacore Fishing Ground in the South Pacific

The albacore fishing grounds were mainly distributed in 10° S–35° S and 160° E–90° W in the South Pacific, with high CPUE in the area of 25° S–35° S and 100° W–130° W (Figure 4). The relationship between monthly average catch and CPUE of South Pacific albacore and time of year (Figure 5) showed that the CPUE reached 41.7 in August, 2021, 43.5 in September, 2020, higher than other months.

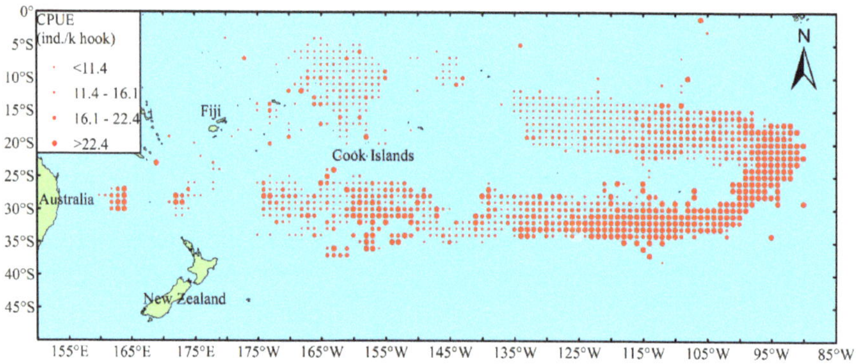

Figure 4. Average albacore CPUE distribution monthly in the South Pacific.

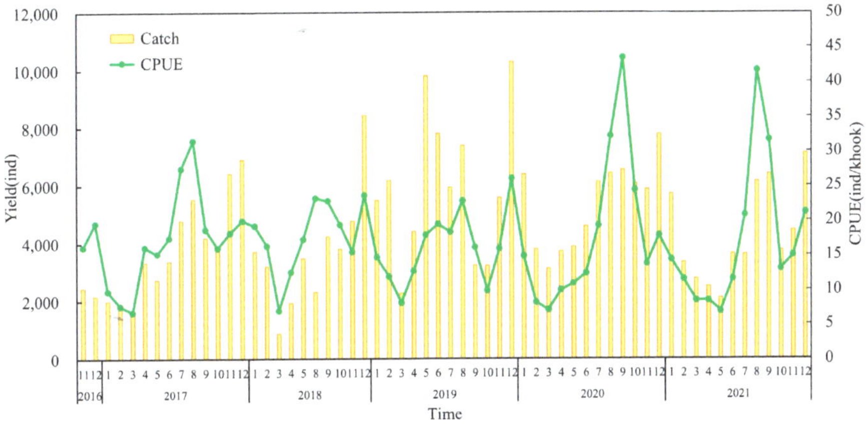

Figure 5. Monthly average catch and CPUE changes of albacore tuna in the South Pacific.

3.2. Correlation between Environmental Variables and CPUE of Albacore

Every environmental factor except year, PDO, and WS had significant correlations with CPUE ($p < 0.001$) (Table 3). SST and DO were the two environmental variables with the highest correlation coefficients. The relationship between SST and CPUE had the strongest negative correlation (−0.48), and the relationship between DO and CPUE presented the strongest positive correlation (0.49).

Table 3. Pearson correlation analysis between environmental factors and CPUE.

Variables	Correlation	p
Year	−0.00	0.81
Month	0.27	<0.05
Lat	−0.37	<0.05
Lon	0.12	<0.05
SOI	0.13	<0.05
PDO	0.01	0.48
SST	**−0.48**	<0.05
CHL	−0.05	<0.05
SSS	−0.17	<0.05
DO	**0.49**	<0.05
SSHA	0.06	<0.05
WS	−0.02	0.28
MLD	0.37	<0.05
PAR	−0.23	<0.05
EKE	−0.06	<0.05
ST50	−0.46	<0.05
SS50	−0.32	<0.05
ST100	−0.39	<0.05
SS100	−0.37	<0.05
ST200	−0.36	<0.05
SS200	−0.42	<0.05

3.3. Comparative Analysis of Binary Classification Models

The abundance of fishing grounds was divided into low-abundance and high-abundance areas through three discretization methods (Table 4). RoC curves (Figure 6) and PR curves (Figure 7) showed that when the sampling ratio was appropriate, the prediction performance of the ensemble learning model was better than that of the single model in scheme C. The AUC of STK, CAT, and RF reached 0.817, 0.817, and 0.814, respectively. The AP values of STK, CAT, and XGB reached 0.906, 0.914, and 0.91, respectively. Prediction performance by STK (Table 5), when the sampling ratio was relatively balanced, showed higher P and R rates of high-abundance fishing grounds. Using the first quartile resulted in the highest $F1$ value, reaching 0.89 with an accuracy rate of 0.823. Thus, although the accuracy was higher (reached 0.976) than other segmentation methods, the A and R were lower (reached 0.417 and 0.192). These results suggested that the quartile method was the most appropriate discretization method for dividing high- and low-abundance fishing grounds.

Table 4. Number and imbalance of different methods for classifying fishing ground abundance levels.

Scheme	Division Basis	Number of High Abundance Fishing Grounds	Number of Low Abundance Fishing Grounds	Imbalance Rate
A	Median	2312	2311	1
B	Third percentile first quantile	3097	1526	2
C	Quartile first quantile	3467	1156	3
D	Two fifths percentile	2774	1849	1.5
E	K-means clustering	650	3973	6
F	1-R partition nodes	102	4521	44

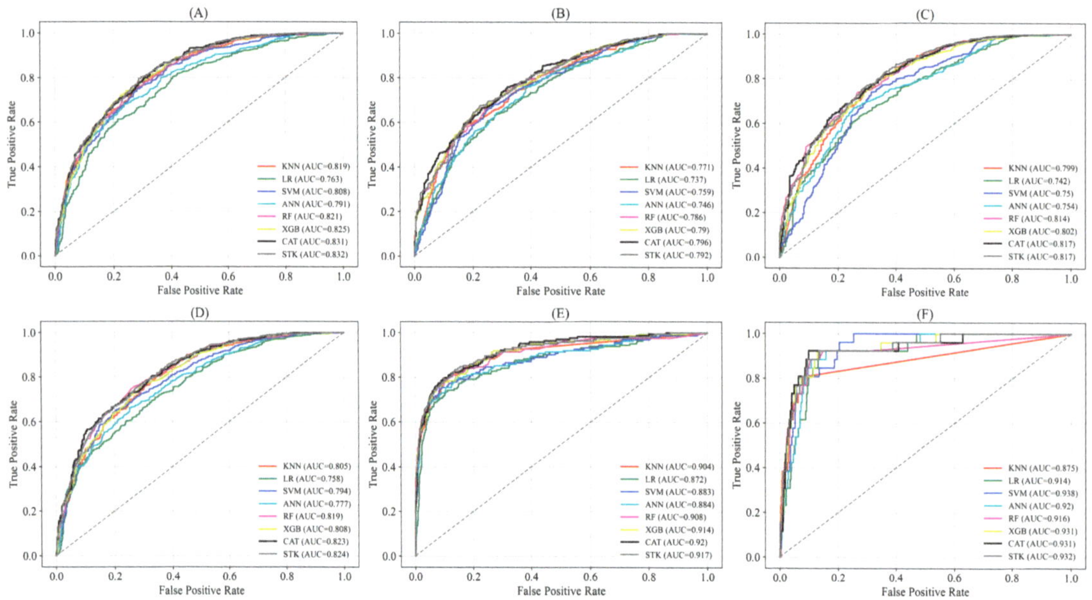

Figure 6. Six discretization methods were used to segment CPUE, and eight learners were used to predict the RoC curve obtained from 25% of the training set. (**A**) Median, (**B**) Third percentile first quantile, (**C**) Quartile first quantile, (**D**) Quintile second quantile, (**E**) K-means clustering, (**F**) 1-R partition nodes.

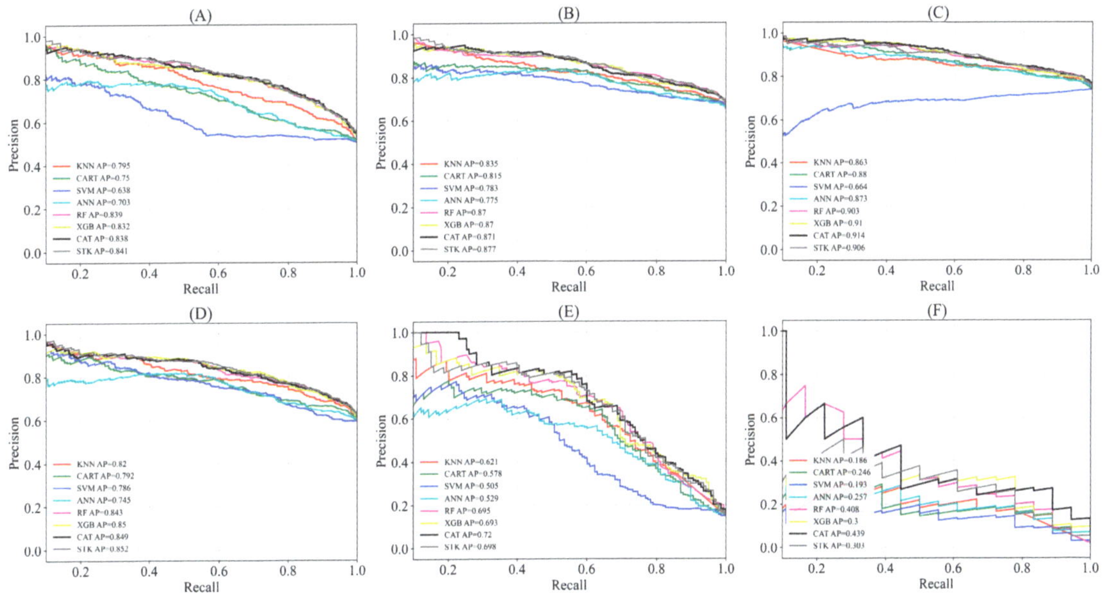

Figure 7. 6 segmentation CPUE methods and 8 binary classifiers were used to predict the relationship between probability, positive class recall, and accuracy. (**A**) Median, (**B**) Third percentile first quantile, (**C**) Quartile first quantile, (**D**) Quintile second quantile, (**E**) K-means clustering, (**F**) 1-R partition nodes.

Table 5. STK was used to predict the A-F group, selecting a classification threshold of 0.5.

Scheme	*A*	*P*	*R*	*F1*
A	0.744	0.743	0.746	0.744
B	0.746	0.773	0.879	0.822
C	0.823	**0.833**	**0.955**	**0.89**
D	0.753	0.771	0.837	0.802
E	0.925	0.828	0.589	0.688
F	**0.976**	0.417	0.192	0.263

3.4. Comparative Analysis of Multiple Classification Models

The PCA results (Figure 8) indicated that there was no significant difference between various fishing grounds (Table 6) in this study. Principal component PC1 explained 46.07% of the variation, while PC2 explained 13.47%. The clustering discretization division of fishing grounds had the highest accuracy in prediction (top-1), reaching 0.741 (Figure 9) for XGB when the abundance level of the fishing ground was divided into 3 levels. When the abundance level of the fishing ground was divided into 5 levels by clustering division, the precision prediction accuracy of STK was the highest, reaching 0.54. The comparison of the top-k accuracy predicted by each learner (Table 7) showed that the prediction performance of the K-means clustering algorithm for dividing the fishing ground level was significantly better than that of the equifrequency division ($p < 0.001$), and the same result was also observed in low fault tolerance prediction ($p < 0.001$).

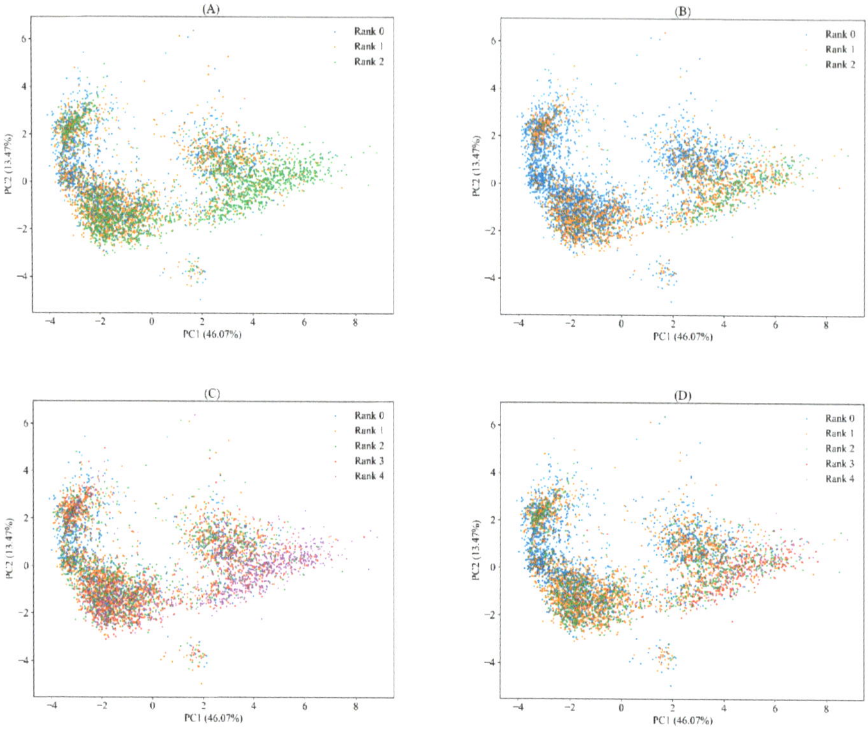

Figure 8. PCA analysis was applied for 18 environmental factors and mapped all abundance levels of fishing grounds to PC1 and PC2. (**A**) Third percentile method. (**B**) K-means clustering for K = 3. (**C**) Fifth percentile method. (**D**) K-means clustering for K = 5.

Table 6. Proportion of fishery abundance samples divided by equal frequency and K-means clustering.

Scheme	Segmentation Method	Parameter	Sample Proportion
A	Third percentile method	-	1: 1: 1
B	K-means clustering	K = 3	2872: 1415: 336
C	Fifth percentile method	-	1: 1: 1: 1: 1
D	K-means clustering	K = 5	1972: 1538: 683: 317: 106

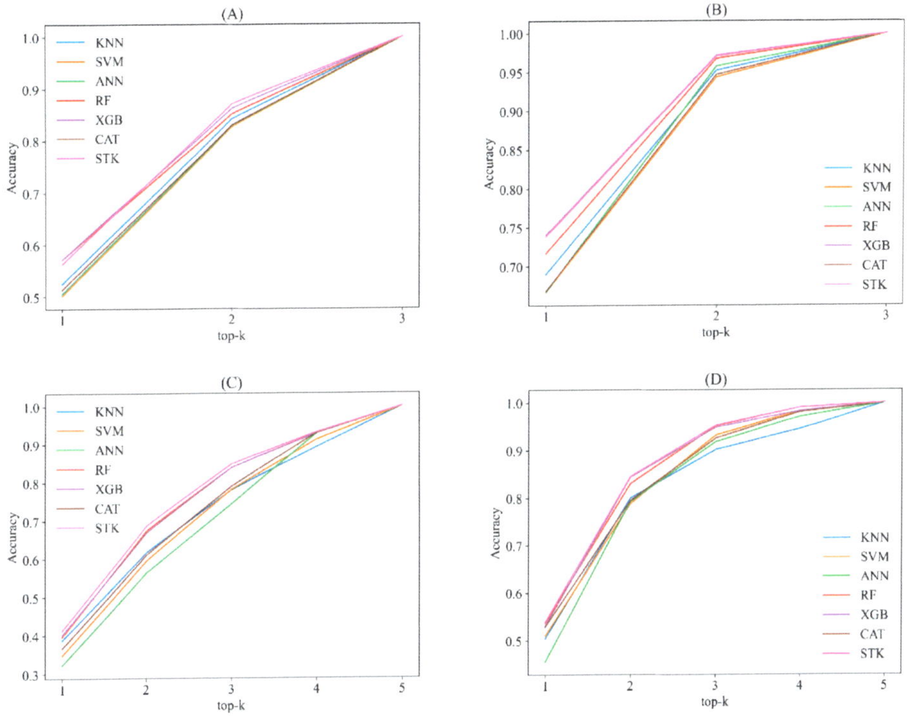

Figure 9. (**A**) Top-3 accuracy of each learner's prediction of fishery abundance after discretizing and segmenting fishery abundance levels using the third percentile method. (**B**) Top-3 accuracy of using the K-means clustering algorithm to divide fishery abundance levels by K = 3; (**C**) Top-5 accuracy predicted by using quintile discretization to segment the abundance level of fishing grounds; (**D**) Top-5 accuracy of using K-means clustering algorithm to classify the abundance level of fishing grounds with K = 5.

Table 7. Differential analysis of Top-k prediction performance using paired sample *t*-test.

Scheme	Scheme	Top-k	t	*p*
A	B	1	−43.392	<0.001
A	B	2	−33.639	<0.001
C	D	1	−20.906	<0.001
C	D	2	−19.472	<0.001
C	D	3	−13.458	<0.001
C	D	4	−16.306	<0.001

3.5. Performance of Fishing Ground Forecast Models

The actual test performance of using the quartile method to classify high- and low-abundance fishing grounds and predicting fishing grounds through STK was shown in Figure 10. Figure 10A shows that the STK model performed well in predicting high-abundance fishing grounds, and the prediction model showed that high-abundance fishing grounds appeared in all survey areas, especially between 15° S–35° S and 90° W–130° W, where a higher density of high-abundance fishing grounds are found. Figure 10B shows that the STK model had certain problems in predicting negative categories, i.e., low-abundance fishing grounds. The model exhibited low recall characteristics for low-abundance fishing grounds, and most of the survey areas of 5° S–20° S, 120° W–180° W were classified as low-abundance fishing grounds, corresponding with the previous result.

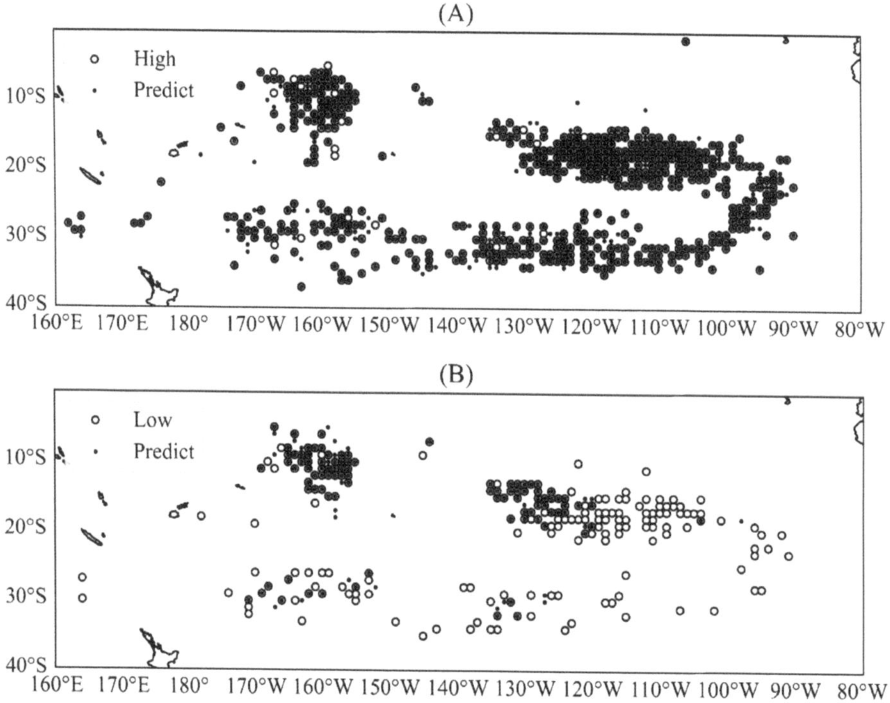

Figure 10. (**A**) The distribution of real and predicted fishing grounds for high abundance fishing grounds in the test set, and (**B**) the distribution of real and predicted fishing grounds for low abundance fishing grounds in the test set.

When CPUE was classified in 3 levels by K-means, the distribution of actual and predicted fishing grounds (Figure 11) showed that high-abundance fishing grounds have a higher probability of occurrence at 90° W–120° W and 25° S–35° S. When CPUE was classified into 5 levels (Figure 12), the distribution of low-abundance and high-abundance fishing grounds predicted by the model was similar to that of low-abundance fishing grounds classified into 3 levels. However, the discrimination performance for high-abundance and low-abundance fishing grounds was poor, because when the sampling ratio difference was too large, the learner was conservative in discriminating minority classes and tended to predict majority classes to achieve the minimum learning loss function.

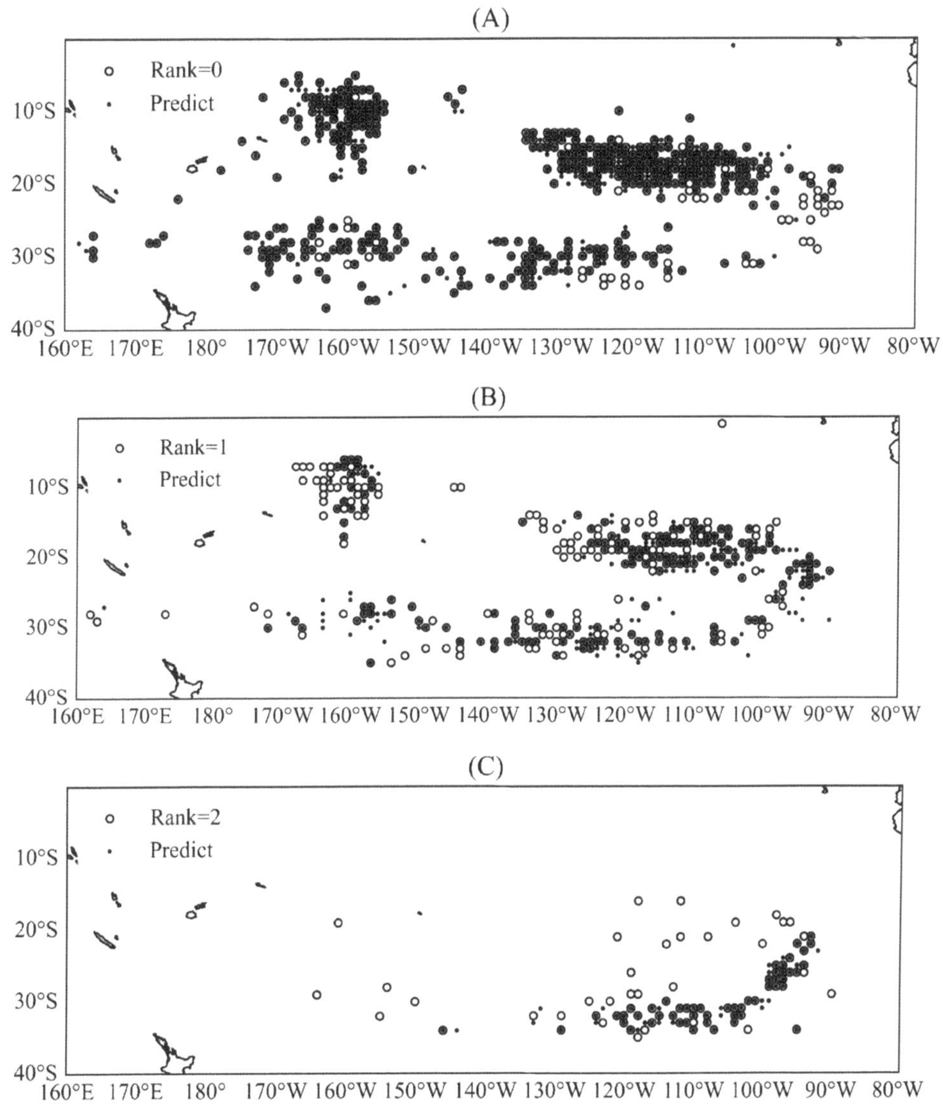

Figure 11. The abundance of fishing grounds was divided into 3 levels by K-means clustering and modeled using the STK methods. (**A**) The actual low abundance fishing grounds and prediction points, (**B**) the actual medium abundance fishing grounds and prediction points, and (**C**) the actual high abundance fishing grounds and prediction points.

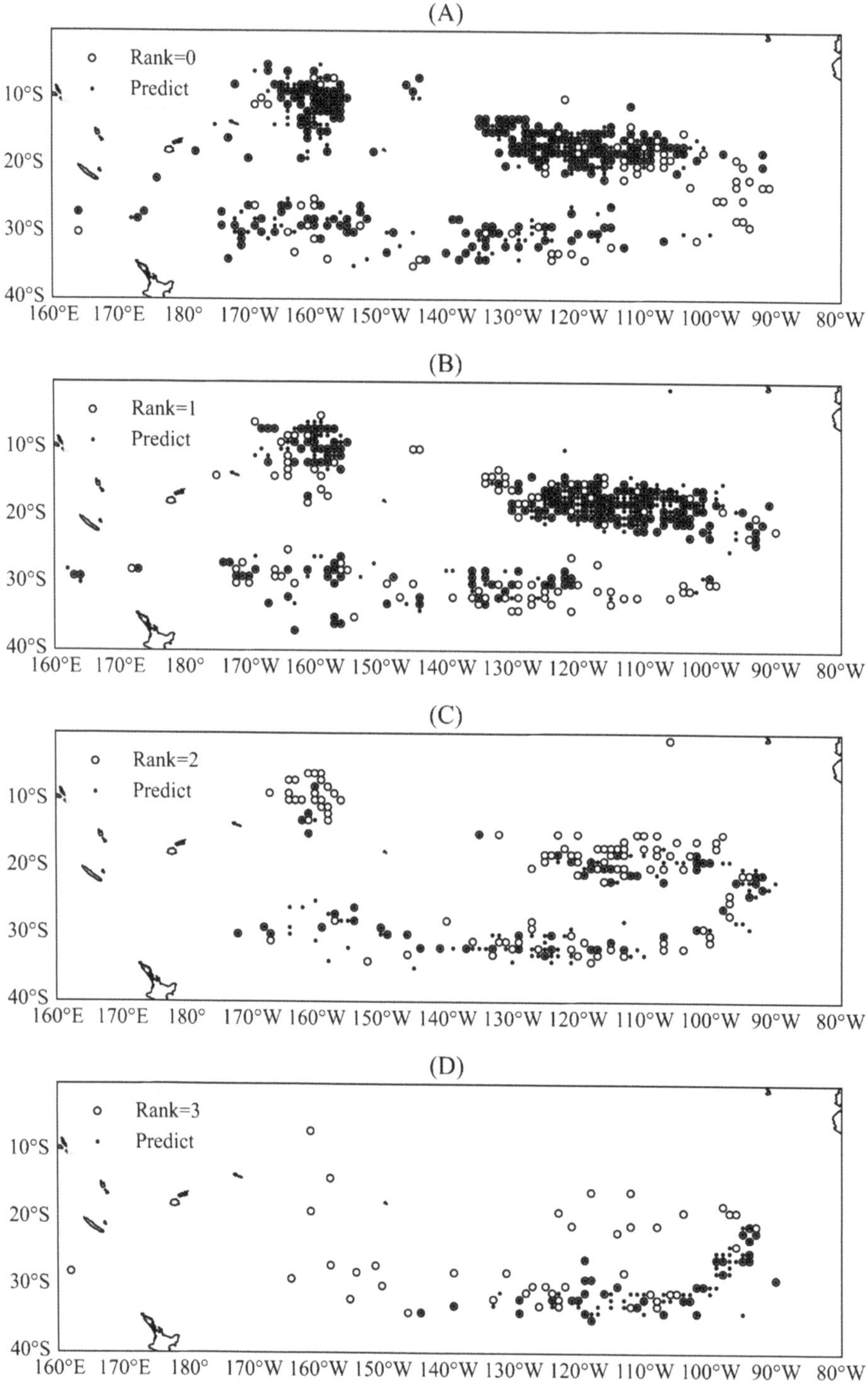

Figure 12. *Cont.*

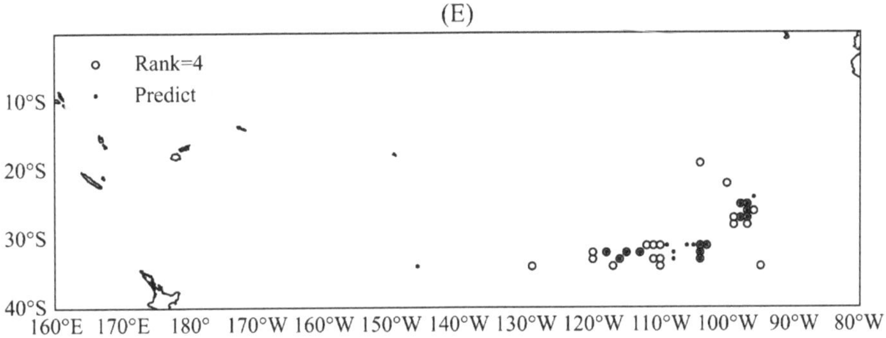

Figure 12. The abundance of fishing grounds was divided into 5 levels by K-means clustering and modeled using the STK method. (**A**) actual lowest abundance fishing grounds and prediction points, (**B**) actual low abundance fishing grounds and prediction points, (**C**) actual medium abundance fishing grounds and prediction points, (**D**) actual high abundance fishing grounds and prediction points, and (**E**) highest abundance fishing grounds and prediction points.

4. Discussion

4.1. Model Comparison and Analysis

Fishing ground forecasting is a technique that predicts the distribution of specific fish species within a specified time and location based on ocean environmental and historical data, and can be expanded to predict expected catch, fishing season, quantity, and quality [13]. Currently, most fishery forecasts have established the assumption that the life history of marine fish is closely related to the physical and chemical conditions of the ocean [14,18]. Traditional fishing ground predictions are built with habitat index models, generalized linear models, and generalized additive models, which are greatly affected by human factors, have weak generalization ability, difficult feature transformation, and insufficient fitting ability [16,19]. Machine learning is a growing method of data mining, which can explore the potential relationship between environmental variables and fishery abundance, establish the relationship between environmental factors and fishery resource abundance through certain rules, and play an important role in fishery forecasting research. Compared to traditional habitat index models and linear models, machine learning algorithms are capable of mining large-scale, complex, variable, and high-dimensional ocean data information, with a stronger ability to fit data and achieve high-precision and high-generalization ability in fishing ground forecasting [18].

A single machine learning model such as KNN, LR, SVM, or ANN can achieve good fishing ground prediction results. Compared with a single model, the ensemble learning algorithm can provide more accurate prediction power and better prediction generalization by building multiple learners, sampling and forecasting random forests through bootstrapping, and selecting prediction results through boosting in XGB and CAT. RF had better performance compared to linear models and ANN, predicting the habitat distribution characteristics of the spring short octopus in Haizhou Bay and its relationships with environmental factors [38]. Additionally, several ensemble algorithms were applied to estimate the biomass of major economically important crabs in Zhoushan fishing grounds, and XGB had the best estimation effect [39]. CAT was uniquely applied to forecast the albacore fishery in the South Pacific Ocean and has shown greater advantages than other single models. STK had a superior performance in predicting potential fishing areas for target fish species in the fishing grounds, thanks to advantages in combating data noise, non-Gaussian distribution, and non-linear relationships compared to other models. It was found that in predicting albacore fishing grounds in the South Pacific with STK, models had higher accuracy compared to other machine learning models. Mugo and Saitohused 2020 applied machine learning models to model the overall habitat of bonito in the western

Pacific Ocean, and the results showed that ensemble learning algorithms have higher generalization ability compared to other models [40].

We propose three main reasons why the STK model had superior performance in predicting fishing grounds. (1) For fishing ground label classification, KNN, RF, and XGB had superior performance and were more stable compared to other models, making overfitting less likely to occur. However, due to insignificant differences between fishing grounds, the performance upper limit of secondary learners is low, and the accuracy of predictions is greatly affected by the distribution of data. The STK model integrates relatively stable and accurate KNN, RF, and XGB, and uses a structurally simple ANN as a meta learner to further prevent model overfitting, resulting in better performance in fishery label classification tasks. (2) The use of the 5-fold cross validation method fully utilizes the entire dataset, reduces the risk of overfitting, and improves the model's generalization ability. (3) Considering that CPUE does not always follow a normal distribution, multiple methods of separating data were tested to avoid various problems caused by imbalanced data during model prediction.

4.2. Impact of Environmental Factors on Model Prediction

The life history of albacore is closely related to ocean temperature, salinity, dissolved oxygen concentration, and sea level height [41]. As large-scale migratory marine bony fish, albacore are usually distributed in tropical and temperate waters, and their abundance is highly correlated with temperature. This may be because warm water is suitable for albacore to lay eggs, and during the spawning period, the development and survival rate of eggs and juveniles greatly depends on the appropriate temperature [42]. Appropriate water temperature also helps to maintain the stability of water quality, providing suitable environmental conditions to support the growth and development of juvenile fish. In addition, warm waters nurture more bait organisms, with short life-cycle cephalopods and small fish serving as prey for albacore. Prey habitats are widely distributed in temperate and tropical waters [43], providing more feeding opportunities for albacore. Albacore prefer high-salinity waters, as high-salinity waters often imply relatively stable water masses and higher biological suitability [44]. As a top predator, albacore require continuous swimming to breathe [4], and albacore caught by longline fishing are rarely found in low dissolved oxygen concentrations [1]. Albacore have a higher CPUE in the range of dissolved oxygen concentrations around 240 mol/L [45]. Albacore prefer waters with higher dissolved oxygen levels than other tuna species [46]. Abnormalities in sea level height are often closely related to changes in ocean circulation and marine environment, and the variation in sea surface height is mainly related to factors such as ocean currents, water masses, and tides as a characterization of ocean hydrodynamics [47]. Abnormal changes in sea level height may lead to corresponding changes in the migration path and distribution range of albacore, which are closely related to the occurrence of mesoscale eddies [48]. Albacore are sensitive to the flow and changes in the marine environment, and find suitable temperature, salinity, and dissolved oxygen conditions through changes in the distribution of ocean circulation.

Chlorophyll-a concentration, photosynthetic effective radiation, and eddy energy have indirect effects on the distribution of albacore. The concentration of chlorophyll-a is directly related to the level of primary productivity [49], but albacore have a high trophic level and do not directly consume primary productivity. However, the distribution of albacore fishing grounds is still indirectly affected by chlorophyll concentration and photosynthetically active radiation. Photosynthetically active radiation (PAR) is the main driving force for the growth and reproduction of phytoplankton in the ocean [8,41], and phytoplankton are an important food source for albacore prey organisms. Appropriate photosynthetic and effective radiation conditions can promote the growth of phytoplankton, thereby providing abundant food resources and attracting the aggregation of albacore. The correlation between eddy energy and CPUE is relatively small. EKE is calculated through ocean current velocity and affects the three-dimensional distribution of circulation, water temperature, and chlorophyll, indirectly affecting the distribution of albacore fishing grounds [10].

Albacore have vertical distribution patterns correlated with vertical temperature, salinity, and mixed layer depth. Changes in temperature and salinity of vertical water layers in the ocean have a direct impact on the ecological processes of growth, development, reproduction, metabolism, and migration distribution of marine organisms [50,51]. Albacore are active in water layers below 100 m, with an optimal water temperature of 20–26 °C and a salinity range of 35.6–35.9. It was reported that environmental factors in different water layers have varying effects on the CPUE of albacore [52]. Depth of the mixed layer (MLD) also has an effect on the vertical distribution of albacore. The hydrological properties of the mixed layer in the same sea area follow the same trends, and environmental factors such as temperature and salinity are similar. In tropical latitudes, the vertical distribution of albacore follows a day-night pattern, in which albacore occupy shallow and warmer waters above MLD at night and deeper and cooler waters below MLD during the day. However, in temperate latitudes, albacore are almost exclusively found in shallow waters above MLD [5].

In recent years, global climate change has significantly impacted marine ecosystems, and typical climate events have altered the habitat distribution of marine organisms to some extent [53]. Climate change has a significant impact on the life history of albacore. Lehodey showed subtropical albacore had low replenishment levels during El Niño and high replenishment levels during La Niña, which is opposite to the patterns of bonito (*Katsuwonus pelmis*) and yellow-finned tuna (*Thunnus albacares*) [54]. The low correlation between CPUE and PDO may have been due to the low number of years collecting longline fishing logs, which was consistent with previous studies and had a high correlation with SOI. Besides, the short time series in this study contains less abnormal climate information, and it is easy to ignore the influence factors of climate change on tuna distribution when training data.

4.3. Impact of Different Partitioning Methods on the Model

The area of research on albacore tuna in the South Pacific is mostly distributed in the southwestern Pacific Ocean west of the Cook Islands [15,18], and there is relatively little research on the distribution of albacore in the southeastern Pacific Ocean. The main research area for this study was the Cook Islands, providing a more comprehensive reference for the prediction and application of albacore tuna fishing grounds in the South Pacific. Overall, there was significant agreement between the predicted and actual locations of most fishing grounds. However, the main reasons for the relatively few deviations in predictions of high abundance fishing grounds may have been (1) long line fishing operations have a large spatial span [19]; (2) additionally, the actual distribution of fishing grounds is relatively discrete and outlier fishing grounds may have not been considered as positive classes by the model; (3) finally, due to the large study area, the difference in environmental factors was not significant, which may lead to a decrease in the classification performance of the model (Figure 9).

The proportion of samples used to generate data labels is one of the main factors influencing the effectiveness of classifying fishing area abundance levels [20,21]. When studying large ocean areas, the effects of marine environment, fishing factors, and human factors are difficult to estimate. Therefore, the definition and discrimination of fishing area levels can offset the influence of these unpredictable factors [22]. When fishing grounds are defined in two levels, the sampling ratio may be imbalanced, and the model may bias the prediction results towards the majority class to improve accuracy. Although STK performed better than other models, its overall generalization ability decreased, and there may be a concurrent decrease in the recall rate of low abundance fishing grounds in actual fishing ground prediction (Figure 11), especially in predicting low-abundance fishing grounds at higher latitudes. Multi-level classification methods can be used to avoid the problem of imbalanced sample partitioning. The forecast results support the conclusion that the prediction effect of multi-class abundance is better than that of binary classification prediction models (Figures 11 and 12). Moreover, this study suggests the top-k indicator is most appropriate for multi-level classification of fishing ground forecasts.

4.4. Outlook

Using methods other than equal frequency partitioning to divide CPUE can cause imbalanced sampling, which can lead to a model leaning towards the majority class prediction and resulting in low recall or accuracy in the prediction results. In addition, the division of spatiotemporal distribution was based on $1^\circ \times 1^\circ$ degree and monthly units, and there was a significant lag in time measuring fishing ground predictions in practice. There was no continuous large-scale data support in the data collection process and the actual operation dates and locations of fishing grounds could not be standardized. Multiple spatial and space resolutions can be used to compare the forecast resolution in future studies. Further, it is not economically feasible to meet the requirements of an ideal survey with standardized sampling times and methods, resulting in unavoidable error. Optimization algorithms such as particle swarm optimization, simulated annealing algorithm, and genetic algorithm can be introduced in subsequent work to achieve rapid model convergence and improve modeling efficiency and robustness. In addition, in the current relevant studies, the tuna forecasting model mostly took a single species as the response variable, but seldom considered the interactions between species. Therefore, joint species distribution models can also be used to explore the relationship between species in subsequent studies.

5. Conclusions

This research used different discretization methods to divide the CPUE of the South Pacific albacore in different ways. These division methods were used to evaluate indicators for imbalanced sampling and the problem of predicting fishery abundance was transformed into a classification problem. Fishery abundance was divided into 2, 3, and 5 levels. Multiple machine learning algorithms were tested to predict fishery abundance levels through remote sensing environmental factors. The accuracy and stability of STK were found to be higher than those of other models. However, due to the inherent circumstances of fishery production data, there are shortcomings in predicting the temporal aspects of fishing grounds. Therefore, in the future, it is necessary to improve the collection and organization of fishery data and further study the environmental attribute values of fishing grounds to reflect the marine environment more accurately. The implementation of these improvements will provide more reliable and real-time fishery and environmental information for fishing ground prediction models, thereby more accurately predicting and guiding fishery production. In addition, climate change challenges for tuna fisheries require more diverse and accurate fishery forecasting techniques. A variety of forecasting methods and technologies can provide suitable fishing programs that will contribute to the sustainable management of tuna fisheries.

Author Contributions: Conceptualization, F.C.; Data curation, Q.D.; Funding acquisition, W.Y. and W.Z. (Weifeng Zhou); Visualization, D.L.; Writing—original draft, J.L.; Writing—review and editing, W.Z. (Wenbin Zhu). All authors have read and agreed to the published version of the manuscript.

Funding: National Key RD Program of China (2023YFD2401303), Pelagic Fishery Resources Monitoring Project of Zhejiang Province (2024HZ003).

Institutional Review Board Statement: This study applied machine learning for tuna fishing ground forecast by the tuna long line fishing log data from long-line vessels in the South Pacific Ocean, there was not any animal experiment in the study.

Informed Consent Statement: Not applicable.

Data Availability Statement: The datasets and codes that support the findings of this study are available from the corresponding author upon reasonable request.

Acknowledgments: The authors wish to thank the commercial fishing company for providing fishing logs of longline vessels in the South Pacific Ocean. This work was funded by the National Key RD Program of China (2023YFD2401303) and routine monitoring program for distant-water fishery resources of Zhejiang province(2023HZ024).

Conflicts of Interest: The authors have no conflicts of interest to declare.

References

1. Nikolic, N.; Morandeau, G.; Hoarau, L.; West, W.; Arrizabalaga, H.; Hoyle, S.; Fonteneau, A. Review of albacore tuna, *Thunnus alalunga*, biology, fisheries and management. *Rev. Fish. Bio. Fish.* **2017**, *27*, 775–810. [CrossRef]
2. Watanabe, H.; Kubodera, T.; Masuda, S.; Kawahara, S. Feeding habits of albacore *Thunnus alalunga* in the transition region of the central North Pacific. *Fish. Sci.* **2004**, *70*, 573–579. [CrossRef]
3. Dragon, A.C.; Senina, I.; Titaud, O.; Calmettes, B.; Conchon, A.; Arrizabalaga, H.; Lehodey, P. An ecosystem-driven model for spatial dynamics and stock assessment of North Atlantic albacore. *Can. J. Fish. Aquat. Sci.* **2015**, *72*, 864–878. [CrossRef]
4. Lehodey, P.; Senina, I.; Nicol, S.; Hampton, J. Modelling the impact of climate change on South Pacific albacore tuna. *Deep-Sea. Res. Pt. II.* **2015**, *113*, 246–259. [CrossRef]
5. Vaihola, S.; Yemane, D.; Kininmonth, S. Spatiotemporal Patterns in the Distribution of Albacore, Bigeye, Skipjack, and Yellowfin Tuna Species within the Exclusive Economic Zones of Tonga for the Years 2002 to 2018. *Diversity.* **2023**, *15*, 1091. [CrossRef]
6. Williams, A.J.; Allain, V.; Nicol, S.J.; Evans, K.J.; Hoyle, S.D.; Dupoux, C.; Vourey, E.; Dubosc, J. Vertical behavior and diet of albacore tuna (*Thunnus alalunga*) vary with latitude in the South Pacific Ocean. *Deep-Sea. Res. Pt. II.* **2015**, *113*, 154–169. [CrossRef]
7. Reglero, P.; Santos, M.; Balbín, R.; Laíz-Carrión, R.; Alvarez-Berastegui, D.; Ciannelli, L.; Jiménez, E.; Alemany, F. Environmental and biological characteristics of Atlantic bluefin tuna and albacore spawning habitats based on their egg distributions. *Deep-Sea. Res. Pt. II.* **2017**, *140*, 105–116. [CrossRef]
8. Zainuddin, M.; Saitoh, K.; SAITOH, S.I. Albacore (*Thunnus alalunga*) fishing ground in relation to oceanographic conditions in the western North Pacific Ocean using remotely sensed satellite data. *Fish. Oceanogr.* **2008**, *17*, 61–73. [CrossRef]
9. Mondal, S.; Lee, M.A. Habitat modeling of mature albacore (*Thunnus alalunga*) tuna in the Indian Ocean. *Front. Mar. Sci.* **2023**, *10*, 1258535. [CrossRef]
10. Tussadiah, A.; Pranowo, W.S.; Syamsuddin, M.L.; Riyantini, I.; Nugraha, B.; Novianto, D. Characteristic of eddies kinetic energy associated with yellowfin tuna in southern Java Indian Ocean. *IOP Conf. Ser. Earth Environ. Sci.* **2018**, *176*, 012004. [CrossRef]
11. Singh, A.A.; Sakuramoto, K.; Suzuki, N. Impact of climatic factors on albacore tuna *Thunnus alalunga* in the South Pacific Ocean. *Amer. J. Clim. Chan.* **2015**, *4*, 295. [CrossRef]
12. Lindegren, M.; Checkley, D.M., Jr.; Koslow, J.A.; Goericke, R.; Ohman, M.D. Climate-mediated changes in marine ecosystem regulation during El Niño. *Glob. Change Biol.* **2018**, *24*, 796–809. [CrossRef] [PubMed]
13. Gao, F.; Chen, X.; Guan, W.; Li, G. A new model to forecast fishing ground of *Scomber japonicus* in the Yellow Sea and East China Sea. *Acta Ocean. Sin.* **2016**, *35*, 74–81. [CrossRef]
14. Cui, X.; Tang, F.; Zhou, W.; Wu, Z.; Yang, S.; Hua, C. Fishing ground forecasting model of *Ommastrephes* bartramii based on support vector machine (SVM) in the Northwest Pacific Ocean. *South China Fish. Sci.* **2016**, *12*, 1–7. (In Chinese with English abstract)
15. Mao, J.; Chen, X.; Yu, J. Forecasting fishing ground of *Thunnus alalunga* based on BP neural network in South Pacific Ocean. *Acta. Ocean. Sin.* **2016**, *10*, 34–43. (In Chinese with English abstract)
16. Chen, X.; Fan, W.; Cui, X.; Zhou, W.; Tang, F. Fishing ground forecasting of *Thunnus alalunga* in Indian Ocean based on random forest. *Acta. Ocean. Sin.* **2013**, *35*, 158–164. (In Chinese with English abstract)
17. Vaihola, S.; Kininmonth, S. Environmental Factors Determine Tuna Fishing Vessels' Behavior in Tonga. *Fishes.* **2023**, *8*, 602. [CrossRef]
18. Hou, J.; Zhou, W.; Fan, W.; Zhang, H. Research on fishing grounds forecasting models of albacore tuna based on ensemble learning in South Pacific. *South China Fish. Sci.* **2020**, *5*, 42–50. (In Chinese with English abstract)
19. Song, L.; Li, T.; Zhang, T.; Sui, H.; Li, B.; Zhang, M. Comparison of machine learning models within different spatial resolutions for predicting the bigeye tuna fishing grounds in tropical waters of the Atlantic Ocean. *Fish. Oceanogr.* **2023**, *32*, 509–526. [CrossRef]
20. Xu, L.; Chi, D. Machine Learning Classification Strategy for Imbalanced Data Sets. *Comput. Eng. Appl.* **2020**, *56*, 12–27. (In Chinese with English abstract)
21. Wardhani, N.W.S.; Rochayani, M.Y.; Iriany, A.; Sulistyono, A.D.; Lestantyo, P. Cross-validation metrics for evaluating classification performance on imbalanced data. In Proceedings of the 2019 International Conference on Computer, Control, Informatics and Its Applications, Tangerang, Indonesia, 23–24 October 2019; pp. 14–18.
22. Shabani, F.; Kumar, L.; Ahmadi, M. A Comparison of Absolute Performance of Different Correlative and Mechanistic Species Distribution Models in an Independent Area. *Ecol. Evol.* **2016**, *6*, 5973–5986. [CrossRef] [PubMed]

23. Feng, Y.; Chen, X.; Gao, F.; Liu, Y. Impacts of changing scale on Getis-Ord Gi hotspots of CPUE: A case study of the neon flying squid (*Ommastrephes bartramii*) in the northwest Pacific Ocean. *Acta. Ocean. Sin.* **2018**, *37*, 67–76. [CrossRef]
24. Krishna, K.; Murty, M.N. Genetic K-means algorithm. *IEEE Trans. Sys. Man. Cyber. Pt. B* **1999**, *29*, 433–439. [CrossRef] [PubMed]
25. Jacod, J.; Protter, P. *Discretization of Processes*; Springer: Berlin, Heidelberg, 2011.
26. Ren, L.; Ma, Y.; Shi, H.; Chen, X. Overview of Machine Learning Algorithms. In *Lecture Notes in Electrical Engineering*; Springer: Singapore, 2020; pp. 672–678.
27. LaValley, M.P. Logistic regression. *Circulation* **2008**, *117*, 2395–2399. [CrossRef] [PubMed]
28. Hearst, M.A.; Dumais, S.T.; Osuna, E.; Platt, J.; Scholkopf, B. Support vector machines. Springer Science+Business Media, LLC: New York, NY, USA, 1998; Volume 13, pp. 18–28.
29. Suryanarayana, I.; Braibanti, A.; Sambasiva Rao, R.; Ramam, V.A.; Sudarsan, D.; Nageswara Rao, G. Neural networks in fisheries research. *Fish. Res.* **2008**, *92*, 115–139. [CrossRef]
30. Breiman, L. Random forests. *Mach. Learn.* **2001**, *45*, 5–32. [CrossRef]
31. Chen, T.; He, T.; Benesty, M.; Khotilovich, V.; Tang, Y.; Cho, H.; Chen, K.; Mitchell, R.; Cano, I.; Zhou, T.; et al. *R Package, version 0.4-2*; Xgboost: Extreme Gradient Boosting; The R Foundation: Vienna, Austria, 2015; Volume 1, pp. 1–4.
32. Prokhorenkova, L.; Gusev, G.; Vorobev, A.; Dorogush, A.V.; Gulin, A. CatBoost: Unbiased boosting with categorical features. *Ad. Neural. Infor. Process. Syst.* **2018**, *31*, 6637–6647.
33. Džeroski, S.; Ženko, B. Is combining classifiers with stacking better than selecting the best one? *Mach. Learn.* **2004**, *54*, 255–273. [CrossRef]
34. Fawcett, T. An introduction to ROC analysis. *Pat. Recog. Let.* **2006**, *27*, 861–874. [CrossRef]
35. Majnik, M.; Bosnić, Z. ROC analysis of classifiers in machine learning: A survey. *Intel. Dt. Analy.* **2013**, *17*, 531–558. [CrossRef]
36. Harding, J.; Shahbaz, M.; Srinivas, S.; Kusiak, A. Data mining in manufacturing: A review. *ASME Trans. J. Manuf. Sci. Eng.* **2006**, *128*, 969–976. [CrossRef]
37. Tan, H.; Wu, Y.; Shen, B.; Jin, P.; Ran, B. Short-term traffic prediction based on dynamic tensor completion. *IEEE Trans. Intell. Transp. Syst.* **2016**, *17*, 2123–2133. [CrossRef]
38. Cui, Y.; Liu, S.; Zhang, Y.; Xu, B.; Ji, Y.; Zhang, C.; Xue, Y. Habitat characteristics of *Octopus ocellatus* and their relationship with environmental factors during spring in Haizhou Bay, China. *Chin. J. Appl. Ecol.* **2022**, *33*, 1686–1692. (In Chinese with English abstract)
39. Yang, C.; Li, X.; Liu, Q.; Wang, Y. Application of machine learning methods for estimating the biomass of economically crabs in the Zhoushan fishery. *Mar. Sci.* **2023**, *9*, 61–70. (In Chinese with English abstract)
40. Mugo, R.; Saitoh, S.I. Ensemble modelling of Skipjack tuna (*Katsuwonus pelamis*) habitats in the western North Pacific using satellite remotely sensed data; a comparative analysis using machine-learning models. *Remote. Sens.* **2020**, *12*, 2591. [CrossRef]
41. Zainuddin, M.; Kiyofuji, H.; Saitoh, K.; Saitoh, S.I. Using multi-sensor satellite remote sensing and catch data to detect ocean hot spots for albacore (*Thunnus alalunga*) in the northwestern North Pacific. *Deep-Sea. Res. Pt. II.* **2006**, *53*, 419–431. [CrossRef]
42. Ashida, H.; Gosho, T.; Watanabe, K.; Okazaki, M.; Tanabe, T.; Uosaki, K. Reproductive traits and seasonal variations in the spawning activity of female albacore, *Thunnus alalunga*, in the subtropical western North Pacific Ocean. *J. Sea. Res.* **2020**, *160*, 101902. [CrossRef]
43. Jackson, G.D.; Meekan, M.G.; Wotherspoon, S.; Jackson, C.H. Distributions of young cephalopods in the tropical waters of Western Australia over two consecutive summers. *ICES. J. Mar. Sci.* **2008**, *65*, 140–147. [CrossRef]
44. Shcherbina, A.Y.; D'Asaro, E.A.; Riser, S.C.; Kessler, W.S. Variability and interleaving of upper-ocean water masses surrounding the North Atlantic salinity maximum. *Oceanography* **2015**, *28*, 106–113. [CrossRef]
45. Mondal, S.; Wang, Y.C.; Lee, M.A.; Weng, J.S.; Mondal, B.K. Ensemble three-dimensional habitat modeling of Indian Ocean immature albacore tuna (*Thunnus alalunga*) using remote sensing data. *Remote. Sens.* **2022**, *14*, 5278. [CrossRef]
46. Arrizabalaga, H.; Dufour, F.; Kell, L.; Merino, G.; Ibaibarriaga, L.; Chust, G.; Irigoien, X.; Santiago, J.; Murua, H.; Fraile, I.; et al. Global habitat preferences of commercially valuable tuna. *Deep-Sea Res. Pt. II.* **2015**, *113*, 102–112. [CrossRef]
47. Kai, E.T.; Marsac, F. Influence of mesoscale eddies on spatial structuring of top predators' communities in the Mozambique Channel. *Prog. Oceanogr.* **2010**, *86*, 214–223.
48. Zhou, C.; He, P.; Xu, L.; Bach, P.; Wang, X.; Wan, R.; Zhang, Y. The effects of mesoscale oceanographic structures and ambient conditions on the catch of albacore tuna in the South Pacific longline fishery. *Fish. Oceanogr.* **2020**, *29*, 238–251. [CrossRef]
49. Iriarte, J.L.; González, H.E.; Liu, K.K.; Rivas, C.; Valenzuela, C. Spatial and temporal variability of chlorophyll and primary productivity in surface waters of southern Chile (41.5–43 S). *Estuarine. Coast. Shelf. Sci.* **2007**, *74*, 471–480. [CrossRef]
50. Lougee, L.A.; Bollens, S.M.; Avent, S.R. The effects of haloclines on the vertical distribution and migration of zooplankton. *J. Exp. Mar. Bio. Eco.* **2002**, *278*, 111–134. [CrossRef]
51. Wu, J.; Jin, L. Exploration of the classification and main characteristics of marine ecosystems. *Inter. J. Mar. Sci.* **2023**, *13*, 1–7.
52. Xu, H.; Song, L.; Shen, J.; Li, Y.; Zhang, M. The relationship between the spatial-temporal distribution of albacore tuna CPUE and the marine environment variables in waters near the Cook Islands based on GAM. *Mar. Sci. Bull.* **2023**, *4*, 444–455. (In Chinese with English abstract)

53. Du Pontavice, H.; Gascuel, D.; Reygondeau, G.; Maureaud, A.; Cheung, W.W. Climate change undermines the global functioning of marine food webs. *Glob. Change Biol.* **2020**, *26*, 1306–1318. [CrossRef]
54. Lehodey, P.; Chai, F.; Hampton, J. Modelling climate-related variability of tuna populations from a coupled ocean–biogeochemical-populations dynamics model. *Fish. Oceanogr.* **2003**, *12*, 483–494. [CrossRef]

fishes

MDPI

Article

A Study on the Impact of Environmental Factors on Chub Mackerel *Scomber japonicus* Fishing Grounds Based on a Linear Mixed Model

Jiasheng Li [1,2], Fenghua Tang [1], Yumei Wu [1], Shengmao Zhang [1], Weifeng Zhou [1] and Xuesen Cui [1,*]

[1] Key and Open Laboratory of Remote Sensing Information Technology in Fishing Resource, East China Sea Fisheries Research Institute, Chinese Academy of Fishery Sciences, Shanghai 200090, China; lijiasheng1984@126.com (J.L.); tangfh@ecsf.ac.cn (F.T.); wuym@ecsf.ac.cn (Y.W.); zhangsm@ecsf.ac.cn (S.Z.); zhouwf@ecsf.ac.cn (W.Z.)

[2] College of Fisheries and Life Science, Shanghai Ocean University, Shanghai 201306, China

* Correspondence: cuixuesen@eastfishery.ac.cn

Abstract: Chub mackerel (*Scomber japonicus*) is a commercially important fish species which are widely distributed in the North Pacific. Based on the fishery data from China's high-sea light-purse seine fishing from 2014 to 2020 and the marine environment factors, a mixed linear model considering the actual spatiotemporal stratification of the catch per unit effort (CPUE) was established to analyze the fixed and random effects of marine environmental factors on the CPUE of chub mackerel and to investigate the relationship between the abundance of chub mackerel resources in the Northwest Pacific and two marine environmental factors: sea surface temperature (SST) and chlorophyll-a concentration (CHL). The results showed that SST had a significant fixed effect on the CPUE. In contrast, the natural logarithm of chlorophyll (logCHL) had no fixed effect on the CPUE. Based on the monthly analysis, random fluctuations were observed in the impact of logCHL on the CPUE. LogCHL and CPUE show a positive correlation during spawning and wintering periods and a negative correlation during the feeding period. The study showed that when fishery sampling data exhibit spatiotemporal stratification, linear mixed models can effectively incorporate both the fixed and random effects of environmental factors on the CPUE of chub mackerel. Linear mixed models can play an important role in analyzing the fluctuations in resource abundance and the mechanisms governing the formation of fishing grounds for chub mackerel in the Northwest Pacific.

Keywords: chub mackerel; linear mixed model; fishing grounds; Northwest Pacific

Key Contribution: A mixed linear model was introduced to analyze the fixed and random effects of marine environmental factors on CPUE of chub mackerel; SST had a significant fixed effect on CPUE, random fluctuations were observed in the impact of logCHL on CPUE.

Citation: Li, J.; Tang, F.; Wu, Y.; Zhang, S.; Zhou, W.; Cui, X. A Study on the Impact of Environmental Factors on Chub Mackerel *Scomber japonicus* Fishing Grounds Based on a Linear Mixed Model. *Fishes* **2024**, *9*, 323. https://doi.org/10.3390/fishes9080323

Academic Editors: Yongjun Tian, Heng Zhang, Ying Xue, Yunzhi Yan and Luoliang Xu

Received: 14 June 2024
Revised: 25 July 2024
Accepted: 9 August 2024
Published: 14 August 2024

1. Introduction

Chub mackerel is a small fish species which inhabit the upper and middle layers of subtropical to temperate waters and is widespread in the Indian Ocean, the Pacific coast of South America, and the Western Pacific [1–3]. Based on the migration patterns and spawning areas, chub mackerel in the Northwest Pacific can be divided into two stocks: the Tsushima Warm Current (TWC) stock, which is distributed in the East China Sea, the Yellow Sea, and the Sea of Japan, and the Pacific stock, which is distributed around the edge of the Kuroshio current [2,4]. Chub mackerel plays an important role in global fisheries, with its stock biomass and catch showing significant year-to-year fluctuations [5]. Chinese fishing vessels have been conducting purse seine fishing operations in the Northwest Pacific since 2014. Catches of chub mackerel peaked at 150,000 tonnes in 2015 and fell to 80,000 tonnes in 2020 [6].

Previous studies have suggested that the marine environment was closely related to the growth rate, distribution of resources, and changes in fishing grounds for chub mackerel. During their early life stages, the growth, migration, and recruitment of chub mackerel are influenced by both biological and abiotic factors [1,3,7–9]. Chub mackerel larvae can be transported to downstream areas of the Kuroshio–Oyashio Transition Area (KOTA) for feeding purposes. The KOTA is characterized by abundant bait resources and plays a role as an important habitat for several small pelagic fish species, including chub mackerel [1,7,10]. In the Northwest Pacific, the concentration of chlorophyll can serve as a reference indicator for the foraging of chub mackerel and can influence the aggregation of chub mackerel on a large scale, especially in the vicinity of the Kuroshio and Oyashio fronts [11,12]. In the southern waters of the East China Sea, the distribution of chub mackerel is influenced by the invasion of the warm Kuroshio current [13]. Sea surface temperature (SST) is considered to be an important environmental factor affecting the spawning and migration patterns of chub mackerel [4,14], as well as affecting the catch per unit effort (CPUE), which is used as an index of relative resource abundance [15,16]. For example, SST had a positive effect on the CPUE between 6 °C and 19 °C and a negative effect between 19 °C and 35 °C for the Korean large purse seine fishery [15]. Within the appropriate temperature and salinity range (suitable water temperature of 20 °C and suitable salinity of 33.75‰), chub mackerel tend to aggregate at the convergence zone between the cold and warm water masses, where significant thermal gradients and salinity gradients exist [17]. Furthermore, the fishing grounds for chub mackerel are affected by changes in the path of the Kuroshio current [17] and climatic regime shifts [18]

To better understand the relationship between the spatiotemporal distribution and resource abundance of chub mackerel and environmental factors, researchers have developed different models using different statistical or machine learning methods, such as GLM and GAM models [15,19,20], minimum volume ellipsoid [3], the habitat suitability index [21,22], and deep learning models [10,23]. These studies suggest that the spatiotemporal distribution of chub mackerel depends on environmental factors, including SST, chlorophyll-a concentration (CHL), sea surface height, and sea surface salinity. However, there exist some differences in the importance of various marine factors among different studies.

However, previous studies on the relationship between chub mackerel resource abundance and the marine environment usually did not take sufficient account of the differences among various spatiotemporal groups of fishery sample data. The age composition of chub mackerel always shows significant variability across different years [6], and the degree of influence of environmental factors on fish at different ages is often different [24,25]. This variability results in the year stratification of data. Furthermore, chub mackerel display obvious migratory behavior, leading to shifts in fishing areas across seasons. The water mass configuration in the fishing ground also varies across different zones and seasons. Consequently, the impact of marine environmental factors on mackerel fluctuates across different months or fishing zones, exhibiting certain patterns within the same month or fishing zone [11,26]. For fishery data with stratification, the different layers can be processed separately according to different months or fishing zones. However, the statistical significance of the results would be lost in a regression analysis due to the insufficient sample size. On the other hand, excessive data subsampling can lead to more complex data pre-processing and analysis as there is excessive stratification (such as an extensive division of fishing zones), which makes the application of the model difficult.

Some studies used a mixed-effects model to analyze the distribution of Georges Bank yellowtail flounder, and the results showed that the mixed-effects model was more parsimonious than the fixed effects model [27]. In this study, in order to enhance our understanding of the impact of marine environment changes on the distribution of mackerel under various spatiotemporal conditions, we analyze the fixed and random effects of SST and CHL on the CPUE of chub mackerel during various month and fishing zone stratifications through the mixed linear model, and investigate the effect of marine environmental changes on chub mackerel CPUE over time and space. Furthermore, this study aims to offer theoretical insights into the sustainable development and management of chub mackerel resources in the Northwest Pacific.

2. Materials and Methods

2.1. Environmental and Fishery Data

Previous studies demonstrated that SST (°C) and CHL (mg/m^3) could have a significant impact on the catch and distribution of chub mackerel resources [3,9,11,28–30]. Therefore, SST and CHL were included in the model of the present study. The monthly averaged SST (Figure 1a) and the monthly averaged CHL (Figure 1b) were acquired from Oregon State University's Ocean Productivity database (http://sites.science.oregonstate.edu/ocean.productivity/, accessed on 1 June 2022), and these data were resampled to a spatial resolution of 0.5° × 0.5°.

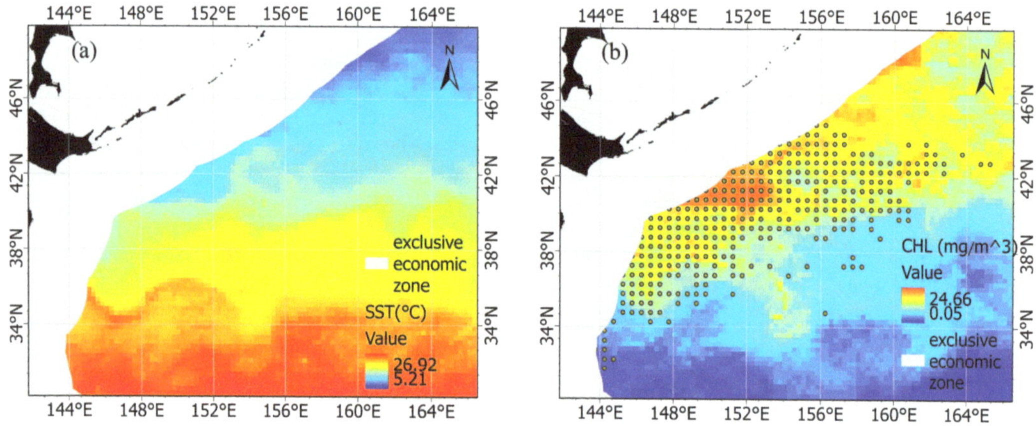

Figure 1. The distribution of (**a**) the sea surface temperature in October 2015, the distribution of (**b**) the chlorophyll-a (CHL) in October 2015, and the fishing area for chub mackerel in the high seas of the Northwest Pacific during 2014–2020 (yellow circles).

The fishing vessel production of China's light-purse seine operations in the high seas of the Northwest Pacific from 2014 to 2020 was collected. The dataset includes the fishery operation dates, locations, net-haul numbers, catches, and so on. The study area covers the traditional fishing grounds of chub mackerel between 140–165° E and 30–50° N in the North Pacific. For each month, we aggregated the fishery data at a spatial resolution of 0.5° × 0.5°. The CPUE (t/net) was calculated with Equation (1).

$$\text{CPUE} = C/N \quad (1)$$

in which C is the total catch (t) in a single fishing zone, and N is the total number of nets hauled (net) in the same zone. To generate the dataset needed for modeling, CPUE was matched with environmental data by the month and zone.

2.2. Normalization and Multicollinearity Detection of Variables

The distribution of CPUE, SST and CHL was shown in Figure 2. We tested the distribution of the CPUE, SST, and CHL to obtain skewness and kurtosis values. The results showed that, except for SST, the skewness of CPUE and CHL was much greater than 0, and the kurtosis was much greater than 3, which indicates a skewed distribution (Table 1, Figure 3). In order to obtain the normally distributed data, the CPUE, SST, and CHL were transformed by taking the natural logarithm or square root of each variable. The natural logarithm is also a method used when dealing with skew data in linear mixed-effects models [27]. The results indicated that the square root of the CPUE (denoted as sqrtCPUE) has a normal distribution (skewness closer to 0, kurtosis closer to 3), and the natural logarithm of CHL (denoted as logCHL) is normally distributed (as shown in Table 1 and Figure 3). The raw SST is close to a normal distribution, so the original SST is directly applied to the model (Table 1).

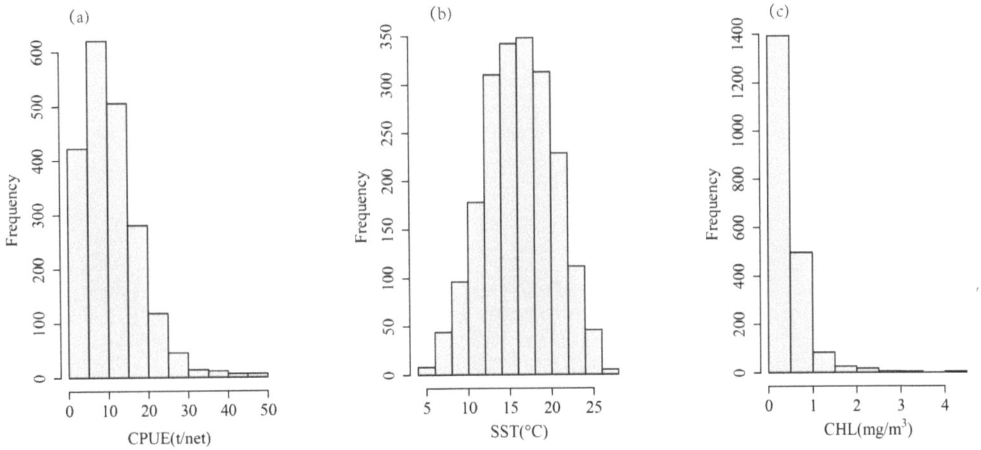

Figure 2. Data distribution of (**a**) CPUE, (**b**) SST, and (**c**) CHL.

Table 1. Skewness and kurtosis of distribution before and after variable transformation.

Variable	Original Distribution		Distribution after Transformation			
			Square Root		Natural logarithm	
	Skewness	Kurtosis	Skewness	Kurtosis	Skewness	Kurtosis
CPUE	1.318	6.337	−0.002	3.459	−1.957	9.027
SST	−0.021	2.610	−0.373	2.952	−0.789	3.842
CHL	4.077	27.165	1.958	9.613	0.511	3.763

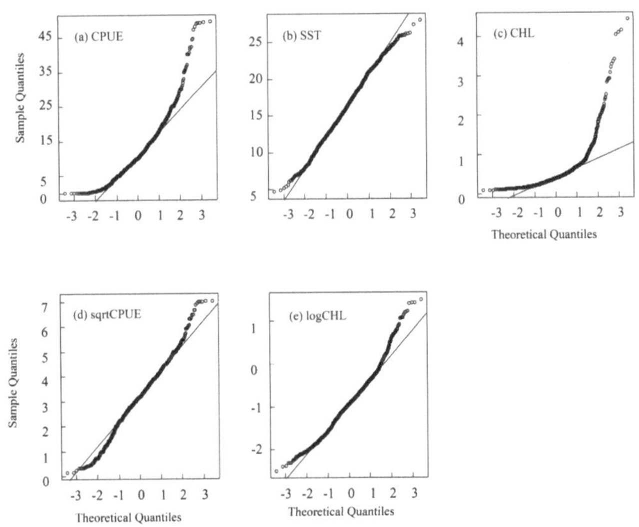

Figure 3. Normal distribution QQ chart of original (**a**) CPUE, (**b**) SST, (**c**) CHL, (**d**) sqrtCPUE, and (**e**) logCHL.

Variance inflation factors (VIF) were used to detect multicollinearity. The variance inflation factor (VIF) of SST and logCHL are both 1.990, which is below the threshold of 5, indicating the absence of multicollinearity between SST and logCHL [31].

2.3. Linear Mixed Model

Mixed-effects models incorporate both fixed effects and random effects [32]. The fixed effects in the model can effectively capture the overall trend of the dependent variable as it changes with the independent variable, while the random effects represent individual characteristics. Linear mixed models (LMMs) are flexible extensions of linear models which incorporate fixed and random effects [33]. Linear mixed-effects models can be mathematically manipulated as matrices as follows [34]:

$$Y = X\beta + Zu + \varepsilon \quad (2)$$

where $Y = (y_1'/y_2', \ldots, y_N')'$ is the $n \times 1$ vector of responses, and $n = n_1 + n_2 + \ldots + n_N$. β is the $p \times 1$ vector of fixed effects, X is the $n \times p$ design matrix for fixed effects, Z is the $n \times (q_1 + q_2 + \ldots + q_N)$ matrix of random effects, b is the $(q_1 + q_2 + \ldots + q_N) \times 1$ vector, and ε is the $n \times 1$ vector of errors, which captures the differences between the observed responses and the predicted responses based on the fixed and random effects in the model.

To determine the effects of SST and chlorophyll concentration on the different levels of CPUE, SST and logCHL were selected as possible fixed effects, and Year, Month, and fishing Zone were considered random effects. We used the lme4 package in the R environment to run the model [32,35]. The model Equation (2) can be written as follows [35]:

sqrtCPUE~1 + SST + logCHL + (1 + SST + logCHL | Year) + (1 + SST + logCHL | Month) + (1 + SST + logCHL | Zone) (3)

in which variables without parentheses represent fixed effects, while Year, Month, and Zone in parentheses denote random effects. The variables to the left of the vertical line in parentheses are covariates; the constant value 1 represents the random intercept.

2.4. Selection of the Optimal Model

Initially, a null model can be developed by considering only the random intercepts, without factoring in the environmental variables. In simpler terms, Equation (3) can be expressed in the following manner:

SqrtCPUE~1 + (1 | Year) + (1 | Month) + (1 | Zone) (4)

By adding the SST and logCHL to the initial model, in fixed effect and random effect terms, 256 candidate models in all can be obtained. In order to choose the optimal model, we fit each candidate model to estimate the model parameters and performed significance tests on both fixed and random terms. Then, each model was assessed via information-based criteria such as AIC, BIC, log likelihood rate (logLik), and residuals. In addition, as strong correlations can affect the estimation of other covariate coefficients, the correlation between slopes and intercept within random effects was checked to ensure that the absolute values of the correlation coefficients between them were less than 0.7 [36–38].

3. Results

3.1. Seasonal Variation and Spatial Distribution of CPUE

The average annual CPUE of chub mackerel showed a trend of decreasing from 2014 to 2020. The CPUE peaked in 2015 and 2018 with values of 15.501 t/net and 11.363 t/net, respectively. In 2020, the CPUE reached its lowest point with a value of 8.355 t/net. In addition, the CPUE showed seasonal variations: The CPUE was relatively lower than 10 t/net from April to July. However, it increased rapidly after August and peaked in November with a value of 14.426 t/net, and then fell rapidly to 5.427 t/net in December.

On account of the spatial distribution patterns of fish, fishing areas with lower CPUE values (less than 5 t/net) were distributed in waters south of 34° N, while fishing areas with higher CPUE values (more than 15 t/net) were located between 148° and 159° E and were relatively dispersed.

As shown in Figure 4, the mean CPUE of chub mackerel varied significantly across different years, months, and spatial locations.

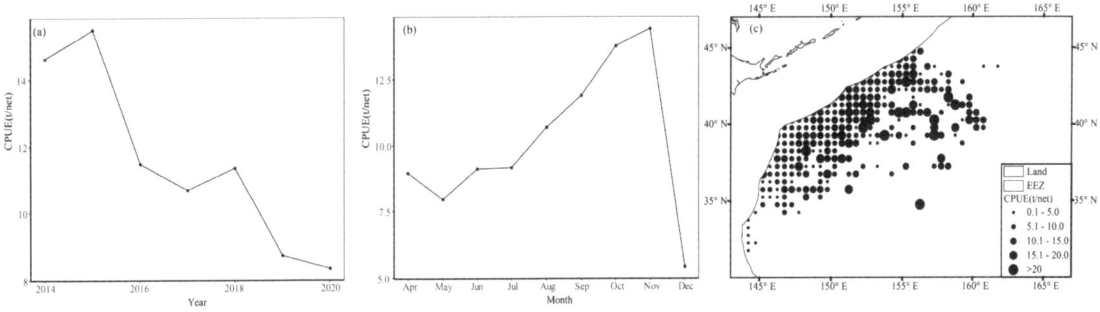

Figure 4. Spatiotemporal variation in (**a**) annual average CPUE, (**b**) average monthly CPUE, and (**c**) spatial distribution of average CPUE for chub mackerel from 2014 to 2019.

3.2. *Optimal Linear Mixed Model*

After calculating the AIC and other information-based criteria for a total of 256 candidate models, the models with fixed effect and random effect items that were statistically significant ($p < 0.05$) and the weak correlation (absolute value was less than 0.7) between the slope and intercept of random effect items were retained. As a result, six candidate models were finally obtained, as shown in Table 2. For comparison, the regression model with only fixed effects is also listed in the table, numbered 0.

Table 2. Six preliminarily selected candidate models.

Scenario	Model Structure	AIC	BIC	logLik	Deviance
0	1 + sst + logCHL	6.265	6.287	−3.128	2.589
1	1 + (1 ǀ Year) + (1 ǀ Month) + (1 ǀ Zone)	6.077	6.105	−3.034	6.067
2	1 + (1 ǀ Year) + (1 + logCHL ǀ Month) + (1 ǀ Zone)	6.055	6.094	−3.020	6.041
3	1 + (1 + logCHL ǀ Year) + (1 + logCHL ǀ Month) + (1 ǀ Zone)	6.056	6.106	−3.019	6.038
4	1 + sst + (1 ǀ Year) + (1 ǀ Month) + (1 ǀ Zone)	6.058	6.091	−3.023	6.045
5	1 + sst + (1 ǀ Year) + (1 + logCHL ǀ Month) + (1 ǀ Zone)	6.035	6.080	−3.010	6.019
6	1 + sst + logCHL + (1 ǀ Year) + (1 ǀ Month) + (1 ǀ Zone)	6.054	6.093	−3.020	6.040

As shown in Table 2, the values of AIC, BIC, and logLik in the scenario 0 model were relatively higher and did not show random effects. Among scenarios 1–6 with random effects, the scenario 5 model exhibited the highest accuracies with the lowest values of AIC, BIC, LogLik, and Deviance; thus, the scenario 5 model was selected as the optimal model, including SST as the fixed effect and three random effects. There was only a random intercept in the random effect zone; meanwhile, there were both an intercept and a slope of logCHL term in the random effect of Month.

3.3. *Mixed Effects of SST and logCHL on CPUE*

3.3.1. Fixed Effect Parameters

The optimal model (Model 5, Table 2) retained SST as a fixed effect. The intercept was positive, and the SST slope value was negative (Table 3). The t-test results for the intercept and slope indicated that both of them were statistically significant ($p < 0.001$).

Table 3. Coefficients table of fixed effects in the model.

Variable	Estimate	Std. Error	Degree of Freedom	T	p
(Intercept)	3.794	0.211	27.275	17.974	0.000 ***
SST	−0.050	0.010	366.871	−4.933	0.000 ***

*** $p < 0.001$.

3.3.2. Random Effect Parameters

The variances and standard deviations (Std. Dev) of the random slope and random intercept coefficients in three random effects are shown in Table 4. In the Month group, the correlation coefficient between the random intercept and the slope of the covariate logCHL was −0.07, indicating a very weak correlation. This suggests that the estimation of model parameters was stable, with relatively low error and variability.

Table 4. Parameter estimation for random effects in the model.

Groups	Coefficient	Variance	Std. Dev.	Correlation
Zone	(Intercept)	0.031	0.175	/
Month	(Intercept)	0.094	0.307	−0.070
	logCHL	0.084	0.289	
Year	(Intercept)	0.093	0.304	/
Residual		1.075	1.037	/

A likelihood ratio test (LRT) was also conducted on random effects. After removing three random effect terms, the accuracy of the model significantly decreased compared to that of the original model (p-values were all less than 0.05), suggesting that retaining the three random effect terms has great effects on the accuracy of the model (Table 5).

Table 5. ANOVA-like table for random effects of the model.

Deleted Single Term	The Number of Parameters	Log-LPT	AIC	LRT	Degree of Freedom	p
None	8	−3009.6	6035.1			
(1 ǀ year)	7	−3055.4	6124.9	91.772	1	0.000 ***
logCHL in (1 + logCHL ǀ m)	6	−3022.8	6057.6	26.424	2	0.000 ***
(1 ǀ zone)	7	−3012.3	6038.6	5.483	1	0.019 *

* $p < 0.05$, *** $p < 0.001$.

The average random intercept values for the Year group are shown in Figure 5a, with the highest value (0.353) observed in 2015 and the lowest value (−0.480) observed in 2019. Their trend over the years is consistent with the annual variation in the CPUE in Figure 4a. For the random effect of the Month, the fluctuation in the logCHL slope and intercept is shown in Figure 5b. The average slopes for April, May, November, and December are positive, while those from June to October are negative. In terms of the average intercept in the Month group, the values are positive from August to November, while they are negative for the remaining months, and the trend in the change is basically consistent with the monthly changes in the CPUE in Figure 4b.

As shown in Figure 6, the random intercept in the Zone group ranged between −0.164 and 0.260 with an average value of 0.000 ± 0.066. It is worth noting that there was a region located between 150–152° E and 39–41°, with relatively high slope values, which was very different from the high CPUE distribution in Figure 4c. The average slope value in the region was 0.098 ± 0.069.

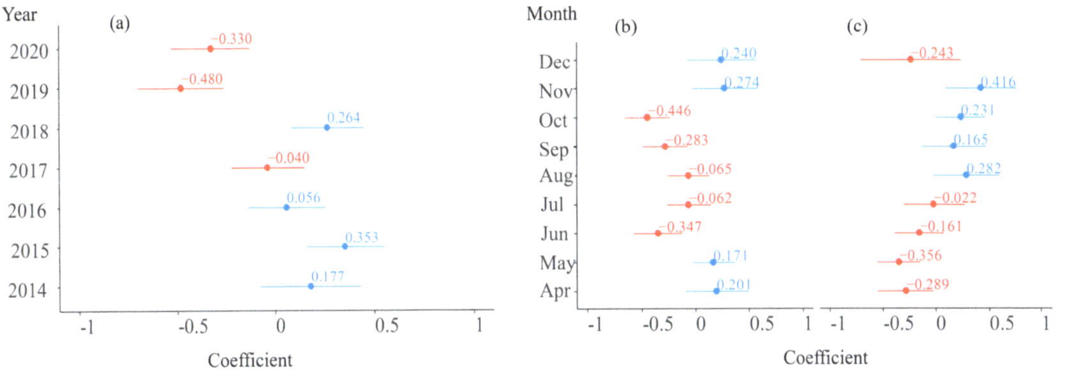

Figure 5. Slope and intercepts in random effects of Year and Month. (**a**) Random intercept in Year group. (**b**) Random slope of logCHL in Month group. (**c**) Random intercept in Month group (Values in blue font denote positive numbers, while those in red font signify negative numbers.).

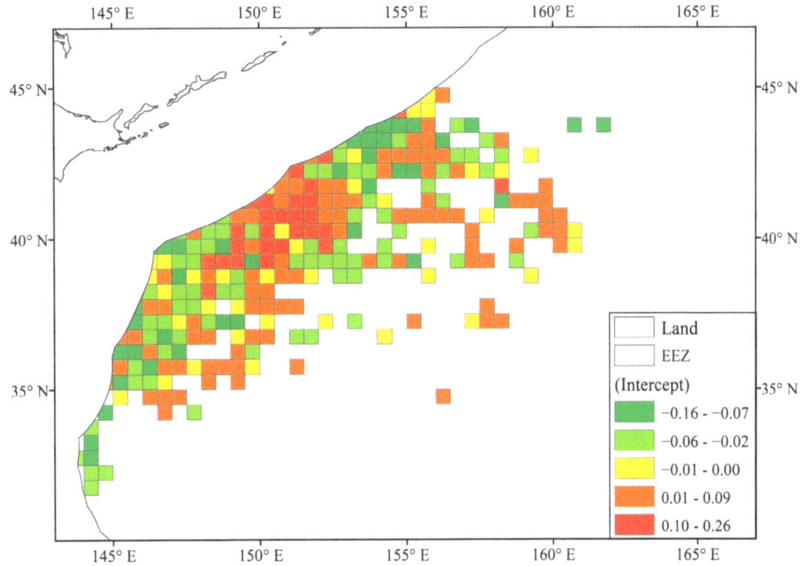

Figure 6. Spatial distribution of random intercepts in the Zone group.

3.4. Residual Test

One of the fundamental assumptions of mixed linear models is that the residuals are normally distributed. The skewness of residuals obtained in the present model was 0.031, which was close to 0 for the normal distribution. The kurtosis value was 3.532, which is close to 3 for the normal distribution. In addition, the points in the Quantile–Quantile Plot (Q-Q Plot) of the residuals were roughly distributed along a straight line (Figure 7). Therefore, it could be concluded that the residuals from the LMM closely approximate a normal distribution.

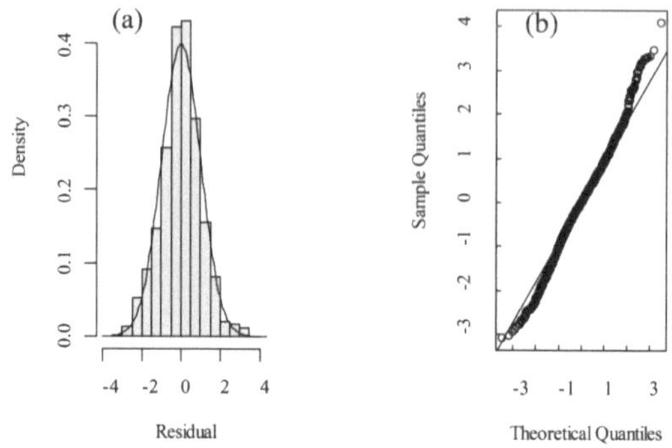

Figure 7. Normality assessment by (**a**) residual distribution and (**b**) Q-Q plot.

4. Discussion

4.1. Influence of SST on Chub Mackerel

The catch and biomass of chub mackerel are closely related to the recruitment stock population [4]. As a poikilotherm, the growth, mortality, and habitat selection of the chub mackerel are affected by seawater temperature and are highly sensitive to changes in seawater temperature [28–30,39]. The SST not only influences the growth rate of chub mackerel [5,7,40] but also has a great impact on the spatial distribution of chub mackerel. Previous studies have suggested that the optimal temperature for the fishing grounds of chub mackerel in the Northwest Pacific is between 14 and 18 °C, and larger catches are easily obtained in waters around 16 °C [11]. Within the appropriate temperature range, the catches will increase with higher water temperature and decrease with lower temperature [41]. However, some studies suggested that a lower SST often corresponds to a higher concentration of chub mackerel in fishing grounds in the East China Sea [9]. In the central Atlantic region of the Middle East, there is a significant negative correlation between chub mackerel catch and SST [42]. In the coastal sea around Japan, an increased SST can result in a decrease in stock size for chub mackerel [43]. Our study showed that the slope of SST in fixed effect exhibited a negative value (Table 3), which is consistent with the above mentioned studies [9,42,43]. It implies that the CPUE of chub mackerel in the Northwest Pacific will increase as the SST generally decreases. In other words, lower water temperatures can lead to higher bait densities, which would facilitate fish growth [7,44].

As is well known, ocean warming has affected the distribution of marine life and global fisheries across time and space [45,46]. Higher water temperatures not only cause short-term changes in the CPUE of chub mackerel, but also have long-term effects on the abundance of the stock. During the spawning period in winter, the effect of seawater temperature suitability on its abundance has a lag effect of 1–2 years (Wang et al., 2022 [14]). Furthermore, the increasing SST with the times has also impacted the habitat location and initiation time of northward migration for chub mackerel. In the 1990s, the chub mackerel spawning ground remained near the Izu Islands until June. However, due to the increase in SST, the species gradually shifts northward after 2000 and begins to move northward in May. As the SST increases, the migration time also advances. On the other hand, the center of gravity of the fishing grounds of the Northwest Pacific light-purse seine showed a gradual trend of northeastward movement. There was a longitude shift of 0.92° to the east and a latitude shift of 1.57° to the north from 2014 to 2019 [47]. As small pelagic fish, chub mackerel are highly sensitive to environmental impacts, and climate change may amplify this sensitivity [44,48]. In this study, the random intercept in the year group and the actual CPUE value both showed a decreasing trend (Figures 4a and 5a). In addition, overfishing and other climate changes have effects on the resource abundance of chub mackerel [5,49]; however, SST is an important factor in the distribution, productivity, and survival of this species.

4.2. The Effect of Chlorophyll-a Concentration on Chub Mackerel

Chub mackerel primarily prey on zooplanktons in the early life stage. These planktonic animals are always abundant in regions with high phytoplankton biomass in the Northwest Pacific [50,51]. Therefore, although chub mackerel are not a direct predator of marine phytoplankton, the organisms that make up the intermediate levels of the food web are affected by primary productivity [52]. Thus, the CPUE of chub mackerel also has a certain correlation with chlorophyll-*a* concentration. Studies have suggested that there is a significant positive correlation between the most suitable ranges of chlorophyll-*a* concentration and the CPUE of chub mackerel in the Northwest Pacific [11]. The optimal range of chlorophyll-*a* for chub mackerel in the East China Sea and North Pacific ranges from 0.3 to 0.8 mg/m^3, which can be used as an indicator for predicting potential fishing grounds of chub mackerel [11,42]. In this study, logCHL did not have a fixed effect on CPUE. However, after grouping by months, the addition of the logCHL random slope significantly improved the fitting accuracy of the model (Table 5, $p < 0.001$). This suggests that the influence of chlorophyll-*a* concentration on CPUE in different months cannot be ignored.

In this study, the coefficient sign of logCHL in different months had both positive and negative signs, which indicated that the effect of logCHL on CPUE in different seasons was not fixed in direction. As shown in Figure 5b, the months with positive logCHL coefficient values were April–May and November–December (data were not available for January to March), which correspond to the wintering and spawning period of chub mackerel [2,4,53]. During this period, the fish forage and grow in the wintering grounds and prepare for the next breeding season. Therefore, the oceanographic conditions of wintering and spawning grounds have a crucial impact on the recruitment and stock of chub mackerel [4]. From June to October, chub mackerel are in the summer feeding period [54,55], and they are often distributed on the high-temperature and high-salinity side of the front zone [12,56]. In this study, the average coefficient of logCHL from June to October was close to or less than 0 (see Figure 5b), which suggested that the increase in chlorophyll-*a* concentration will not result in an increased CPUE during the summer; in fact, the CPUE even decreases. This phenomenon not only exists in the CPUE of chub mackerel; studies have also shown that there is a negative correlation between the CPUE of neon flying squid (*Ommastrephes bartramii*) and chlorophyll-*a* concentration in the Northwest Pacific from June to September. This may be attributed to the predominant location of fishing grounds distributed along the Kuroshio front during this season, where the chlorophyll-*a* concentration is relatively lower than that on the Oyashio side [57,58]. The habitat characteristics and migration routes of chub mackerel are similar to those of neon flying squid, leading to a significant overlap in their fishing grounds. Therefore, the resource abundance of the two species and the chlorophyll-*a* concentration both showed a negative correlation.

4.3. Implications of Intercepts in Random Effects of Year and Zone

The random intercepts under Year and Zone in the model were significant (Table 5), indicating spatiotemporal differences in CPUE in different years and fishing zones. It can also be said that they are the inherent level of CPUE when the other explanatory variables (SST and logCHL) are constant. The abundance of chub mackerel resources is influenced by large-scale and mesoscale marine climate phenomena such as El Niño [3,8,59], and fishing pressure can also cause changes in the stock structure of chub mackerel [5,43]. Therefore, the stock of chub mackerel shows significant fluctuations in the Northwest Pacific in different years [5,54]. The fluctuation always has a large timescale and cannot be correlated with real-time SST and chlorophyll concentration in the model. However, it can be utilized as a random intercept for grouping years in the model. Furthermore, chub mackerel fishing grounds are located in the northwest Pacific Ocean, where the Oyashio cold current and the Kuroshio warm current intersect [60,61], resulting in variations in the spatial distribution of chub mackerel resources. The water areas with high random intercepts (150–152° E, 39–41° N) mentioned in Section 3.3.2 have a high catch rate of chub mackerel [39] and the highest fishing density [11,47]. The model successfully captured this spatial distribution feature through Zone groups.

Therefore, temporal trends or spatial distributions expressed by the random intercept, although similar to the features in Figure 4a,c, actually have significant differences from them. The random intercepts in the model were stable features of the spatiotemporal distribution of chub mackerel CPUE, as they resulted from removing the influence of environmental changes.

4.4. Scalability of Linear Mixed-Effects Regression

Firstly, linear mixed-effects regression, as a traditional statistical method, is not based on deep learning (DL) techniques. Instead, it falls under the umbrella of classical statistical modeling approaches [33,34]. Therefore, it can be computationally intensive, especially for large datasets with numerous observations and complex random effects structures. Secondly, when the relationship of the data is not linear, a nonlinear mixed-effect model can be used [62]. As the extension of linear mixed-effects regression, such a model can handle more complex data relationships. By considering factors such as computational complexity and the extensibility of the model structure, we can confirm that linear mixed-effects models have strong scalability in fishery research.

4.5. Limitations

The first problem is the complexity of interpreting the results, especially when dealing with multiple random effects and interactions between fixed and random effects. In the present study, the levels of the grouping variables, including Year, Month and Zone, crossed each other, which can have an adverse effect on the interpretation of the model. However, some nonlinear models, such as the Gaussian mixture model [63], can be interpreted well for each Gaussian distribution corresponding to a subgroup of data.

The second problem is the high degree of parametrization, involving many free parameters (Figures 5 and 6). While these parameters were capable of capturing intricate relationships and variations between the CPUE and environmental and spatiotemporal factors, highly parametric models pose certain risks. On the one hand, they may exhibit overfitting tendencies, wherein noise in the data is captured instead of the genuine underlying pattern. Consequently, when these parameters are utilized to forecast fishing grounds, significant discrepancies between the predicted and actual locations may arise. On the other hand, such models can complicate interpretation due to their excessive complexity, hindering the extraction of meaningful insights. Therefore, the decision to employ a linear mixed-effect regression model should consider the balance between model flexibility and complexity.

The third problem is the incomplete distribution of data samples in time and space. This research utilizes commercial fishing data solely from mainland China, and there is no data coverage within the EEZ of Japan and Russia. In fact, the spatial habitat range of chub mackerel also includes the inner EEZ, especially near the spawning grounds close to the Izu Islands in Japan [5,28]. The model did not include data from January to March due to the lack of fishing operations. The absence of data in both time and space can also be viewed as an imbalance in sampling, which will inevitably result in a reduction in model accuracy.

5. Conclusions

The traditional linear model is a simple and intuitive method for data analysis. However, when applied to fishery production or survey data, it often encounters the problem of multiple samplings in the same fishing zone, season, or year, easily leading to data stratification and the contradiction of assumptions of traditional linear models. In the present study, we utilized a linear mixed model to analyze the relationship between the CPUE of chub mackerel and environmental factors, including SST and chlorophyll concentration. The findings of this study are as follows:

(1) SST is a fixed effect and there exists a negative correlation between SST and CPUE;

(2) The effect of chlorophyll concentration (logCHL) on the CPUE does not have a fixed effect, but the coefficient of the covariate (logCHL) for the Month group shows significance, and it will have different effects on CPUE with seasonal changes. Chlorophyll concentration is positively correlated with CPUE during wintering and spawning periods and is negatively correlated with CPUE during feeding periods;
(3) The inherent differences in CPUE in time and space can be captured through random intercepts, thereby enhancing the accuracy and generalization ability of the model. This flexibility enables the model to more comprehensively describe the structure and variability in the CPUE, helping us better reveal the complex relationships and patterns behind the changes in chub mackerel CPUE.

Author Contributions: Conceptualization, X.C. and F.T.; methodology, X.C.; writing—original draft preparation, J.L.; writing—review and editing, F.T. and S.Z.; supervision, X.C.; project administration, Y.W. and W.Z. All authors have read and agreed to the published version of the manuscript.

Funding: This research was funded by the National Key R&D Program of China, grant number 2023YFD2401303, and the APC was funded by the National Key R&D Program of China.

Institutional Review Board Statement: Not applicable.

Informed Consent Statement: Not applicable.

Data Availability Statement: The environmental data used in this study are available on the Ocean Productivity database of Oregon State University: http://sites.science.oregonstate.edu/ocean.productivity/ (accessed on 14 January 2024).

Conflicts of Interest: The authors declare that they have no competing interests.

References

1. Guo, C.; Ito, S.; Kamimura, Y.; Xiu, P. Evaluating the Influence of Environmental Factors on the Early Life History Growth of Chub Mackerel (*Scomber japonicus*) Using a Growth and Migration Model. *Prog. Oceanogr.* **2022**, *206*, 102821. [CrossRef]
2. Shiraishi, T.; Okamoto, K.; Yoneda, M.; Sakai, T.; Ohshimo, S.; Onoe, S.; Yamaguchi, A.; Matsuyama, M. Age Validation, Growth and Annual Reproductive Cycle of Chub Mackerel *Scomber japonicus* off the Waters of Northern Kyushu and in the East China Sea. *Fish. Sci.* **2008**, *74*, 947–954. [CrossRef]
3. Torrejón-Magallanes, J.; Ángeles-González, L.E.; Csirke, J.; Bouchon, M.; Morales-Bojórquez, E.; Arreguín-Sánchez, F. Modeling the Pacific Chub Mackerel (*Scomber japonicus*) Ecological Niche and Future Scenarios in the Northern Peruvian Current System. *Prog. Oceanogr.* **2021**, *197*, 102672. [CrossRef]
4. Wang, L.; Ma, S.; Liu, Y.; Li, J.; Liu, S.; Lin, L.; Tian, Y. Fluctuations in the Abundance of Chub Mackerel in Relation to Climatic/Oceanic Regime Shifts in the Northwest Pacific Ocean since the 1970s. *J. Mar. Syst.* **2021**, *218*, 103541. [CrossRef]
5. Wang, Z.; Ito, S.; Yabe, I.; Guo, C. Development of a Bioenergetics and Population Dynamics Coupled Model: A Case Study of Chub Mackerel. *Front. Mar. Sci.* **2023**, *10*, 1142899. [CrossRef]
6. Yukami, R.; Nishijima, S.; Kamimura, Y.; Furuichi, S. Stock Assessment and Evaluation for Chub Mackerel (Fiscal Year 2022). In *Marine Fisheries Stock Assessment and Evaluation for Japanese Waters*; Japan Fisheries Agency and Japan Fisheries Research and Education Agency: Tokyo, Japan, 2022; pp. 39–42.
7. Taga, M.; Kamimura, Y.; Yamashita, Y. Effects of Water Temperature and Prey Density on Recent Growth of Chub Mackerel *Scomber japonicus* Larvae and Juveniles along the Pacific Coast of Boso–Kashimanada. *Fish. Sci.* **2019**, *85*, 931–942. [CrossRef]
8. Cui, K.; Chen, X. Study of the relationships between SST and mackerel abundances in the Yellow and East China Seas. *South China Fish. Sci.* **2007**, *3*, 20–25.
9. Guan, W.; Chen, X.; Gao, F.; Li, G. Environmental Effects on Fishing Efficiency of *Scomber japonicus* for Chinese Large Lighting Purse Seine Fishery in the Yellow and East China Seas. *J. Fish. Sci. China* **2009**, *16*, 949–958.
10. Han, H.; Yang, C.; Jiang, B.; Shang, C.; Sun, Y.; Zhao, X.; Xiang, D.; Zhang, H.; Shi, Y. Construction of Chub Mackerel (*Scomber japonicus*) Fishing Ground Prediction Model in the Northwestern Pacific Ocean Based on Deep Learning and Marine Environmental Variables. *Mar. Pollut. Bull.* **2023**, *193*, 115158. [CrossRef]
11. Dai, S.; Tang, F.; Fan, W.; Zhang, H.; Cui, X. Distribution of Resource and Environment Characteristics of Fishing Ground of Scomber Japonicas in the North Pacific High Seas. *Mar. Fish.* **2017**, *39*, 372–382.
12. Miao, Z. Relation between Pneumatophorus and Carangidae Fishing Grounds in the Summer-Autumn and Ocean Hydrologic Environment in the Northern Part of the East China Sea. *J. Zhejiang Coll. Fish.* **1993**, *12*, 32–39.
13. Sassa, C.; Tsukamoto, Y. Distribution and Growth of *Scomber japonicus* and S. Australasicus Larvae in the Southern East China Sea in Response to Oceanographic Conditions. *Mar. Ecol. Prog. Ser.* **2010**, *419*, 185–199. [CrossRef]

14. Wang, L.; Ma, S.; Liu, Y.; Li, J.; Sun, D.; Tian, Y. Climate-Induced Variation in a Temperature Suitability Index of Chub Mackerel in the Spawning Season and Its Effect on the Abundance. *Front. Mar. Sci.* **2022**, *9*, 996626. [CrossRef]
15. Owiredu, S.A.; Onyango, S.O.; Song, E.-A.; Kim, K.-I.; Kim, B.-Y.; Lee, K.-H. Enhancing Chub Mackerel Catch Per Unit Effort (CPUE) Standardization through High-Resolution Analysis of Korean Large Purse Seine Catch and Effort Using AIS Data. *Sustainability* **2024**, *16*, 1307. [CrossRef]
16. Shi, Y.; Zhang, X.; Yang, S.; Dai, Y.; Cui, X.; Wu, Y.; Zhang, S.; Fan, W.; Han, H.; Zhang, H.; et al. Construction of CPUE Standardization Model and Its Simulation Testing for Chub Mackerel (*Scomber japonicus*) in the Northwest Pacific Ocean. *Ecol. Indic.* **2023**, *155*, 111022. [CrossRef]
17. Li, Y.; Pan, L.; Yan, L.; Chen, X. Individual-based model study on the fishing ground of chub mackerel in the East China Sea. *Acta Oceanol. Sin.* **2014**, *36*, 67–74. [CrossRef]
18. Chen, L.F.; Zhu, G.P. Analysis of Influence on Spatial Distribution of Fishing Ground for Antarctic Krill Fishery in the Northern South Shetland Islands Based on GWR Model. *Ying Yong Sheng Tai Xue Bao* **2018**, *29*, 938–944. [CrossRef] [PubMed]
19. Hashimoto, M.; Nishijima, S.; Yukami, R.; Watanabe, C.; Kamimura, Y.; Furuichi, S.; Ichinokawa, M.; Okamura, H. Spatiotemporal Dynamics of the Pacific Chub Mackerel Revealed by Standardized Abundance Indices. *Fish. Res.* **2019**, *219*, 105315. [CrossRef]
20. Shi, Y.; Zhang, H.; Ma, Q.; Yang, C.; Zhao, G. *Standardized CPUE of Chub Mackerel (Scomber japonicas) Caught by the China's Lighting Purse Seine Fishery up to 2019*; NPFC-2021-TWG CMSA04-WP17; East China Sea Fisheries Research Institute, Chinese Academy of Fishery Sciences: Shanghai, China, 2021.
21. Chen, X.; Li, G.; Feng, B.; Tian, S. Habitat Suitability Index of Chub Mackerel (*Scomber japonicus*) from July to September in the East China Sea. *J. Oceanogr.* **2009**, *65*, 93–102. [CrossRef]
22. Lee, D.; Son, S.; Kim, W.; Park, J.M.; Joo, H.; Lee, S.H. Spatio-Temporal Variability of the Habitat Suitability Index for Chub Mackerel (*Scomber japonicus*) in the East/Japan Sea and the South Sea of South Korea. *Remote Sens.* **2018**, *10*, 938. [CrossRef]
23. Xiao, G. Construction and Comparison of Fishing Ground Forecast Model of Chub Mackerel (*Scomber japonicus*) in Pacific Northwest (Master). Master's Thesis, Shanghai Ocean University, Shanghai, China, 2022.
24. Gilbert, C.S.; Gentleman, W.C.; Johnson, C.L.; DiBacco, C.; Pringle, J.M.; Chen, C. Modeling Dispersal of Sea Scallop (*Placopecten Magellanicus*) Larvae on Georges Bank: The Influence of Depth-Distribution, Planktonic Duration and Spawning Seasonality. *Prog. Oceanogr.* **2010**, *87*, 37–48. [CrossRef]
25. Brochier, T.; Colas, F.; Lett, C.; Echevin, V.; Cubillos, L.A.; Tam, J.; Chlaida, M.; Mullon, C.; Freon, P. Small Pelagic Fish Reproductive Strategies in Upwelling Systems: A Natal Homing Evolutionary Model to Study Environmental Constraints. *Prog. Oceanogr.* **2009**, *83*, 261–269. [CrossRef]
26. Li, J.; Cui, X.; Tang, F.; Fan, W.; Han, Z.; Wu, Z. Spatiotemporal Analysis of Marine Environmental Influence on the Distribution of Chub Mackerel in the Northwest Pacific Ocean Based on Geographical and Temporal Weighted Regression. *J. Sea Res.* **2024**, *200*, 102514. [CrossRef]
27. Hyun, S.-Y.; Cadrin, S.X.; Roman, S. Fixed and Mixed Effect Models for Fishery Data on Depth Distribution of Georges Bank Yellowtail Flounder. *Fish. Res.* **2014**, *157*, 180–186. [CrossRef]
28. Kanamori, Y.; Takasuka, A.; Nishijima, S.; Okamura, H. Climate Change Shifts the Spawning Ground Northward and Extends the Spawning Period of Chub Mackerel in the Western North Pacific. *Mar. Ecol. Prog. Ser.* **2019**, *624*, 155–166. [CrossRef]
29. Wootton, R.J. *Ecology of Teleost Fishes*; Springer: Dordrecht, The Netherlands, 1989; ISBN 978-94-010-6859-8.
30. Yasuda, T.; Kawabe, R.; Takahashi, T.; Murata, H.; Kurita, Y.; Nakatsuka, N.; Arai, N. Habitat Shifts in Relation to the Reproduction of Japanese Flounder *Paralichthys Olivaceus* Revealed by a Depth-Temperature Data Logger. *J. Exp. Mar. Biol. Ecol.* **2010**, *385*, 50–58. [CrossRef]
31. Akinwande, M.O.; Dikko, H.G.; Samson, A. Variance Inflation Factor: As a Condition for the Inclusion of Suppressor Variable(s) in Regression Analysis. *Open J. Stat.* **2015**, *5*, 754–767. [CrossRef]
32. Pinheiro, J.C.; Bates, D.M. (Eds.) Linear Mixed-Effects Models: Basic Concepts and Examples. In *Mixed-Effects Models in S and S-PLUS*; Springer: New York, NY, USA, 2000; pp. 3–56, ISBN 978-0-387-22747-4.
33. Montesinos López, O.A.; Montesinos López, A.; Crossa, J. Linear Mixed Models. In *Multivariate Statistical Machine Learning Methods for Genomic Prediction*; Montesinos López, O.A., Montesinos López, A., Crossa, J., Eds.; Springer International Publishing: Cham, Switzerland, 2022; pp. 141–170, ISBN 978-3-030-89010-0.
34. Chen, Z.; Zhu, S.; Niu, Q.; Zuo, T. Knowledge Discovery and Recommendation with Linear Mixed Model. *IEEE Access* **2020**, *8*, 38304–38317. [CrossRef]
35. Bates, D.; Mächler, M.; Bolker, B.; Walker, S. Fitting Linear Mixed-Effects Models Using Lme4. *J. Stat. Softw.* **2015**, *67*, 1–48. [CrossRef]
36. Akoglu, H. User's Guide to Correlation Coefficients. *Turk. J. Emerg. Med.* **2018**, *18*, 91–93. [CrossRef]
37. Bates, M.D.; Castellano, K.E.; Rabe-Hesketh, S.; Skrondal, A. Handling Correlations Between Covariates and Random Slopes in Multilevel Models. *J. Educ. Behav. Stat.* **2014**, *39*, 524–549. [CrossRef]
38. Chung, Y.; Gelman, A.; Rabe-Hesketh, S.; Liu, J.; Dorie, V. Weakly Informative Prior for Point Estimation of Covariance Matrices in Hierarchical Models. *J. Educ. Behav. Stat.* **2015**, *40*, 136–157. [CrossRef]
39. Wang, L.; Li, Y.; Zhang, R.; Tian, Y.; Zhang, J.; Lin, L. Relationship between the Resource Distribution of *Scomber japonicus* and Seawater Temperature Vertical Structure of Northwestern Pacific Ocean. *Period. Ocean. Univ. China* **2019**, *49*, 29–38. [CrossRef]

40. Kamimura, Y.; Takahashi, M.; Yamashita, N.; Watanabe, C.; Kawabata, A. Larval and Juvenile Growth of Chub Mackerel *Scomber japonicus* in Relation to Recruitment in the Western North Pacific. *Fish. Sci.* **2015**, *81*, 505–513. [CrossRef]
41. Cui, G.; Zhu, W.; Dai, Q. Temporal and Spatial Distribution of the Mackerel Fishing Ground in the Northwest Pacific and Its Relationship with Sea Surface Temperature and Chlorophyll Concentration. *Ocean. Dev. Manag.* **2021**, *38*, 95–99.
42. Li, X.; Pang, Z.; Zhu, J.; Ying, Y.; Sun, S. Spatial-Temporal Patterns in Fishing Ground of *Scomber japonicus* in Center Eastern Central Atlantic Ocean. *Fish. Sci.* **2018**, *37*, 31–37. [CrossRef]
43. Suda, M.; Watanabe, C.; Akamine, T. Two-Species Population Dynamics Model for Japanese Sardine *Sardinops Melanostictus* and Chub Mackerel *Scomber japonicus* off the Pacific Coast of Japan. *Fish. Res.* **2008**, *94*, 18–25. [CrossRef]
44. Yatsu, A.; Watanabe, T.; Ishida, M.; Sugisaki, H.; Jacobson, L.D. Environmental Effects on Recruitment and Productivity of Japanese Sardine Sardinops Melanostictus and Chub Mackerel *Scomber japonicus* with Recommendations for Management. *Fish. Oceanogr.* **2005**, *14*, 263–278. [CrossRef]
45. Cheung, W.W.L.; Watson, R.; Pauly, D. Signature of Ocean Warming in Global Fisheries Catch. *Nature* **2013**, *497*, 365–368. [CrossRef] [PubMed]
46. Ma, Z.; Xu, Z.; Zhou, J. Effect of Global Warming on the Distribution of Lucifer Intermedius and *L. hanseni* (Decapoda) in the Changjiang Estuary. *Prog. Nat. Sci.* **2009**, *19*, 1389–1395. [CrossRef]
47. Zhao, G.; Wu, Z.; CUI, X.; FAN, W.; Shi, Y.; Xiao, G.; Tang, F. Spatial Temporal Patterns of Chub Mackerel Fishing Ground in the Northwest Pacific Based on Spatial Autocorrelation Model. *Haiyang Xuebao* **2022**, *44*, 22–35. [CrossRef]
48. Caramantin-Soriano, H.; Vega-Pérez, L.A.; Ñiquen, M. The Influence of the 1992–1993 El Niño on the Reproductive Biology of *Scomber japonicus* Peruanus (Jordán & Hubb, 1925). *Braz. J. Oceanogr.* **2009**, *57*, 263–272. [CrossRef]
49. Yu, W.; Wen, J.; Chen, X.; Li, G.; Li, Y.; Zhang, Z. Effects of Climate Variability on Habitat Range and Distribution of Chub Mackerel in the East China Sea. *J. Ocean. Univ. China* **2021**, *20*, 1483–1494. [CrossRef]
50. Lin, P.; Ma, J.; Chai, F.; Xiu, P.; Liu, H. Decadal Variability of Nutrients and Biomass in the Southern Region of Kuroshio Extension. *Prog. Oceanogr.* **2020**, *188*, 102441. [CrossRef]
51. Ozawa, T.; Kawai, K.; Uotani, I. Stomach Content Analysis of Chub Mackerel *Scomber japonicus* Larvae by Quantification I Method. *Nippon Suisan Gakkaishi* **1991**, *57*, 1241–1245. [CrossRef]
52. Smith, R.C.; Dustan, P.; Au, D.; Baker, K.S.; Dunlap, E.A. Distribution of Cetaceans and Sea-Surface Chlorophyll Concentrations in the California Current. *Mar. Biol.* **1986**, *91*, 385–402. [CrossRef]
53. Watanabe, C.; Yatsu, A. Long-Term Changes in Maturity at Age of Chub Mackerel (*Scomber japonicus*) in Relation to Population Declines in the Waters off Northeastern Japan. *Fish. Res.* **2006**, *78*, 323–332. [CrossRef]
54. Watanabe, C.; Nishida, H. Development of Assessment Techniques for Pelagic Fish Stochs: Applications of Daily Egg Production Method and Pelagic Trawl in the Northestern Pacific Ocean. *Fish. Sci.* **2002**, *68*, 97–100. [CrossRef] [PubMed]
55. Watanabe, T. Morphology and Ecology of Early Stages of Life in Japanese Common Mackerel, *Scomber japonicus* Houttuyn, with Special Reference to Fluctuation of Population. *Bull. Tokai Reg. Fish. Res. Lab.* **1970**, *62*, 1–283.
56. Shen, H.; Hong, M. Marine Environment and Distribution of Major Pelagic Fishes in the Northwest Pacific Ocean. *Morden Fish. Inf.* **1986**, *2*, 4–8.
57. Fan, W.; Cui, X.; Yumei, W.U. The Study on Fishing Ground of Neon Flying Squid, Ommastrephes Bartrami, and Ocean Environment Based on Remote Sensing Data in the Northwest Pacific Ocean. *Chin. J. Oceanol. Limnol.* **2009**, *27*, 408–414. [CrossRef]
58. Wang, Y.; Bi, R.; Zhang, J.; Gao, J.; Takeda, S.; Kondo, Y.; Chen, F.; Jin, G.; Sachs, J.P.; Zhao, M. Phytoplankton Distributions in the Kuroshio-Oyashio Region of the Northwest Pacific Ocean: Implications for Marine Ecology and Carbon Cycle. *Front. Mar. Sci.* **2022**, *9*, 865142. [CrossRef]
59. Guo, A.; Zhang, Y.; Yu, W.; Chen, X.J.; Qian, W. Influence of El Niño and La Niña with Different Intensity on Habitat Variation of Chub Mackerel Scomber Japonicas in the Coastal Waters of China. *Haiyang Xuebao* **2018**, *40*, 58–67.
60. Kuroda, H.; Suyama, S.; Miyamoto, H.; Setou, T.; Nakanowatari, T. Interdecadal Variability of the Western Subarctic Gyre in the North Pacific Ocean. *Deep Sea Res. Part I Oceanogr. Res. Pap.* **2021**, *169*, 103461. [CrossRef]
61. Qiu, B. Kuroshio and Oyashio Currents. In *Encyclopedia of Ocean Sciences*; Elsevier: Amsterdam, The Netherlands, 2019; pp. 384–394, ISBN 978-0-12-813082-7.
62. Abbas-Aghababazadeh, F.; Lu, P.; Fridley, B.L. Nonlinear Mixed-Effects Models for Modeling in Vitro Drug Response Data to Determine Problematic Cancer Cell Lines. *Sci. Rep.* **2019**, *9*, 14421. [CrossRef]
63. Aiadi, O.; Kherfi, M.L. A New Method for Automatic Date Fruit Classification. *Int. J. Comput. Vis. Robot.* **2017**, *7*, 692. [CrossRef]

fishes

MDPI

Article

Community Structure Characteristics and Changes in Fish Species at Poyang Lake after the Yangtze River Fishing Ban

Yanping Zhang [1], Haixin Zhang [1], Zijun Wu [1], Mingguang Zhao [2] and Guangpeng Feng [1,2,*]

[1] Jiangxi Institute for Fisheries Sciences, Poyang Lake Fisheries Research Centre of Jiangxi, Nanchang 330039, China; zhangyanpingxie@163.com (Y.Z.); zhang73860@126.com (H.Z.); wuzijun0911@outlook.com (Z.W.)

[2] East China Sea Fisheries Research Institute, Chinese Academy of Fishery Sciences, Shanghai 200090, China; 13320732837@163.com

* Correspondence: fenggp@ecsf.ac.cn

Abstract: To understand the fish community structure characteristics and changes in Poyang Lake after the Yangtze River fishing ban in 2021, seven sampling stations were established, namely in Hukou, Lushan, Duchang, Yongxiu, Nanjishan, Ruihong, and Poyang, for fishery resource surveys between 2020 and 2021. The results showed that 93 fish species were identified, belonging to 11 orders, 20 families, and 61 genera, which was an increase of 4.5% from before the fishing ban. Of these, 52 Cyprinidae species were identified, comprising the highest proportion (55.91%). Four invasive alien species (hybrid sturgeon, *Cirrhina mrigala*, *Piaractus brachypomus*, and *Mugil cephalus*) were identified, and the number of alien species in Poyang Lake was higher than before the fishing ban. Carnivorous and mid-lower-level fish showed a significant increase, accounting for 47.31% and 38.71% of the total species, respectively. Compared with the Yangtze River before the fishing ban, the body length and body weight of the main economic fish in Poyang Lake increased by 6.10–61.26% and by 15.14–291.57%, respectively. In terms of age structure, the proportion of major economically important fish aged 1 or 2 years decreased significantly, while the proportion of older fish increased substantially. There was little difference in the diversity of fish communities at different stations in Poyang Lake. In terms of biodiversity, the Shannon–Wiener index ranged from 2.158 to 2.909, with Poyang having the highest value and Nanjishan the lowest. Margalef's index ranged from 4.265 to 6.459, with Lushan having the highest value and Nanjishan the lowest. Pielou's index ranged from 0.617 to 0.822, with Duchang having the highest value and Nanjishan the lowest. Hence, the Yangtze river fishing ban has played an important and positive role in the restoration of fishery resources in Poyang Lake. However, long-term tracking and monitoring are needed to provide information to comprehensively evaluate the ecological impact of the Poyang Lake fishing ban.

Keywords: Poyang Lake; fish resources; Yangtze River fishing ban

Key Contribution: The Yangtze River fishing ban has played an important and positive role in the restoration of fishery resources in Poyang Lake.

Citation: Zhang, Y.; Zhang, H.; Wu, Z.; Zhao, M.; Feng, G. Community Structure Characteristics and Changes in Fish Species at Poyang Lake after the Yangtze River Fishing Ban. *Fishes* **2024**, *9*, 281. https://doi.org/10.3390/fishes9070281

Academic Editors: Paulo Branco and Marko Ćaleta

Received: 22 April 2024
Revised: 5 July 2024
Accepted: 8 July 2024
Published: 15 July 2024

1. Introduction

As the longest river in China, the Yangtze River has a diverse ecological environment, is one of the richest biodiversity areas in the world, and is one of the 35 priority ecoregions for conservation by the World Wide Fund for Nature (WWF). Poyang Lake is the largest freshwater lake in China and is one of the two existing large lakes connected to the Yangtze River [1]. The lake is a key water body involved in the ten-year fishing ban on the Yangtze River. Starting from 1 January 2018, a phased implementation of the fishing ban was implemented on the 332 aquatic life reserves (including 53 aquatic flora and fauna nature reserves and 279 aquatic germplasm resources reserves) in the Yangtze River Basin. The four aquatic life reserves (three aquatic flora and fauna nature reserves and one aquatic germplasm resources reserve) in Poyang Lake were included, with a total area of 1314.23 km^2 and spanning eight counties (cities and districts) in the vicinity of the lake. On 1 January 2020, a complete fishing ban was implemented for the Jiangxi section of the Yangtze River and the Poyang Lake Aquatic Reserve. Since 1 March 2020, all commercial fishing operations and any other activities that might harm the fish resources or fishery conditions of Poyang Lake have been prohibited. As of 31 October 2020, all retired fishing boats and nets were put under unified management. On 1 January 2021, a comprehensive fishing ban was implemented in Poyang Lake.

Poyang Lake is a key natural lake that connects with the mid-lower Yangtze River. It is an important feeding and fattening ground for fish that migrate to the Yangtze River, as well as a breeding ground for fish that migrate across the mouth of the Yangtze River [2,3]. Thus, Poyang Lake occupies a special place in the protection of Yangtze River fish germplasm resources and fish population conservation. As a typical floodplain lake at the mid-lower Yangtze River, the complex hydrological dynamics of Poyang Lake promote the spatiotemporal heterogeneity of the landscape and facilitate the material, energy, and biological exchange among various landscape elements while influencing the succession of various communities. In this sense, Poyang Lake contains rich and diverse biological resources. Field surveys and monitoring before the Yangtze River fishing ban recorded a total of 136 fish species in Poyang Lake between 1955 and 2000, and 101 fish species between 1997 and 2000 [2,4]. In 2010, 72 fish species were recorded in Poyang Lake; 89 fish species were recorded between 2012 and 2013; 62 fish species were recorded between 2018 and 2019; and 52 fish species were detected in 2019 [5–7]. Overall, a total of 212 fish species were recorded from 1980 to 2000. However, the number of fish species decreased to 174 for the period 2000–2017 [8]. In 2018, 106 fish species were observed in Poyang Lake and its tributaries during water restoration projects [9]. At that time, 37.7% of the indigenous fish species and 36.7% of the migratory fish species were not detected, indicating a significant change in the fish species composition [8].

To address these emerging problems, investigations have been performed on the community structure of fish in Poyang Lake, with sampling taking place before the comprehensive ban on commercial productive fishing in 2020. These studies showed that during their wintering period, fish in the Yangtze River-connecting waterways of Poyang Lake largely inhabited the three water areas north of Hukou County, Xieshan Mountain, and near Pingfeng Mountain. There has been relatively little research on the community structure of fish in Poyang Lake after the Yangtze River fishing ban. This study investigated fishery resources in Poyang Lake, in the Jiangxi section of the mainstream of the Yangtze River, and at the estuaries of the "Five Rivers" that flow into Poyang Lake in the period 2020–2021 after the fishing ban came into effect. The composition and biodiversity of fish species were analyzed, and the characteristics and changes of fish community structure were explored. The aim of this study was to provide a reference for the comprehensive evaluation of the ecological impact of the fishing ban on Poyang Lake and fish diversity conservation and biological integrity assessment.

2. Materials and Methods

2.1. Study Area and Sampling Sites

Situated in northern Jiangxi Province and on the south bank of the mid-lower Yangtze River, Poyang Lake is the largest inland freshwater lake in China and an internationally renowned wetland. Covering the area between 115°47′ E and 116°45′ E, and 28°22′ N and 29°45′ N, the water level of the lake is affected by the surface runoff of the "Five Rivers" (the Ganjiang River, the Fuhe River, the Xinjiang River, the Raohe River, and the Xiuhe River) as well as incoming water from the Yangtze River. Poyang Lake plays an important role in regulating and storing water before it is discharged into the Yangtze River at the outlet. The Ganjiang River gradually branches out into its tributaries south of Nanchang, with the main branch being joined by the Xiuhu River and flowing into Poyang Lake at

Wucheng. The segment of the Ganjiang River from Wucheng to Zhuxikou is designated as the western waterway. The main branch is joined by the Fuhe River and the Xinjiang River as it enters Poyang Lake and flows to Ruihong Town, Yugan County. The segment from Ruihong Town to Zhuxikou is designated as the eastern waterway. The two waterways merge at Zhuxikou and enter the Yangtze River at Hukou, known as the waterway to the Yangtze River. The Nanjishan of Poyang Lake is a national-level nature reserve. Occupying a large area of sparsely populated land with abundant natural resources, it is an ideal wintering and stopover site for migratory water birds and comprises an apt representation of the Poyang Lake ecosystem. Between 2020 and 2021, seven sampling stations were set up in Hukou, Lushan, Duchang, Yongxiu, Nanjishan, Ruihong, and Poyang for fishery resource surveys. The position, number and investigation frequency of sampling stations are listed in Table 1 and the map of the Poyang Lake area and 7 sampling stations are shown in Figure 1.

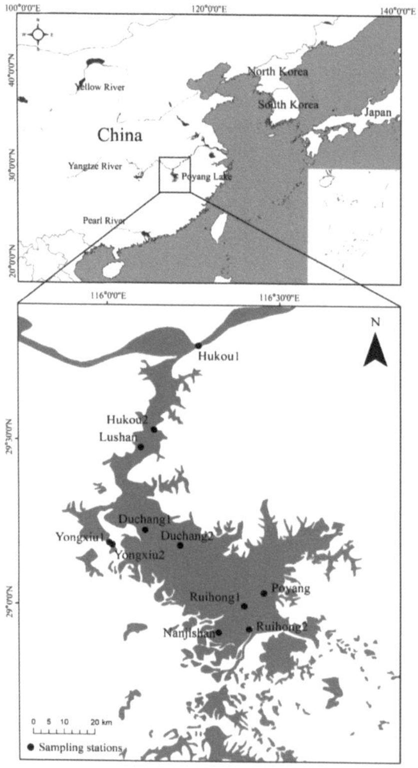

Figure 1. The map of Poyang Lake area and 7 sampling stations.

Table 1. The location and sampling profile of fisheries survey stations in Poyang Lake.

Position	Number of Sampling Sites (Water Areas)	Investigation Time	Total Investigation Frequency
Hukou	2 (Bali River: the west part of the main stream of the Yangtze River; Pingfeng: the watercourse between Poyang Lake and the Yangtze River)	April–June, July–August, September–October, November–December	16
Lushan	1 (the watercourse between Poyang Lake and the Yangtze River)	April–June, July–August, September–October, November–December	8

Table 1. *Cont.*

Position	Number of Sampling Sites (Water Areas)	Investigation Time	Total Investigation Frequency
Duchang	2 (Zhuxikou; Poyang Lake District)	April–June, July–August, September–October, November–December	16
Yongxiu	2 (the mouth of the Ganjiang River; the mouth of the Xiuhe River)	April–June, July–August, September–October, November–December	16
Nanjishan	1 (Poyang Lake District)	July–August, November–December	4
Ruihong	2 (the mouth of the Xinjiang River and Fuhe River; Poyang Lake District)	April–June, July–August, September–October, November–December	16
Poyang	1 (Longkou: the mouth of the Raohe River)	April–June, July–August, September–October, November–December	8

2.2. *Investigation Method*

Fish caught 10 km up- and downstream of the sampling station were collected. Fishing was carried out via active collection by research boats equipped with gillnets (mesh sizes: 2.5 cm, 6 cm, 8 cm, 10 cm, 12 cm, each of 100 m) and fishing baskets (length × height × width = 18 m × 0.33 m × 0.45 m, mesh = 0.8 cm). Collection locations and vessel numbers were recorded. The total length (to the nearest 0.1 cm), body length (to the nearest 0.1 cm), and weight (to the nearest 0.1 g) of the fish were measured. Fish that were difficult to identify on site were fixed in 10% formaldehyde solution after measuring their total length, body length, and weight and were taken back to the laboratory for further assessment.

Classification Method

The classification, Chinese names, and scientific names of the fish were determined using the latest taxonomic data, with the "Chinese Freshwater Fish", "Fauna Sinica-Cypriniformes, Osteichthyes", "Fauna Sinica-Siluriformes, Osteichthytes", and "Latin-Chinese Dictionary of Fish Names by Classification System" as the primary references [10–13]. The sample collection time at each sampling station was 3–5 days.

2.3. *Data Analysis and Statistics*

2.3.1. Historical Data

Fish data from 2002 to 2019 were historical data collected by the authors' research group with the same survey methodology as this study.

2.3.2. Fish Species Classification

Fish were classified with reference to the existing literature. The fish in Poyang Lake were divided into three categories according to their feeding habits: (1) carnivores feeding on vertebrates or invertebrates; (2) herbivores feeding on aquatic vascular plants or algae; (3) omnivores with both carnivorous and herbivorous feeding. Another way of classification was by the preferred habitats in water: (1) top-level fish: those often living in the upper water layer; (2) mid-lower-level fish: those often living in the middle and lower layers of the water; (3) bottom-level fish: those often living in the bottom water layer.

2.3.3. Catch per Unit Effort (CPUE)

Catch per unit effort refers to the average weight or quantity of fish caught per day within a period of time. This study used the catch per vessel per day as a measurement. All research fishing boats fished for the same 10 h per day, and the number of people on board each boat was usually two.

2.3.4. Dominant Fish Species

The Index of Relative Importance (IRI) [14] was used to evaluate the importance of a fish species in the community. The equation is:

$$IRI = (PN + PW) \times F$$

where PN is the number percentage of a species; PW is the weight percentage of a species, and F is its frequency of occurrence across all sampling periods. Species with IRI >1000 were designated as dominant species, and those with $100 \leq IRI < 1000$ were important species. Dominant and important species were collectively referred to as main species.

2.3.5. Biodiversity

The diversity of fish in Poyang Lake was measured using several biodiversity indices: the Shannon–Wiener diversity index (H'), Margalef's richness index (D), and Pielou's evenness index (J'). The equations are:

$$\text{Shannon-Wiener diversity index}: H' = -\sum_{i=1}^{S} P_i \ln P_i$$

$$\text{Margalef's richness index}: D = (S - 1)/\ln N,$$

$$\text{Pielou's evenness index}: J' = H'/\ln S,$$

where n_i is the number of fish species i; N is the total number of fish; P_i is the number percentage of fish species i in the total number of fish N, $P_i = n_i/N$ and S are the total number of fish species at the sampling site.

3. Results

3.1. Fish Community Composition

A total of 14,333 fish were collected during the period 2020–2022. The catch weighed 10,300.048 kg in total. Ninety-three fish species were identified, belonging to 11 orders, 20 families, and 61 genera (Table 2). Of these, 52 Cyprinidae species were identified, comprising the highest proportion (55.91%), followed by nine Bagridae species at 9.68%. There were 54 fish species collected from the Bali River section of the Yangtze River and the mouth of the Raohe River, respectively, 84 species from Poyang Lake, 56 species at the mouth of the Ganjiang River, 54 species at, and 60 species at the mouth of the Xinjiang River and Fuhe River.

Table 2. Trends of fish species composition in Poyang Lake during the past 40 years.

Year	Number of Species Collected	Exotic Species	Feeding Habit			Habitat Water Horizon		
			Carnivores	Herbivores	Omnivores	Top-Level	Mid-Lower	Bottom-Level
Before 1980	117	1	55	4	58	25	41	51
1982–1990	103	1	52	4	47	24	35	44
1997–2000	101	1	49	4	48	25	37	39
2012–2013	89	1	40	4	45	20	33	36
2020–2021	93	4	44	4	45	20	37	36

A total of 44 carnivorous fish species, 45 omnivorous fish species, and four herbivorous fish species were collected (Table 3). Twenty species of upper-level fish, 36 species of mid-lower-level fish, and 37 species of bottom-level fish were collected. Four species of migratory fish were collected, for 4.30% of the total number of fish species. These were *Acipenser sinensis*, *Anguilla japonica*, *Coilia ectenes*, and *Mugil cephalus*. A total of four protected fish species were collected, comprising 4.30% of the total number of fish species. These were *A. sinensis*, *A. japonica*, *Myxocyprinus asiaticus*, and *Leiocassis longirostris*. Four alien species were identified, comprising 4.30% of the total number of fish species. These were hybrid sturgeon (*Acipenser* × *Huso dauricus*), *Cirrhina mrigala*, *Piaractus brachypomus*, and *Mugil cephalus*.

Table 3. The composition, ecological types and distribution of fish species in Poyang Lake.

Species	Ecotype	The Bali River Section of the Yangtze River	Poyang Lake District	The Mouth of the Ganjiang River	The Mouth of the Xiuhe River	The Mouth of the Raohe River	The Mouth of the Xinjiang River and Fuhe River.
Acipenseridae							
1. *Acipenser sinensis*	C; B		△				
2. Hybrid sturgeon (*Acipenser* × *Huso dauricus*) *	C; B		△				
Engraulidae							
3. *Coilia ectenes*	C; U	△	△	△	△	△	△
4. *Coilia brachygnathus*	C; U	△	△	△	△	△	△
Salangidae							
5. *Neosalanx jordani*	C; U		△				
6. *Protosalanx hyalocranius*	C; U		△				△
Anguillidae							
7. *Anguilla japonica*	C; L	△				△	
Catostomidae							
8. *Myxocyprinus asiaticus*	O; L	△	△				△
Cyprinidae							
Danioninae							
9. *Zacco platypus*	O; U			△			
10. *Opsariichthys bidens*	O; U	△	△	△			
Leuciscinae							
11. *Mylopharyngodon piceus*	C; B	△	△		△		△
12. *Ctenopharyngodon idella*	H; L	△	△	△	△	△	△
13. *Squaliobarbus curriculus*	O; L		△	△			△
14. *Ochetobius elongatus*	C; L		△	△			
15. *Elopichthys bambusa*	C; U		△	△			△
Culterinae							
16. *Sinibrama macrops*	O; L	△					
17. *Pseudolaubuca sinensis*	O; U		△	△		△	△
18. *Pseudolaubuca engraulis*	O; U						
19. *Toxabramis swinhonis*	C; U		△	△		△	
20. *Hemiculter leucisculus*	O; U		△	△		△	△
21. *Hemiculter bleekeri*	O; U		△	△	△	△	△
22. *Culterichthys erythropterus*	C; U		△	△	△	△	
23. *Culter alburnus*	C; U	△	△	△	△	△	△
24. *Culter mongolicus*	C; U	△	△	△	△	△	△
25. *Culter dabryi*	C; U	△	△	△	△	△	△

Table 3. *Cont.*

Species	Ecotype	The Bali River Section of the Yangtze River	Poyang Lake District	The Mouth of the Ganjiang River	The Mouth of the Xiuhe River	The Mouth of the Raohe River	The Mouth of the Xinjiang River and Fuhe River.
26. *Culter oxycephaloides*	C; U						△
27. *Parabramis pekinensis*	H; L	△	△	△	△	△	△
28. *Megalobrama mantschuricus*	O; L	△	△	△	△	△	△
29. *Megalobrama amblycephala*	H; L		△	△	△	△	△
Xenocyprinae							
30. *Xenocypris argentea*	O; L	△	△	△	△		△
31. *Xenocypris davidi*	O; L		△	△			△
32. *Xenocypris microlepis*	O; L		△				△
33. *Pseudobrama simoni*	O; L		△	△	△	△	△
Hypophthalmichthyinae							
34. *Hypophthalmichthys molitrix*	H; U	△	△	△	△	△	△
35. *Hypophthalmichthys nobilis*	C; U	△	△	△	△	△	△
Gobioninae							
36. *Hemibarbus labeo*	C; B	△	△				
37. *Hemibarbus maculatus*	C; B	△	△	△	△	△	△
38. *Paracanthobrama guichenoti*	O; L		△	△	△	△	
39. *Pseudorasbora parva*	O; L		△		△		△
40. *Sarcocheilichthys sinensis*	O; L	△	△	△	△	△	
41. *Sarcocheilichthys kiangsiensis*	O; L	△	△	△	△		
42. *Sarcocheilichthys nigripinnis*	O; L	△	△	△		△	△
43. *Squalidus argentatus*	O; L	△	△	△	△	△	△
44. *Squalidus nitens*	O; L	△	△				
45. *Coreius heterodon*	O; B		△	△	△		
46. *Rhinogobio typus*	O; B	△	△	△	△	△	△
47. *Abbottina rivularis*	O; B	△	△	△	△	△	△
48. *Microphysogobio tungtingensis*	C; B		△				
49. *Saurogobio dabryi*	O; B	△	△	△	△	△	△
Acheilognathinae							
50. *Acheilognathus gracilis*	O; L		△				
51. *Acheilognathus macropterus*	O; L	△	△	△	△	△	△
52. *Acheilognathus chankaensis*	O; L		△		△	△	△
53. *Acheilognathus taenianalis*	O; L		△	△	△	△	△
54. *Rhodeus ocellatus*	O; L		△			△	△
55. *Rhodeus sinensis*	O; L		△		△		

Table 3. *Cont.*

Species	Ecotype	The Bali River Section of the Yangtze River	Poyang Lake District	The Mouth of the Ganjiang River	The Mouth of the Xiuhe River	The Mouth of the Raohe River	The Mouth of the Xinjiang River and Fuhe River.
56. *Paracheilognathus imberbis*	C; L		△			△	△
57. *Paracheilognathus himantegus*	O; L		△			△	△
Labeoninae							
58. *Cirrhina mrigala* *	O; B		△				
Cyprininae							
59. *Cyprinus carpio*	O; B	△	△	△	△	△	△
60. *Carassius auratus*	O; L	△	△	△	△	△	△
Cobitidae							
61. *Parabotia fasciata*	O; B	△	△	△		△	△
62. *Cobitis sinensis*	O; B	△	△				
63. *Cobitis macrostigma*	O; B	△	△				△
64. *Misgurnus anguillicaudatus*	O; B	△	△	△	△	△	△
65. *Paramisgurnus dabryanus*	O; B	△	△			△	
Siluridae							
66. *Silurus asotus*	C; B	△	△	△	△	△	△
67. *Silurus meridionalis*	C; B	△	△			△	△
Bagridae							
68. *Pelteobagrus fulvidraco*	C; B	△	△	△	△	△	△
69. *Pelteobagrus eupogon*	C; B	△	△	△	△	△	△
70. *Pelteobagrus vachelli*	C; B	△	△	△	△	△	△
71. *Pelteobagrus nitidus*	C; B	△	△	△	△	△	△
72. *Leiocassis longirostris*	C; B	△					
73. *Leiocassis crassirostris*	C; B	△	△	△		△	△
74. *Pseudobagrus pratti*	C; B		△				
75. *Pseudobagrus albomarginatus*	C; B	△	△	△		△	△
76. *Hemibagrus macropterus*	C; B	△	△				△
77. *Clarias fuscus*	C; B		△		△	△	△
Hemiramphidae							
78. *Hyporhamphus intermedius*	O; U	△	△			△	
Synbranchidae							
79. *Monopterus albus*	C; B	△	△	△	△	△	△
Serranidae							
80. *Siniperca chuatsi*	C; L	△	△	△	△	△	△
81. *Siniperca kneri*	C; L	△	△	△		△	△
82. *Siniperca scherzeri*	C; L		△		△	△	△

Table 3. *Cont.*

Species	Ecotype	The Bali River Section of the Yangtze River	Poyang Lake District	The Mouth of the Ganjiang River	The Mouth of the Xiuhe River	The Mouth of the Raohe River	The Mouth of the Xinjiang River and Fuhe River.
83. *Siniperca roulei*	C; L	△	△				
Eleotridae							
84. *Odontobutis obscura*	C; B	△	△			△	△
Gobiidae							
85. *Rhinogobius giuroiuns*	C; B	△	△		△		△
Belontidae							
86. *Macropodus chinensis*	O; L		△	△	△		△
87. *Macropodus opercularis*	O; L			△			
Channidae							
88. *Channa argus*	C; B	△	△	△	△	△	△
89. *Channa asiatica*	C; B	△					
Mastacembelidae							
90. *Macrognathus aculeatus*	C; B	△					
91. *Macrognathus sinensis*	C; B		△			△	
Mugilidae							
92. *Mugil cephalus* *	O; B		△				
Characidae							
93. *Piaractus brachypomus* *	O; L		△				

Where H: herbivorous fish; O: omnivorous fish; C: carnivorous fish; U: upper Layer fish; L: lower-middle layer fish; B: bottom layer fish; △ collected during this surveillance; * alien species.

3.2. Fish Community Composition and Variation in Poyang Lake

Previous fish resource surveys in Poyang Lake have recorded a total of 134 fish species (the same species registered under different names was counted as one species) from 26 families and 12 orders (Table 4). Of these, 71 species belonged to the family Cyprinidae, comprising 53.0%; 12 species (9.0%) belonged to the family Bagridae; eight species (6.0%) belonged to the family Cobitidae, and five species (3.7%) belonged to the family Serranidae. There were also four species in the family Salangidae and four species of the family Amblycipitidae, each comprising 3.0% of the total. Less than three species were identified for all other families. In the years leading to 1980, 117 fish species were identified in Poyang Lake, and 103 fish species were recorded between 1982 and 1990, including one alien species. Between 1997 and 2000, 101 fish species and one alien species were recorded, and 89 fish species and one alien species were identified between 2012 and 2013. In this survey, 93 fish species were recorded, including four alien species, hybrid sturgeon (*Acipenser* × *H. dauricus*), *C. mrigala*, *P. brachypomus*, and *M. cephalus*. The number of alien species in Poyang Lake has increased, while *Psephurus gladius*, *Tenualosa reevesii*, *Cynoglossus joyneri*, *Takifugu ocellatus*, and *Takifugu obscurus* have not been detected from 2001 to 2021. *P. gladius* and *T. reevesii* are considered extinct. Neither the order Cyprinodontiformes nor Tetraodontiformes have been observed for the past decade.

3.3. Community Structure Variation in Commercially Important Fish Species

The variation in body length was 6.10–61.26% between 2012 and 2021, with *Mylopharyngodon piceus* showing the smallest increase in body length and *Siniperca chuatsi* showing the largest increase (Table 5). In terms of weight, the corresponding change was 15.14–291.57%, with *S. chuatsi* showing the largest increase and *Cyprinus carpio* showing the smallest increase (Table 6). In terms of age structure, which was judged by the annual lines on the scales of the fishes, the proportion of major economically important fish aged 1 or 2 years at maturity decreased significantly in 2021 compared to 2012, while the proportion of more mature fish increased substantially (Tables 7 and 8).

3.4. Community Structure Characteristics of Fish Species

The average catch per unit effort for Poyang Lake was 58.22 kg/boat/day. The highest CPUE was 204.5 kg/boat/day at Hukou, and the lowest was 17.10 kg/boat/day at Nanjishan (Table 9). The biodiversity indices used in this study were the Shannon–Wiener index, Margalef's richness index, and Pielou's evenness index (Table 8). The average value of the Shannon–Wiener index was 2.64, with the highest number being 2.901 at Poyang and the lowest being 2.158 at Nanjishan. The average value of Margalef's richness index was 5.38. The highest value was 6.459 at Lushan, and the lowest was 4.265 at Nanjishan. The average value of Pielou's evenness index was 0.72, with the highest value being 0.822 at Duchang City, and the lowest being 0.617 at Nanjishan. The dominant species at the Hukou, Lushan, Duchang, Yongxiu, Nanjishan, Ruihong, and Poyang sampling sites were, respectively, *Cyprinus carpio* and *Megalobrama terminalis*; *Hypophthalmichthys nobilis*, *Coilia brachygnathus*, and *M. terminalis*; *H. nobilis*, *Ctenopharyngodon idella*, and *Hypophthalmichthys molitrix*; *H. molitrix*, *H. nobilis*, and *M. terminalis*; *Carassius auratus*, *H. nobilis*, and *Paracanthobrama guichenoti*; *H. nobilis* and *H. molitrix*; *H. molitrix*, *C. idella*, and *M. terminalis*. Species with the highest IRI were *C. carpio*, *H. nobilis*, *H. molitrix*, *C. acratus*, and *H. molitrix* in the Hukou, Lushan, Duchang, Yongxiu, Nanjishan, Ruihong, and Poyang sampling sites, respectively (Table 10).

Table 4. The dynamics of fish fauna in Poyang Lake in the last 40 years.

Year	Acipenseriformes	Anguilliformes	Clupeiformes	Salmoniformes	Cypriniformes	Siluriformes	Cyprinodontiformes	Beloniformes	Synbranchiformes	Perciformes	Pleuronectiformes	Tetraodontiformes	Mugiliformes	Characoidei
Before 1980	2	1	3	3	70	17	1	1	1	14	2	2	0	0
1982–1990	0	1	3	4	61	18	1	1	1	12	1	1	0	0
1997–2000	0	1	3	4	62	14	1	1	1	14	0	0	0	0
2012–2013	1	1	2	3	56	10	1	1	1	12	1	0	0	0
2020–2021	2	1	2	2	58	12	0	1	1	12	0	0	1	1

Table 5. The body length composition of major economic fishes in Poyang Lake.

Species of Fish	The Body Length in 2012 (cm)			The Body Length in 2021 (cm)			The Amplitude of Variation (%)
	Range	Average Value	Sample Size	Range	Average Value	Sample Size	
Mylopharyngodon piceus	11–82	24.6	101	9.4–80	26.1	58	6.1
Ctenopharyngodon idella	8.7–70	29.5	108	11.3–82	44.7	331	51.53
Hypophthalmichthys molitrix	11.5–74	28.4	173	18.6–64.3	39.9	434	40.49
Hypophthalmichthys nobilis	12.2–77.5	32.9	85	16.2–86	52.3	240	58.97
Cyprinus carpio	8.1–94.5	21.4	185	5.8–56.3	29.0	126	35.51
Carassius auratus	5.9–17.6	10	118	5.8–26.2	12.9	629	29
Siniperca chuatsi	8–46.4	19.1	109	11.5–51	30.8	224	61.26
Silurus asotus	15.2–107.5	30.1	92	12.1–135	37.2	43	23.59
Culter alburnus	12.1–62.3	25.1	99	8.9–67.5	32.4	113	29.08

Table 6. The body weight composition of major economic fishes in Poyang Lake.

Species of Fish	The Body Weight in 2012 (g)			The Body Weight in 2021 (g)			The Amplitude of Variation (%)
	Range	Average Value	Sample Size	Range	Average Value	Sample Size	
Mylopharyngodon piceus	20–12,500	533	101	17.3–8725	714	58	33.96
Ctenopharyngodon idella	29–6500	810	108	31.3–12,800	2183	331	169.51
Hypophthalmichthys molitrix	27–6250	744	173	8.7–5730	1385	434	86.16
Hypophthalmichthys nobilis	40–10,500	1086	85	85.6–12,830	3335	240	207.09
Cyprinus carpio	15–17,500	687	185	5.4–4032	791	126	15.14
Carassius auratus	7–219	37	118	4.8–435	79	629	113.51
Siniperca chuatsi	10–2508	261	109	28–6540	1022	224	291.57
Silurus asotus	30–10,500	512	92	18.5–20,300	1377	43	168.95
Culter alburnus	16–2650	266	99	7–3800	694	113	160.9

Table 7. Percentage age composition of major fish harvesting in Poyang Lake in 2012.

	1+	**2+**	**3+**	**4+**	**5+**
Culter alburnus	66%	27%	6%	1%	/
Siniperca chuatsi	80%	17%	/	/	/
Silurus asotus	49%	37%	7%	4%	3%
Carassius auratus	100%	/	/	/	/
Cyprinus carpio	86%	7%	3%	1%	3%
Hypophthalmichthys nobilis	71%	21%	5%	2%	1%
Hypophthalmichthys molitrix	82%	11%	6%	/	1%
Ctenopharyngodon idella	54%	19%	15%	11%	1%
Mylopharyngodon piceus	56%	28%	13%	2%	1%

Table 8. Percentage age composition of major fish harvesting in Poyang Lake in 2021.

	1+	**2+**	**3+**	**4+**	**5+**
Culter alburnus	42%	10%	26%	21%	1%
Siniperca chuatsi	36%	20%	22%	21%	1%
Silurus asotus	47%	21%	14%	14%	5%
Carassius auratus	99%	1%	/	/	/
Cyprinus carpio	76%	12%	12%	/	/
Hypophthalmichthys nobilis	27%	41%	28%	3%	1%
Hypophthalmichthys molitrix	67%	32%	1%	/	/
Ctenopharyngodon idella	35%	34%	26%	5%	1%
Mylopharyngodon piceus	62%	29%	7%	/	2%

Table 9. Catch per unit of fishing effort (CPUE) of different fishing gears and fish diversity in each section of Poyang Lake.

Position	Netting Gear	Day	CPUE (kg/Ship/Day)	Shannon–Wiener Diversity Index	Margalef's Richness Index	Pielou's Evenness Index
Hukou	Drift net, fish cage	20	204.5	2.586	6.355	0.692
Lushan	Trammel net, fish cage	20	27.65	2.724	6.459	0.696
Duchang	Trammel net, fish cage	20	44.60	2.900	5.225	0.822
Yongxiu	Trammel net, fish cage	20	32.67	2.825	5.482	0.761
Nanjishan	Trammel net, fish cage	10	17.10	2.158	4.265	0.617
Ruihong	Trammel net, fish cage	20	57.34	2.375	4.859	0.653
Poyang	Trammel net, fish cage	20	23.67	2.909	5.015	0.812

Table 10. Dominant species and fish harvesting composition in each section of Poyang Lake.

Position/ Species		*Parabramis pekinensis*	*Ctenopharyngodon idella*	*Culter dabryi*	*Coilia brachygnathus*	*Siniperca chuats*	*Carassius auratus*	*Cyprinus carpio*	*Hypophthalmichthys molitrix*	*Culter mongolicus*	*Culter alburnus*	*Megalobrama terminalis*	*Megalobrama amblycephala*	*Coilia ectenes*	*Hypophthalmichthys nobilis*	*Leiocassis longirostris*	*Paracanthobrama guichenoti*
Hukou	N (%)	1.58	0.32	0.32	2.21	6.15	1.10	9.15	2.68	1.56	0.47	28.39	-	12.46	7.73	11.04	-
	W (%)	0.11	0.13	0.00	0.02	0.40	0.03	82.14	2.18	0.18	0.08	3.21	-	0.13	6.32	2.20	-
	IRI	<300	<300	<300	<300	<300	<300	4057	<300	<300	<300	316	-	699	624	736	-
Lushan	N (%)	0.86	0.20	8.47	29.98	2.18	6.04	0.66	2.94	4.16	2.49	12.48	0.20	2.69	2.74	-	0.25
	W (%)	0.67	1.06	3.75	3.71	4.40	1.67	1.90	22.69	1.32	2.08	14.65	0.43	0.65	37.03	-	0.03
	IRI	<300	<300	1222	3369	658	771	<300	2563	548	456	2679	<300	334	3977	-	<300
Duchang	N (%)	6.55	6.55	0.25	11.09	1.18	13.36	1.93	5.04	1.34	0.59	6.81	0.42	3.19	6.55	-	13.86
	W (%)	39.43	26.85	0.02	0.50	2.43	0.54	1.67	14.28	0.07	0.42	8.12	0.42	0.28	39.43	-	0.45
	IRI	<300	3341	<300	1159	<300	1391	<300	1932	<300	<300	14.92	<300	348	4598	-	1431
Yongxiu	N (%)	4.95	4.61	4.07	7.39	5.97	16.81	2.37	9.97	1.97	1.02	15.12	1.08	1.83	4.95	-	4.47
	W (%)	26.41	13.79	1.62	0.52	11.28	2.64	4.83	22.26	2.41	1.55	6.45	1.33	0.35	26.41	-	0.15
	IRI	429	1840	569	791	1725	1945	720	3222	437	<300	2157	<300	<300	3137	-	463
Nanjishan	N (%)	0.77	0.88	2.87	1.93	0.28	23.28	0.33	0.17	1.16	0.39	4.08	0.55	-	0.77	-	26.14
	W (%)	27.85	17.64	1.48	0.55	1.11	9.06	0.55	1.55	6.47	0.98	7.19	1.87	-	27.85	-	1.70
	IRI	<300	1852	435	<300	<300	3234	<300	<300	763	<300	1127	<300	-	2862	-	2784
Ruihong	N (%)	1.53	1.78	0.74	2.32	2.81	6.16	1.13	8.04	0.69	0.89	5.42	3.55	1.68	1.53	-	1.13
	W (%)	78.23	3.18	0.28	0.94	3.55	0.14	1.10	6.93	0.65	0.08	1.11	1.34	0.16	78.23	-	0.02
	IRI	<300	496	<300	325	318	630	<300	1497	<300	<300	654	489	<300	5982	-	<300
Poyang	N (%)	1.68	4.19	9.78	12.66	0.28	13.69	3.82	5.40	1.40	1.86	12.94	1.68	0.74	1.68	-	0.47
	W (%)	10.81	17.22	6.96	1.95	0.99	2.85	8.69	23.77	2.39	4.57	5.04	1.74	0.17	10.81	-	0.14
	IRI	638	2141	1674	1461	386	1654	1251	2917	<300	510	1798	<300	<300	937	-	<300

4. Discussion

4.1. Post-Fishing Ban Variation in the Fish Community Composition

The prevalent fish species in Poyang Lake belong to the families Cypriniformes and Cyprinidae, exhibiting the compositional characteristics of fish communities in East Asian rivers and lakes [15]. In recent years, the fish community composition has been affected by factors such as flooding, water temperature, water turbidity, and the fishing ban. The 1996 and 1998 Yangtze River floods brought large numbers of farmed fish and migratory fish into Poyang Lake, resulting in a significant change in the species composition of the lake [16]. Since the implementation of the Yangtze River fishing ban, it has had a positive effect on the recovery of fish communities in the lakes of the middle and lower reaches of the Yangtze River. Hu et al. performed a quantitative study of the characteristics of water and sediment movement in the lake and the changes in the lake basin from the influence of sedimentation and erosion as well as the hydrological and ecological effects, using measured data of sediment transport in and out of the lake from 1956 to 2020 [17]. They found that before 2000 Poyang Lake was largely dominated by sedimentation and, from 2001 to 2010, both sedimentation and erosion were present in the lake basin. From 2011 to 2020, the lake basin was entirely under the influence of erosion. The water temperature in Poyang Lake showed a lateral trend of gradual increase from north to south, and from west to east. This is because the northern part of the lake is a waterway to the Yangtze River, where water runs deep and flows fast, resulting in strong heat dissipation. Compared to the west bank, the east bank of the main lake has a lower flow velocity, and the flow is affected by the bay. This region thus features low wind–wave mixed energy, and heat is not transferred effectively in the lateral direction. The water is emerald-green in color, and the sunlight penetrates relatively deep. From 2010 to 2019, although the distribution of fish species changed along environmental gradients, the similarity among communities increased, and there was a tendency towards uniformity [7]. In the years leading up to 1980, 117 species of fish were recorded in Poyang Lake. Between 1982 and 1990, 103 species of fish were recorded, including one alien species. From 1997 to 2000, 101 fish species were recorded, including one alien species. Between 2012 and 2013, 89 fish species were recorded, including one alien species. In this study, 93 fish species were recorded, including four alien species. The composition of fish species in Poyang Lake is thus constantly changing, although less than 90 species play a dominant role in the structure of the fish community. The fish population in Poyang Lake reached a peak in 1998. However, the variety of fish species in the lake showed a declining trend in 2012–2013. The increase in the number of species in 2020–2021 following the Yangtze River fishing ban is small, while the increase in exotic species is obvious.

Further analysis showed that the species identified at various stations in Poyang Lake and the Jiangxi section of the Yangtze River mainstream were similar, consisting largely of Cyprinidae, particularly omnivorous and bottom-level species. The fish in Poyang Lake are primarily bottom-level and omnivorous types. However, after the comprehensive ban on commercial fishing in 2020, there were significant increases in carnivorous and mid-lower-level fish. It has been speculated that, after the ban, carnivorous fish had abundant food and were not threatened by fishing, and this benefited their population recovery. In contrast, carnivorous and mid-lower-level fish are the main targets of commercial fishing. After the fishing ban, it became easier to collect samples of these species for research purposes and monitoring, as manifested by the increase in the observed number of species.

Strong spatial distribution characteristics were also seen in the species composition, with the highest number of fish species observed in Poyang Lake. Significant differences in fish species exist among the sampling sites at river mouths. Sixty species of fish were identified at the mouths of the Xinjiang and Fuhe Rivers, while only 45 species were found at the Xiuhe River mouth. For other rivers, at least 54 species of fish species were identified at their river mouths. Among the five rivers, Ganjiang River contributes the highest average flow into Poyang Lake per year, about 54.2% of the total influx into the lake. The Raohe River contributes the smallest flow volume into Poyang Lake (9.4%), followed by the 9.8%

of the Xiushui River, while the Xinjiang River and the Fuhe River contribute 14.5% and 12.1%, respectively [18,19]. In summary, there may not be a close relationship between the number of fish species identified at a river mouth and the average influx into the lake at that location. The Xiushui River is located in the mountains of northwestern Jiangxi Province, with its mouth also in the northwest. We thus speculate that the habitat characteristics of the Xiuhe River and the gradual increase in water temperature of Poyang Lake in the north–south and west–east directions is one reason for the relatively low number of fish species found at the mouth of the Xiuhe River. Zhang et al. classified 24 sampling sites into the Xiushui group and the Poyang Lake group [15]. In this study, the IRI values of Hukou (*C. carpio* and *M. terminalis*), Lushan (*H. nobilis*, *C. auratus*, *M. terminalis*, *H. molitrix*, and *C. dabryi*), Duchang (*H. nobilis*, *C. idella*, *H. molitrix*, *M. terminalis*, *P. guichenoti*, *C. acratus*, and *C. brachygnathus*), Yongxiu (*H. molitrix*, *H. nobilis*, *M. terminalis*, *C. idella*, and *S. chuatsi*), Nanjishan (*C. acratus*, *H. nobilis*, *P. guichenoti*, *C. idella*, and *M. terminalis*), Ruihong (*H. nobilis* and *H. molitrix*), and Poyang (*H. molitrix*, *C. idellus*, *M. terminalis*, *C. dabryi*, *C. acratus*, *C. brachygnathus*, and *C. carpio*) were $\geq$1000, making them the dominant species. Poyang Lake attracts fish with different movement and migration habits by providing targeted spawning grounds and foraging conditions. Fish that lay buoyant eggs periodically choose river and lake habitats for spawning and fattening [20], while fish that produce sticky eggs choose coastal areas with dense water grass for spawning and open water for wintering and seeking refuge [21]. Hukou is well connected with the mainstream of the Yangtze River and tributaries such as the Ruihong and Xinjiang Rivers. Hukou features rich ecological niches and a complex hydrological and sediment environment that enable it to support greater biodiversity [22]. The higher hydrologic connectivity facilitates greater ecological niche breadth, accommodating varied fish species [23]. Strong interspecific competition may have resulted in the relatively low number of dominant species at this location.

4.2. Post-Fishing Ban Variation in the Community Diversity

The habitat environment and the activities of fish induce changes in their community structure and diversity [24,25]. For the fish community in Poyang Lake, the key area for fishery resources has changed from water areas around the three mountains, Poyang County, and Yugan County in 2010 to Hukou County, Xingzi County, and Xinjian County in 2019. In other words, the key areas moved from the lake center to the Yangtze River waterway and closer to the mainstream of the Yangtze River [7]. This study produced similar findings. Generally speaking, there was little difference in the diversity of fish communities at different stations in Poyang Lake. In terms of biodiversity, the Shannon–Wiener index ranged from 2.158 to 2.909, with Poyang having the highest value and Nanjishan the lowest. Margalef's index ranged from 4.265 to 6.459, with Lushan having the highest value and Nanjishan the lowest. Pielou's index ranged from 0.617 to 0.822, with Duchang having the highest value and Nanjishan the lowest. The diversity of the fish community depends on two factors: the number of species and the evenness of their distribution. When the number of species is constant, a more even distribution of species gives rise to higher diversity [26]. The sampling frequency at Nanjishan was only half that of other stations, which could be the reason for its relatively low biodiversity indices. It should be noted that research on the Yangtze River finless porpoises (*Neophocaena asiaeorientalis*) in Poyang Lake shows that at mid-low water levels, the species is most frequently spotted in water bodies around the three mountains, Longkou, and between Kangshan and the mouths of the three rivers. At high water levels, porpoises are most frequently seen in water bodies around Duchang County and the three mountains. There is partial overlap between the areas with the highest occurrence of Yangtze River finless porpoises and those with high fish diversity [27].

The number of fish species in Poyang Lake increased after the fishing ban in the Yangtze River compared with the number before the fishing ban. The number of fish species collected in this study increased to 93 from the 89 species identified in the 2012–2013 study. Alien species showed the most significant change in species, and the trend of premature maturity, small size, and low quality among major economically important fish species was effectively curbed. Only one alien species was identified in the 2012–2013 study, while four were found in this study, namely hybrid sturgeon (*Acipenser* × *Huso dauricus*), *Cirrhina mrigala*, *Piaractus brachypomus*, and *Mugil cephalus*. Some researchers have pointed out that research on the fishing industry of Poyang Lake should consider the invasion of alien species [7], as this is thought to be an important reason for the decline of indigenous fish resources. The presence of some exotic fish has been recorded in Poyang Lake. These fish species may enter Poyang Lake through escaping farming facilities, artificial introductions, or natural dispersal with currents. The Poyang Lake Basin is an area heavily and adversely affected by invasive fish species in Jiangxi Province. Alien species such as hybrid sturgeon, *C. mrigala*, *Piaractus brachypomum*, and *M. cephalus* are found extensively in the lake, the mainstream and tributaries of the Five Rivers, and their associated water bodies, posing a danger to the fish population, the aquatic ecosystem, and the biodiversity of indigenous fish in the invaded water bodies. There is currently a lack of effective measures for the prevention and control of these invasive species. We recommend undertaking a systematic study on the type, distribution, transmission pathway, invasion mechanism, and ecological impact of these species in the Poyang Lake Basin to identify areas for improvement in existing prevention and control measures and thereby protect the indigenous fish population and the local aquatic system.

4.3. Impact of the Yangtze River Fishing Ban on the Restoration of Fishery Resources in Poyang Lake

The ban on fishing in the Yangtze River has replenished the fish resources of Poyang Lake, especially the resources of the "four major carp" (herring, grass carp, silver carp and bighead carp) have increased significantly. Compared to 2012, increases in body length and weight were observed among the major economically important fish species in 2021, with the increases in the carnivorous fish species *Siniperca chuatsi* being the most prominent. In addition, a significant increase in proportion of fish in more mature age groups was found. The Yangtze River fishing ban policy has had a positive impact on the changes in body length, body weight, and age of major economic fish in Poyang Lake. Studies on the effect of the Yangtze River fishing ban on Coilia ectenes of Poyang Lake, a species that migrates upstream, also indicated that the two years of fishing ban have so far benefitted the population recovery of this migratory fish [28]. Our two-year study also showed that among all seven stations, Hukou scored the highest catch per unit effort, followed by Ruihong and Duchang. It should be noted that higher values of catch per unit effort were observed at the mouths of the Fuhe and Xinjiang rivers near Ruihong and at Zhuxikou near Duchang compared to sampling sites in the lake. Nanjishan, as a representation of the Poyang Lake ecosystem, had the lowest catch per unit effort of 17.10 kg/boat/day, revealing the massing of fish in the channels connecting the Yangtze River and Poyang Lake, and Poyang Lake and other rivers. Significant differences in community composition and biodiversity remain in different regions, possibly due to the interconnectivity between Poyang Lake and Yangtze River, and the effective replenishment of lake fish by the "Five Rivers". Hence, we recommend the continued and increased monitoring and management of key water areas such as the Yangtze River waterway, Poyang Lake waterway, and the mouths of the "Five Rivers" to fully realize their potential in supporting and maintaining the fish diversity of Poyang Lake.

However, from the perspective of available fishery resources, compared to historical peaks, the amount of fishery resources at Poyang Lake has shown a trend of increase followed by decrease. The fishing yield between 1949 and 1980 was 3.72×10^4 t. In the 1990s, the amount of fishery resources increased continuously, reaching an average annual supply of 4.26×10^4 t. Since the end of the 20th century, the supply of fishery resources has declined to an average annual volume of 2.94×10^4 tons (a 30.8% drop). In 1998, the supply of fishery resources reached a peak of 7.19×10^4 t. The lowest supply occurred in 2017, with only 27.0% of the volume in 1998. Jiang et al. suggested that there was a significant change in the fish community structure in Poyang Lake during 2010–2019 [29]. While small fish remained the dominant species, a significant miniaturization was observed in several fish species, possibly due to overfishing. At the same time, interference from human activities has increased in Poyang Lake, especially in areas farther from the mainstream of the Yangtze River such as Poyang County, Xinjian County, and Yugan County. However, will the ten-year fishing ban on the Yangtze River significantly help the restoration of fishery resources in Poyang Lake? As suggested by Yang et al. [30], the discovery of Mugil cephalus, a euryhaline nearshore fish, at two sampling sites, one at the center of the main lake (S9) and one in the waterway to Yangtze River (S20), in October 2020 could be the result of the fishing ban. However, enhanced restoration of natural habitats is critical to the life cycle of various fish species, including spawning and nursery habitats. Also, stricter management of pollutants entering the lake and management of water levels to mimic natural fluctuations will help accelerate the recovery of fish resources in Poyang Lake. They concluded that the fishing ban eliminated the pressure on the ecosystem from fishing and promoted the recovery of fishery resources and the upstream movement of fish, and the conclusions of this study also indicate that the effects of fishing closures on fish stocks are positive. Hence, we believe that the fishing ban on the Yangtze River has played an important and positive role in the restoration of fishery resources in Poyang Lake. However, the restoration of fishery resources in Poyang Lake is a long-term and complex process that requires continuous conservation and management measures. The implementation of the fishing ban policy is a positive start, but more scientific monitoring and reasonable ecological regulation are needed in the future to ensure the sustainable use of fishery resources and the long-term health of the ecological environment in Poyang Lake.

5. Conclusions

In this study, a total of 93 species of fish were found in seven stations of Poyang Lake, of which 55.91% were Cyprinidae species. Meanwhile, the number of alien species (hybrid sturgeon, *Cirrhina mrigala*, *Piaractus brachypomum*, and *Mugil cephalus*) increased after the fishing ban. The ecological types of fish in Poyang Lake showed that 47.31% and 38.71% of the total species were carnivorous and mid-lower-level fishes. For the main economic fish in Poyang Lake, the body length and body weight increased, and the proportion of older fish increased substantially. Moreover, the diversity of fish communities at different stations in Poyang Lake showed no significant difference. Generally, the Yangtze River fishing ban has played an important and positive role in the restoration of fishery resources in Poyang Lake.

Author Contributions: Conceptualization, Y.Z. and G.F.; data curation, Y.Z. and H.Z.; formal analysis, Y.Z. and Z.W.; funding acquisition, G.F.; project administration, G.F.; visualization, H.Z. and Z.W.; writing—original draft, Y.Z., M.Z. and G.F.; writing—review and editing, Y.Z. and G.F. All authors have read and agreed to the published version of the manuscript.

Funding: This research was funded by the Jiangsu Innovation and Entrepreneurship Team Project, grant number JSSCTD202120, and the Breeding Research Project of Jiangxi Province, grant number 2023yyyzgg-02. The APC was funded by Guangpeng Feng.

Institutional Review Board Statement: The study was conducted in accordance with the Declaration of Helsinki and approved by the Committee on Laboratory Animal Welfare and Ethics of Jiangxi Institute for Fisheries Sciences and the Poyang Lake Fisheries Research Centre of Jiangxi (Approval code: jxifs 2023-10, date: 30 October 2023).

Informed Consent Statement: Not applicable.

Data Availability Statement: The data presented in this study are available in the article. Further information is available upon request from the corresponding author.

Conflicts of Interest: The authors declare no conflicts of interest.

References

1. Tian, B.; Gao, P.; Mu, X.; Zhao, G. Water Area Variation and River–Lake Interactions in the Poyang Lake from 1977–2021. *Remote Sens.* **2023**, *15*, 600. [CrossRef]
2. Zhou, C.; Chen, J.; Guo, T.; Ouyang, S.; Wu, X. Environmental DNA metabarcoding reveals the biological community structure in Poyang Lake, China. *Conserv. Genet. Resour.* **2022**, *14*, 437–448. [CrossRef]
3. Wang, Z.; Wu, Q.; Zhou, J.; Ye, Y.; Tong, J. Silver carp, *Hypophthalmichthys molitrix*, in the Poyang Lake belong to the Ganjiang River population rather than the Changjiang River population. *Environ. Biol. Fishes* **2003**, *68*, 261–267. [CrossRef]
4. Huang, L.L.; Wu, Z.Q.; Li, J.H. Fish fauna, biogeography and conservation of freshwater fish in Poyang Lake Basin, China. *Environ. Biol. Fishes* **2013**, *96*, 1229–1243. [CrossRef]
5. Xiong, W.; Xie, D.; Wang, Q.; Wang, H. Non-native species in Poyang Lake Basin: Status, threats and management. *Aquat. Invasions* **2023**, *18*, 119–134. [CrossRef]
6. Wu, S.H. Functional Traits Relationships of Fish Assemblages in Sub-Lakes of Poyang Lake Floodplain. Master's Thesis, Nanchang University, Jiangxi, China, 2020. (In Chinese)
7. Jiang, X.L.; Li, M.Z.; Yang, S.R.; Wang, C.; Zhang, C.; Gao, X. Temporal variation of fish metacommunity structure in Poyang Lake. *Resour. Environ. Yangtze River* **2022**, *31*, 588–601. (In Chinese)
8. Liu, X.; Qin, J.; Xu, Y.; Zhou, M.; Wu, X.; Ouyang, S. Biodiversity pattern of fish assemblages in Poyang Lake Basin: Threat and conservation. *Ecol. Evol.* **2019**, *9*, 11672–11683. [CrossRef]
9. Zhang, X.; Yang, T.Y.; Luo, X.H.; Yuan, C.; Liu, H. Fish phylogenetic community structure in the Poyang Lake and its tributary the Xiushui River in summer. *Acta Hydrobiol. Sin.* **2020**, *44*, 1297–1312. (In Chinese)
10. Chen, Y.Y. *Zoology of China Osteichthys Cypriniformes*; Science Press: Beijing, China, 1998. (In Chinese)
11. Le, P.Q. *Zoology of China Osteichthyes Cypriniformes Volume 3*; Science Press: Beijing, China, 2000. (In Chinese)
12. Chu, X.L.; Zhen, B.S.; Dai, D.Y. *Zoology of China Osteichthyes Siluriformes Volume 2*; Science Press: Beijing, China, 1999. (In Chinese)
13. Wu, H.L.; Shao, G.Z.; Lai, C.F. *Latin-Chinese Dictionary of Fishes Names*; Fisheries publication: Taiwan, China, 1999. (In Chinese)
14. Pinkas, L. Food habits of albacore, bluefin tuna, and bonito in California waters. *Fish Bull.* **1971**, *152*, 100–105.
15. Jin, B.S.; Winemiller, K.O.; Shao, B.; Si, J.; Jin, J.; Ge, G. Fish assemblage structure in relation to seasonal environmental variation in sub-lakes of the Poyang Lake floodplain, China. *Fish. Manag. Ecol.* **2019**, *26*, 131–140. [CrossRef]
16. Yin, C.; Huang, L.; Xu, L.; Huang, J.; Gao, M. Fish diversity in nature reserves of Jiangxi Province, China. *Eco. Mont.* **2016**, *8*, 33–42. [CrossRef]
17. Hu, Z.P.; Wang, S.G. Evolution of scour and sedimentation and its hydrological and ecological effects in Poyang Lake. *Water Resour. Hydropower Eng.* **2022**, *53*, 66–78. (In Chinese)
18. Mo, L. Pollutants Total Maximun Dalily Load Estimation and Distribution of Poyang Lake. Master's Thesis, Nanchang University, Jiangxi, China, 2009. (In Chinese)
19. Huang, A.; Liu, X.; Peng, W.; Dong, F.; Ma, B.; Li, J.; Wang, W. Spatiotemporal characteristics, influencing factors and evolution laws of water exchange capacity of Poyang Lake. *J. Hydrol.* **2022**, *609*, 127717. [CrossRef]
20. Li, H.; Zhang, H.; Yu, L.; Cao, K.; Wang, D.; Duan, X.; Ding, F.; Mao, Z.; Wang, K.; Liu, S.; et al. Managing water level for Large migratory fish at the poyang lake outlet: Implications based on habitat suitability and connectivity. *Water* **2022**, *14*, 2076. [CrossRef]
21. Ding, F. Study on Spawning Field of Fish Spawning on Plants in Dahuchi Lake Using Remote Sensing Technique. Master's Thesis, Shanghai Ocean University, Shanghai, China, 2017. (In Chinese)
22. Naus, C.J.; Reid Adams, S. Fish nursery habitat function of the main channel, floodplain tributaries and oxbow lakes of a medium-sized river. *Ecol. Freshw. Fish* **2017**, *27*, 4–18. [CrossRef]
23. Couto, T.B.d.; Zuanon, J.; Olden, J.D.; Ferraz, G. Longitudinal variability in lateral hydrologic connectivity shapes fish occurrence in temporary floodplain ponds. *Can. J. Fish. Aquat. Sci.* **2018**, *75*, 319–328. [CrossRef]
24. Huang, W.; Hu, T.; Mao, J.; Montzka, C.; Bol, R.; Wan, S.; Li, J.; Yue, J.; Dai, H. Hydrological Drivers for the Spatial Distribution of Wetland Herbaceous Communities in Poyang Lake. *Remote Sens.* **2022**, *14*, 4870. [CrossRef]
25. Mao, Z.G.; Gu, X.H.; Zeng, Q.F. The structure of fish community and changes of fishery resources in Lake Hulun. *J. Lake Sci.* **2016**, *28*, 387–394.
26. Pielou, E.C. The measurement of diversity in different types of biological collections. *J. Theor. Biol.* **1966**, *13*, 131–144. [CrossRef]
27. Han, Y.; Sun, Z.; Fang, H.; Bai, S.; Huang, L.; He, G. Habitat succession of the Yangtze finless porpoise in Poyang Lake under the changing hydrodynamic and feeding environment. *Ecol. Model.* **2020**, *424*, 109035. [CrossRef]
28. Jiang, T.; Yang, J.; Xuan, Z.Y.; Chen, X.; Liu, H. Preliminary report on the effects of resource recovery on anadromous *Coilia nasus* in Poyang Lake under the National 10-Year Fishing Ban. *Prog. Fish. Sci.* **2022**, *43*, 24–30. (In Chinese)
29. Jiang, X.L.; Li, M.Z.; Yang, S.R.; Wang, C.; Zhang, C.; Gao, X. Temporal variation of fish biodiversity in Poyang Lake. *Acta Hydrobiol. Sin.* **2023**, *47*, 376–388. (In Chinese)
30. Yang, M.; Sheng, P.; Zhang, Y.P.; Guo, L.; Wang, H.; Gao, X.; Chen, Y. Characteristics of fish assemblages of Poyang Lake at the initial stage of the fishing ban. *Acta Hydrobiol. Sin.* **2022**, *46*, 1569–1579. (In Chinese)

fishes

Article

Functional Alpha and Beta Diversity of Fish Communities and Their Relationship with Environmental Factors in the Huanghe River (Yellow River) Estuary and Adjacent Seas, China

Shaowen Li [1], Yufang Huang [2], Fan Li [1,*], Yanfen Liu [3,*], Hongliang Ma [2], Xiaomin Zhang [1], Xiuxia Wang [1], Wei Chen [1], Guangxin Cui [1] and Tiantian Wang [4]

1 Shandong Provincial Key Laboratory of Restoration for Marine Ecology, Observation and Research Station of Laizhou Bay Marine Ecosystem, MNR, Shandong Marine Resource and Environment Research Institute, Yantai 264006, China; lishaowen@shandong.cn (S.L.); zhangxiaomin@shandong.cn (X.Z.); wangxiuxia@shandong.cn (X.W.); chenwei@shandong.cn (W.C.); cuiguangxin@shandong.cn (G.C.)
2 Institute of Yellow River Water Resources Protection, Zhengzhou 450004, China; yfhuang@yrwrp.cn (Y.H.); hlma@yrwrp.cn (H.M.)
3 Observation and Research Station of Bohai Strait Eco-Corridor, MNR, Dongying Marine Development Research Institute, Dongying 257000, China
4 Yantai Marine Economic Research Institute, Yantai 264000, China; ytsblz@yt.shandong.cn
* Correspondence: lifan@shandong.cn (F.L.); yfliu@yic.ac.cn (Y.L.)

Abstract: Fisheries and bottom seawater data were collected in July in 2022 and 2023 from 15 sampling stations in the Huanghe River estuary and adjacent seas in China. The patterns of functional alpha and beta diversity of fish communities in this area and their relationships with environmental factors were studied using trait-based analysis. Five functional classifications, which included 16 functional traits, four functional alpha diversity indices (functional richness, functional evenness, functional divergence, and community-weighted mean), three functional beta diversity indices (functional beta diversity and its turnover and nestedness components), and 14 bottom environmental factors were considered. The dominant traits of fish communities were the following: benthivorous feeding habits, small and front-facing mouths, high trophic level, demersal, anguilliform, low growth coefficient, medium resilience, low vulnerability, and adapted to warm temperate conditions. The dominant migration traits and types of fish eggs varied by year, and fish abundance was the main factor affecting the dominant traits of the communities. The results of multiple regression on distance matrices and variance partitioning analysis indicated that ammonia nitrogen content, total phosphorus content, and pH were the main environmental factors that affected fish functional diversity. The N/P ratio had a bottom-up control effect on fish functional diversity. Our findings also revealed that high pH gradients and distant geographical distances can inhibit trait turnover in fish communities.

Keywords: traits; functional diversity; Huanghe River estuary; bottom-water environment; turnover; nestedness

Key Contribution: This study analyzed the functional alpha and beta diversity patterns of fish communities in the Huanghe River estuary and adjacent seas and their relationships with environmental factors based on bottom trawl survey data from July 2022 and 2023. The results of this study enhance our understanding of biodiversity in and near estuaries and will contribute to the conservation and maintenance of these fishes.

Citation: Li, S.; Huang, Y.; Li, F.; Liu, Y.; Ma, H.; Zhang, X.; Wang, X.; Chen, W.; Cui, G.; Wang, T. Functional Alpha and Beta Diversity of Fish Communities and Their Relationship with Environmental Factors in the Huanghe River (Yellow River) Estuary and Adjacent Seas, China. *Fishes* **2024**, *9*, 222. https://doi.org/10.3390/fishes9060222

Academic Editor: Pedro Morais

Received: 7 May 2024
Revised: 10 June 2024
Accepted: 11 June 2024
Published: 12 June 2024

1. Introduction

Determining biodiversity distribution patterns, maintenance mechanisms, and response mechanisms to environmental change are core issues in community ecology [1–3]. In the 1960s, Whittaker divided biodiversity patterns into three scales: alpha diversity (α), beta diversity (β), and gamma diversity (γ) [4,5]. Alpha and gamma diversity share the same characteristics but are applied at different scales. The alpha diversity scale is the smallest and describes the species diversity of a quadrat. The gamma diversity scale is the largest and describes the diversity of species in an entire ecosystem or study area. Beta diversity describes differences in community composition between habitats or the rate at which species are replaced along environmental gradients. Understanding these three types of diversity patterns can solve many ecological and evolutionary biology problems, and they are extremely important for understanding ecosystem functioning and effective biodiversity conservation, as well as ecosystem management [6]. However, for a long time, scholars primarily focused on alpha and gamma diversity. The beta diversity decomposition method was proposed in 2010, and beta diversity-related studies rapidly increased [7].

In terms of diversity research, traditional community ecology mainly focuses on species diversity, such as species composition and quantity. In fact, communities with the same number of species are likely to have substantial differences in living habits, forms, and other aspects [8]. Therefore, traditional taxonomic diversity does not truly reflect the status of species in ecosystems and their impact on ecosystem functions [9,10]. In recent years, ecologists have found that functional diversity based on the functional traits of species is more closely related to the environment than taxonomic diversity. This provides a more scientific basis for species conservation and management [11,12]. An increasing number of scholars have proposed that functional diversity should be used instead of species diversity to study communities.

From the perspective of functional diversity, functional alpha diversity reflects the difference or diversity of functional traits of species within a community. Mason et al. [13] divided functional alpha diversity into functional richness, functional evenness, and functional dispersion, and they proposed a calculation method. Recent studies on the functional alpha diversity of fishes in China has mainly focused on freshwater fishes, and there have been few studies on marine fish communities, with studies only having been conducted in Haizhou Bay [14], Pipa Island Ocean Ranch [15], and Changdao Island [16] in Shandong, China. The functional traits commonly used in the above studies include feeding, movement, and morphological characteristics, such as feeding, mouth position, habitat position, and migration type.

Unlike functional alpha diversity, functional beta diversity describes the changes in the composition of functional diversity between communities under different environmental gradients [17,18]. Recent studies indicated that differing functional beta diversity between communities is mainly due to two processes—species increase or loss and species replacement or turnover. The former leads to differences in species richness between communities, whereas the latter leads to a decrease in common species and an increase in endemic species between communities [19–21]. When a community with fewer species is a subset of a community with more species, it appears as a nested pattern [22,23]. Therefore, functional beta diversity can be decomposed into the form of the addition of the turnover and nested components [24–26]. In China, studies have only been conducted on the functional beta diversity of marine fishes in Haizhou Bay in Shandong Province [27,28] and offshore of Yantai and Weihai [29].

The Huanghe River is the second largest river in China, and its mouth is located in Bohai Bay, China. Here, water of the river and ocean converge to form a sensitive and complex ecosystem at the Huanghe River estuary [30,31]. The average water depth of the Huanghe River estuary and adjacent seas is 11.2 m, with a maximum depth of 18 m and a minimum depth of 2.8 m. The channel width of the Huanghe River at Lijin station (79 km from the estuary) is 0.58 km. In the past 20 years, the annual average runoff at Lijin Station was 168.56×10^2 hm^3, of which the runoff during the flood season was 90.52×10^2 hm^3 and 78.04×10^2 hm^3 during the non-flood season. The minimum value was 41.90×10^2 hm^3 in 2002, and the maximum value was 359.79×10^2 hm^3 in 2020 [32,33]. The influx of freshwater creates a low-salinity zone in the estuary. This area provides a suitable shelter for marine life to reproduce and survive. Additionally, the freshwater and sediment of the Huanghe River transport a lot of nutrients to the coastal area of the estuary. For example, the concentration of dissolved organic carbon from the estuary to offshore of the Huanghe River during the flood season can reach 2.6–3.1 mg L^{-1}, which increases primary productivity [34,35]. Therefore, this area is an important spawning, breeding, and feeding ground for fishes in the Yellow and Bohai Seas of China [36]. The unique environmental gradient is an important reason for the high level of biodiversity in the Huanghe River estuary. According to historical records, there were more than 600 marine species in the Huanghe River estuary [37].

The fish community is an important part of the biodiversity in the Huanghe River estuary. Most studies on fish communities focus on biological characteristics, resource density, species diversity, and community structure, but there are no reports on functional diversity. Comparative studies of alpha and beta diversity on a regional scale represent important research in ecology [38]. The main objectives of this study were to (1) describe the functional alpha and beta diversity patterns of fish communities in the Huanghe River estuary and (2) identify the relationship between their biodiversity patterns and environmental gradients in the estuary. These results can improve our understanding of estuarine and marine biodiversity and will contribute to their conservation and maintenance.

2. Materials and Methods

2.1. Data Source

We conducted bottom trawl surveys of fishery resources in July 2022 and 2023 in offshore areas of the Huanghe River estuary in China. The survey range was 119°00′–119°08′ E, 37°02′–38°03′ N. Samples were collected from 15 sampling stations (37°20′–38°09′ N, 119°05′–119°40′ E), which were divided into three groups according to the distance from the Huanghe River mouth. Sections A, B, and C were located 15–20, 25–35, and 40–55 km from the estuary mouth, respectively (Figure 1).

The fishing vessel used for surveys was "Luchangyu 60003". This vessel has a Yuchai MKF56K00018 engine (Guangxi Yuchai Machinery Group Co., Ltd., Yulin, China), which has a power of 260 kW. The vessel is 24.4 m long and 6.0 m wide and weighs 68 t. A single bottom trawl was used. The trawl mouth has a 30.6 m circumference, 20 mm bag mesh, and ~8 m diameter when towing. The designed average towing speed at each station was 2 kn, and the trawling time was 1 h. The catches were identified on site and counted, and biological parameters (e.g., body length and weight, accurate to 1 mm) measured. The bottom-water temperature (TEM), salinity (SAL), depth (DEP), pH, and dissolved oxygen (DO) were recorded using a profiler (YSI EXO Handheld; Yellow Springs Instrument Co., Inc., Yellow Springs, OH, USA). Bottom chemical oxygen demand (COD) was determined using the basic potassium permanganate method. Bottom dissolved inorganic phosphate (DIP) and nitrogen (DIN, including NO_3–N, NO_2–N, and NH_4–N) were analyzed using segmented flow analysis. Bottom total nitrogen (TN) and total phosphorus (TP) were measured using the potassium persulfate oxidation method. Chlorophyll a (Chla) was measured using a fluorescence spectrophotometer. Suspended solids (SSs) were measured using the membrane filtration method. Biological and environmental samples were collected simultaneously. All indicators were collected and measured according to the Specification for Oceanographic Survey Part of the National Standards of PRC (GB/T 12763–2007).

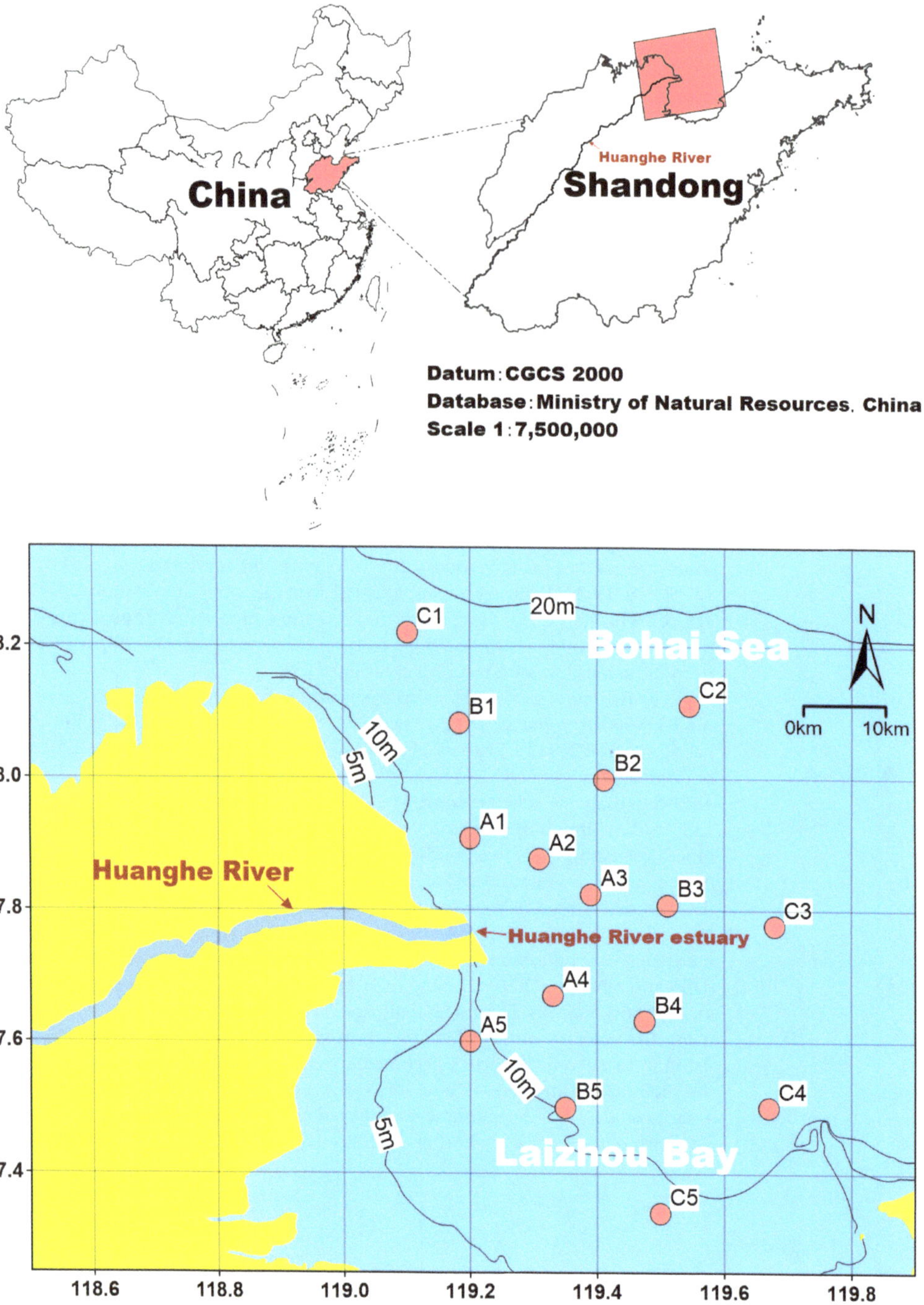

Figure 1. Map showing the study area and locations of the fish sampling stations in the Huanghe River estuary and adjacent seas in China. The labels of stations represent the distance from the estuary. A is for 15–20 km from the estuary mouth, B for 25–35 km, and C for 40–55 km.

2.2. Methods

2.2.1. Functional Traits

Sixteen functional traits related to fish feeding, motility, reproduction, population dynamics, and ecological adaptation were selected, including 9 categorical traits and 7 continuous traits (Table 1). The values or types of functional traits were mainly obtained from the World Fish Database (FishBase) [39] and previous papers [14,15,40–47]. The "species–traits" matrix of fishes is detailed in Appendix A Table A1.

Table 1. Classification of functional traits of fishes in the Huanghe River Estuary and adjacent seas.

Classification of Function	Type of Functional Traits	Variable Type	Functional Trait
Food acquisition	Feeding habits	Ordered categorical variables	Herbivorous (H), Planktivorous (P), Planktivorous and Benthivorous (P–B), Benthivorous (B), Benthivorous and Nektivorous (B–N), Nektivorous (N)
	Mouth position	Ordered categorical variables	Lower, Front, Upper
	Relative mouth size	Ordered categorical variables	Small (S), Comparatively small (CS), Medium (M), Comparatively big (CB), Big (B)
	Trophic level	Continuous variable	2.5~4.8
Locomotion	Habitat	Unordered categorical variable	Demersal (De), Pelagic (Pe)
	Migration type	Ordered categorical variables	Settlement type (ST), Coastal migratory (CM), Offshore migratory (OM), Distantly migrating (DM)
	Body shape	Unordered categorical variable	Fusiform (1), Compressiform (2), Depressiform (3), Anguilliform (4), Asymmetry (5), Sub-cylinder (6), Slightness (7), Cylindrical anterior part and compressed posterior part (8), Band shape (9), Depressed anterior part and compressed posterior part (10), Sub-cylindrical anterior part and compressed posterior part (11)
	Maximum body length (cm)	Continuous variable	9.0~113.0
Reproduction	Body length at first sexual maturity (cm)	Continuous variable	6.8~52.7
	Age at first sexual maturity (a)	Continuous variable	0.8~3.8
	Generation time (a)	Continuous variable	1.0~8.5
	Ecological type of eggs	Unordered categorical variable	Pelagic eggs (PE), Adhesive eggs (AE), Demersal adhesive eggs (DAE), Pelagic adhesive eggs (PAE), Ovoviviparous (O)
Population dynamics	Growth coefficient (k)	Continuous variable	0.18~0.80
	Resilience	Ordered categorical variables	Low resilience (Lr), Medium resilience (Mr), High resilience (Hr)
	Vulnerability	Continuous variable	10~60
Ecological adaptation	Thermophily	Unordered categorical variable	Cold temperate (CT), Warm temperate (T), Warm water (WW)

2.2.2. Functional Alpha Diversity

In this study, four functional diversity indices were used to measure the spatio-temporal variation of functional alpha diversity in fish communities [48,49]:

1. Functional richness, *FRic*, quantifies the size of the ecological space occupied by a species within a community, thus reflecting the stability of the community and its ability to buffer environmental disturbances and resist ecological invasion [50]. *FRic* was calculated as follows. First, a species with an extreme character value

was identified and used as the end point of the smallest convex shape in the N-dimensional character space. Then, the endpoints were connected to form a minimum convex polygon. Finally, the area or volume of the minimum convex polygon was calculated [48].

2. Functional evenness, *FEve*, quantifies the evenness of the abundance of functional traits of species within a community that is distributed in functional space, thus reflecting the overall utilization of resources by species [50]. *FEve* was calculated as follows:

$$dist(i,j) = \sqrt{(a_i - a_j)^2 + (b_i - b_j)^2 + \cdots + (m_i - m_j)^2} \quad (1)$$

$$EW_l = dist(i,j) / (w_i + w_j) \quad (2)$$

$$PEW_l = EW_l / \sum_{l=1}^{S-1} EW_l \quad (3)$$

$$FE_{ve} = \frac{\sum_{l=1}^{S-1} min\left(PEW_l, \frac{1}{S-1}\right) - \frac{1}{S-1}}{1 - \frac{1}{S-1}} \quad (4)$$

where a to m represent the 16 functional traits of species i and j in a multidimensional trait space, $dist(i, j)$ is the Euclidean distance between species i and j, EW_l is the branch length, w_i and w_j are the relative abundances of species i and j, PEW_l is the branch length weight, and S is the number of species.

3. Functional divergence, *FDiv*, quantifies the dispersion of species functional trait abundance within the community in the functional space, thus reflecting the degree of niche differentiation and resource competition among species within the community [51]. *FDiv* was calculated as follows:

$$g_k = \frac{1}{S} \sum_{i=1}^{S} x_{ik} \quad (5)$$

$$dG_i = \sqrt{\sum_{k=1}^{T} (x_{ik} - g_k)^2} \quad (6)$$

$$\overline{dG} = \frac{1}{S} \sum_{i=1}^{S} dG_i \quad (7)$$

$$\Delta d = \sum_{i=1}^{S} w_i \times \left(dG_i - \overline{dG}\right) \quad (8)$$

$$\Delta|d| = \sum_{i=1}^{S} w_i \times \left|dG_i - \overline{dG}\right| \quad (9)$$

$$FD_{iv} = \frac{\Delta d + \overline{dG}}{\Delta|d| + \overline{dG}} \quad (10)$$

where x_{ik} is the value of trait k for species I, g_k is the centroid of trait k, S is the number of species, T is the number of traits, $\overline{dG}$ is the average distance between species i and the centroid, Δd is the dispersion weighted by abundance, and w_i is the relative abundance of species i. The usage conditions were $S > T$.

4. Community-weighted mean, *CWM*, quantifies the weighted average of functional traits of species within a community, thus reflecting the changes in dominant trait values of fish communities. *CWM* is very important for evaluating community dynamics [10]. *CWM* was calculated as follows:

$$WM = \sum_{i=1}^{S} p_i \times trait_i \quad (11)$$

where p_i represents the relative abundance of species i, $trait_i$ is the trait value of species i, and S is the number of species in the community.

2.2.3. Functional Beta Diversity

Species occurrence and non-occurrence data (i.e., occurrence is 1, non-occurrence is 0) were used to calculate the functional beta diversity of fish communities. Before calculation, a Gower distance transformation was performed on the species–traits matrix, and the trait distance was calculated between species. Then, principal coordinate analysis was used to reduce dimensionality. The first four axes (cumulative explanation rate > 90%) were selected as input data for calculating the beta diversity of functional traits [52]. Using the additive decomposition method proposed by Baselga [24], the Sørensen dissimilarity index and multi-dimensional functional space were used to calculate the functional beta diversity of fish communities. This method decomposes functional beta diversity into the sum of turnover and nested components [48]. Measure the contribution of turnover components to functional beta diversity by their proportion to functional beta diversity. A ratio greater than 0.5 indicates that beta diversity is mainly dominated by turnover components, whereas a ratio less than 0.5 indicates that beta diversity is mainly dominated by nested components [18]. Functional beta diversity was calculated as follows:

$$^F\beta_{sor} = {^F\beta_{sim}} + {^F\beta_{sne}} \tag{12}$$

$$^F\beta_{ratio} = {^F\beta_{sim}} / {^F\beta_{sne}} \tag{13}$$

where $^F\beta_{sor}$ represents the functional beta diversity, $^F\beta_{sim}$ is the turnover component of functional beta diversity, $^F\beta_{sne}$ is the nestedness component of functional beta diversity, and $^F\beta_{ratio}$ is the proportion of the turnover component to functional beta diversity.

2.2.4. Data Analysis

The functional alpha diversity indices were calculated using the "FD" package. The functional beta diversity and its components were calculated using the "betapart" package. Multiple regression on distance matrices (MRM) analysis was used to explore the relationship between spatial distance and environmental factors matrices with functional diversity and its components. Environmental factors were tested for correlation using a collinearity test, with correlation indicated by Spearman $\varrho^2 < 0.7$. MRM analysis was performed in the "ecodist" package, and all *p*-values were obtained by 9999 Mantel tests. The distances between two stations and between each station and river mouth were calculated using the "distm" function of the "geosphere" package based on latitude and longitude. The Euclidean distance matrix of environmental factors was calculated using the "vegdist" function of the "vegan" package. A collinearity test on environmental factors was performed using the "varclus" function of the "Hmisc" package, and highly correlated (Spearman $\varrho^2 > 0.7$) environmental factors were excluded. To compare the relative importance of each factor, the "stdize" function in the "MuMIn" package was performed to standardize the Euclidean distance matrix of each factor before running the MRM analysis [52]. Variance partitioning analysis (VPA) was performed to evaluate the contribution of geographic distance and environmental factors to the variance in functional diversity. VPA was performed using the "vegan" package [53].

All statistical analyses were conducted in R 4.2.2 and SPSS 17.0. Surfer 14, and Origin 2022 was also used to plot data.

3. Results

3.1. Fish Species Composition in the Huanghe River Estuary and Adjacent Seas

In the two cruises, a total of 30 sampling stations were surveyed. The species accumulation curve showed a stable upward trend, and the number of samples showed a gradual flattening (Appendix B Figure A1). This indicated that the sampling in this study was adequate. In 2022 and 2023, a total of 37 fish species were caught by bottom trawl surveys in the Huanghe River estuary and adjacent seas; these species belong to 35 genera, 21 families, nine orders, and one class (Appendix A Table A1). Among these species, Perciformes accounted for the highest number of species (19). In 2022, the average species count per station was 12 ± 3, and the dominant species were *Chaeturichthys stigmatias*, *Cynoglossus joyeri*, and *Lateolabrax maculatus*. In 2023, the average species count per station was 17 ± 2, and the dominant species were *C. stigmatias*, *C. joyeri*, *Konosirus punctatus*, *Setipinna taty*, *Thrissa kammalensis*, and *Scomberomorus niphonius*. The species count at each station in both years followed a normal distribution. A paired *t*-test revealed a significant difference in species count between the two years ($p < 0.001$). Stacked plots of species abundance percentages (Figure 2) showed that Gobiidae consistently accounted for the highest proportion of species at each station across two years; in 2022 and 2023, it constituted 73.55% and 35.57%, respectively, of the total species abundance. Furthermore, the proportion of Gobiidae was unaffected by the geographical distance between the stations and the estuary.

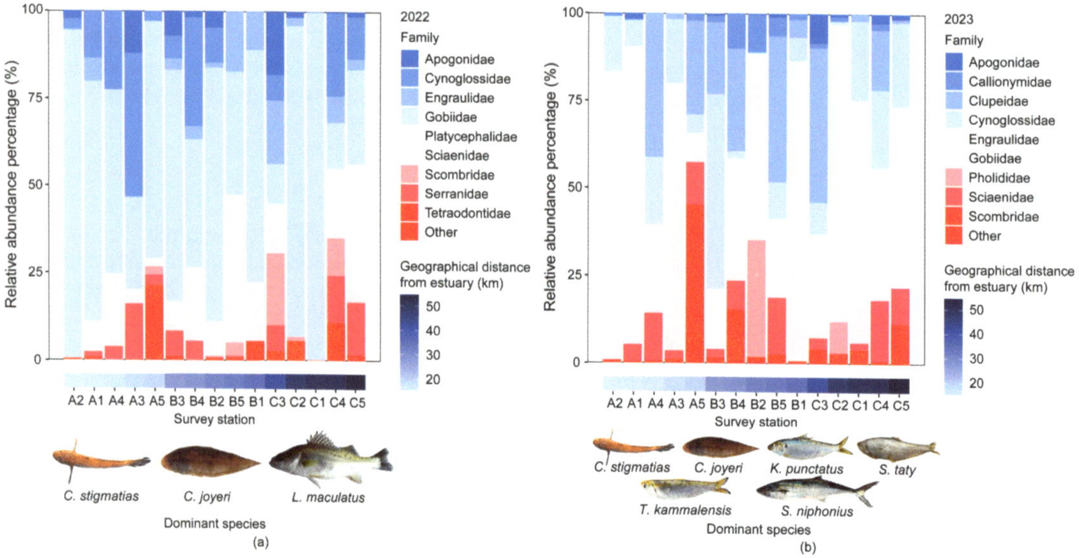

Figure 2. Stacking plots of fish species and dominant species in the Huanghe River estuary seas in 2022 (**a**) and 2023 (**b**).

3.2. Functional Alpha Diversity and Dominant Community Trait Composition of Fishes in the Huanghe River Estuary and Adjacent Seas

Alpha diversity analysis of fish communities in the Huanghe River estuary (Figure 3) revealed that the average *FRic* index values for 2022, 2023, and the two-year period were 0.140 ± 0.117, 0.307 ± 0.175, and 0.223 ± 0.169, respectively. This indicated that the number of species occupying ecological niches in fish communities significantly increased over the two-year period (one-way ANOVA $p < 0.05$). The average *FEve* index values were 0.514 ± 0.166, 0.408 ± 0.104, and 0.461 ± 0.147, respectively, which indicated that the evenness of functional trait abundance distribution among fish species in the multi-dimensional trait space decreased over the two-year period ($p < 0.05$). The average *FDiv* index values were 0.812 ± 0.155, 0.798 ± 0.075, and 0.805 ± 0.120, respectively, but the differences were not significant ($p > 0.05$). This indicated that the dominant species in the fish community deviated from the center of gravity in the multi-dimensional trait space. The abundance distribution of the species' functional traits exhibited a high level of dispersion during the survey period. There was not a significant difference in the functional alpha diversity indices among different sections across survey years. This indicated that the distribution pattern of the functional alpha diversity was not affected by the distance gradient between sampling stations and the river mouth.

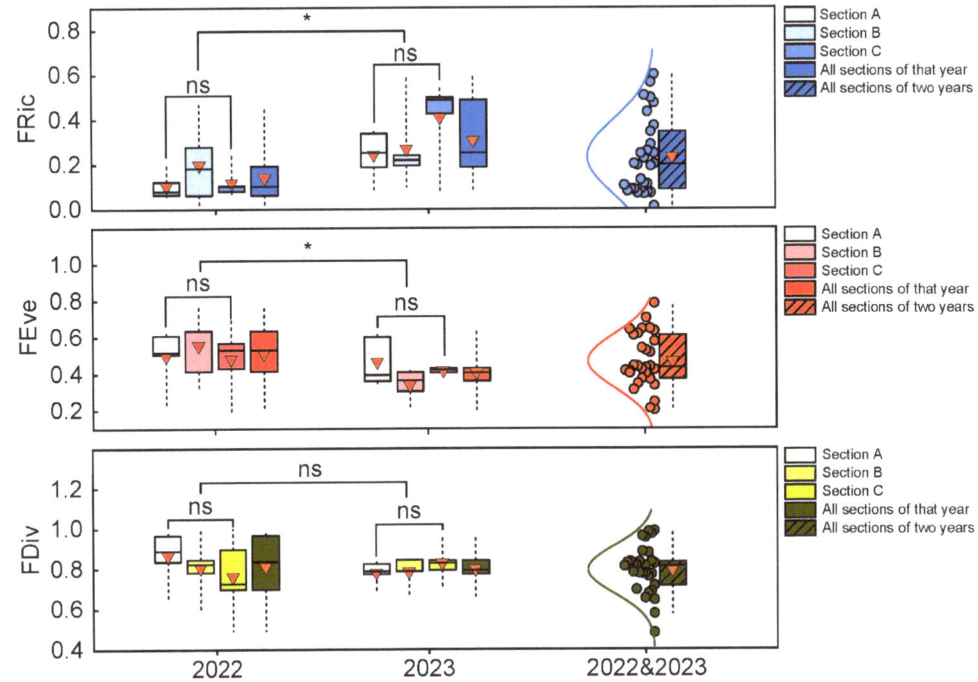

Figure 3. Functional alpha diversity of fishes in the Huanghe River estuary and adjacent seas in 2022 and 2023. Asterisks denote a significant difference (* $p < 0.05$); ns indicates no significant difference.

The *CWM* reflected the dominant trait composition pattern of fish communities in the Huanghe River estuary and adjacent seas (Figure 4). In terms of food acquisition, species with benthivorous feeding habits; small, front-facing mouths; and higher trophic levels were dominant during the survey period. Feeding habits shifted from benthivorous species in 2022 to benthivorous, herbivorous, planktivorous, and nektivorous species in 2023. The trophic levels of dominant species in 2022 and 2023 were 3.876 ± 0.172 and 3.689 ± 0.181, respectively, with significant differences between the two years ($p < 0.05$).

Of the locomotion traits, species that are demersal, anguilliform, and have a maximum body length 30–40 cm were dominant. The dominant migration types shifted from settlement type in 2022 to offshore migratory in 2023. The proportion of species inhabiting pelagic layers also increased. The maximum body lengths of dominant fishes in 2022 and 2023 were 35.21 ± 6.86 cm and 30.70 ± 10.02 cm, respectively, with no significant difference between the two years ($p > 0.05$).

Of the fish reproduction traits, the dominant ecological type of eggs shifted from adhesive eggs in 2022 to pelagic eggs in 2023. The dominant body lengths at first sexual maturity in 2022 and 2023 were 15.09 ± 3.42 cm and 15.09 ± 3.45 cm, respectively. The ages at first sexual maturity were 2.35 ± 0.44 years and 2.46 ± 0.35 years, respectively. The generation times were 2.97 ± 0.62 years and 3.38 ± 0.65 years, respectively. During the survey period, there were no significant differences in body lengths, ages at first sexual maturity, or generation times ($p > 0.05$).

The dominant population dynamics of fish were low growth coefficient, moderate resilience, and low vulnerability during the survey period. The growth coefficients for 2022 and 2023 were 0.423 ± 0.044 and 0.476 ± 0.060, respectively, with a highly significant difference between the two years ($p < 0.01$). The dominant resilience of the fish populations shifted from moderate resilience to coexistence of species with moderate and high resilience. The population vulnerabilities were 26.39 ± 5.75 and 27.14 ± 5.98, respectively, with no significant difference between the two years ($p > 0.05$). In terms of thermophily, warm temperate was the dominant trait during the survey period.

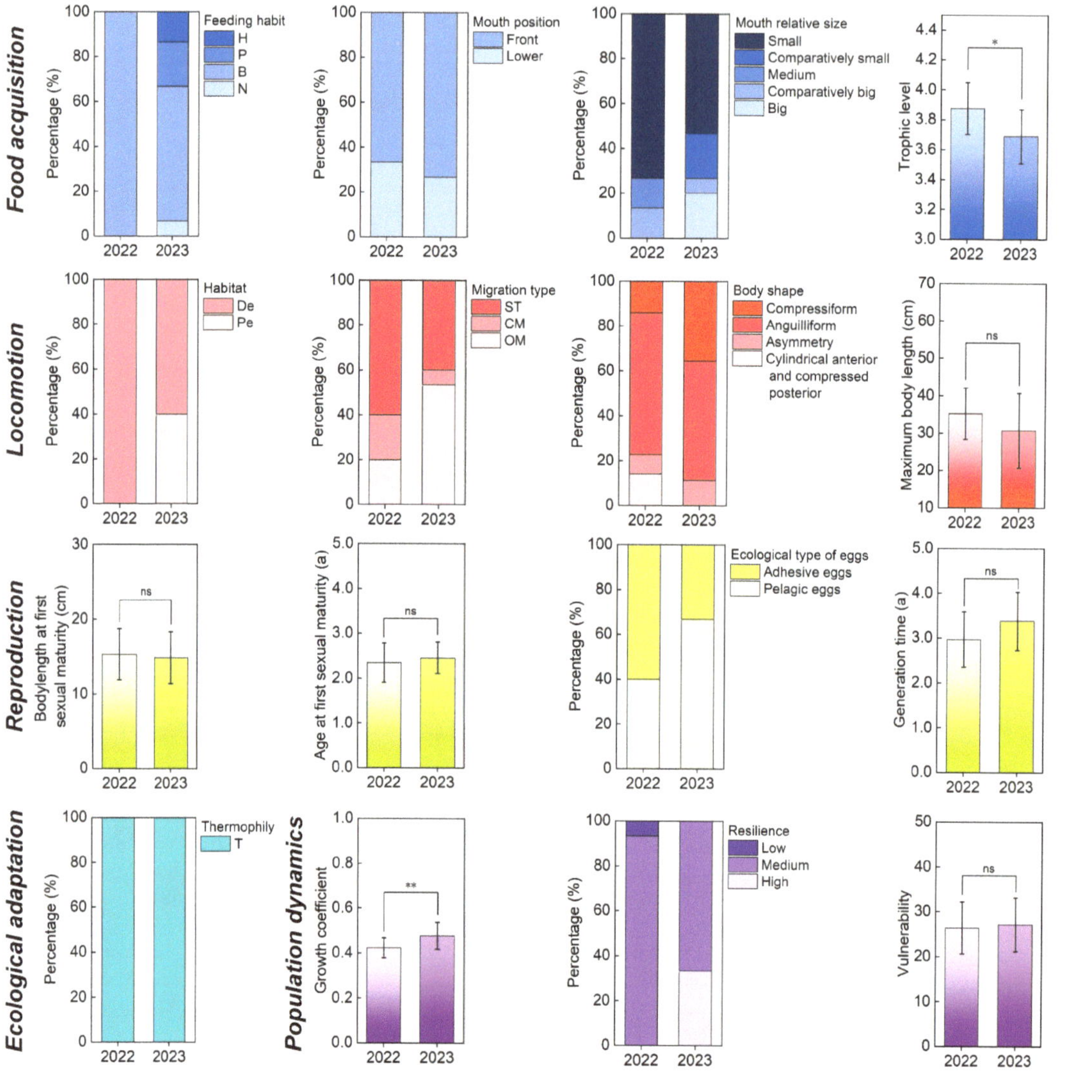

Figure 4. Fish community *CWM* and variation in the Huanghe River estuary and adjacent seas. Asterisks denote significant difference (* $p < 0.05$) or highly significant difference (** $p < 0.01$), and ns indicates no significant difference.

3.3. Functional Beta Diversity of Fishes in the Huanghe River Estuary and Adjacent Seas

In 2022, the mean values of ${}^{F}\beta_{sor}$, ${}^{F}\beta_{sim}$, and ${}^{F}\beta_{sne}$ of the fish communities in the study area were 0.661 ± 0.180, 0.391 ± 0.221, and 0.271 ± 0.224, respectively. The turnover component contributed 59.15% to the functional beta diversity. In 2023, the mean values of the three components were 0.661 ± 0.180, 0.391 ± 0.221, and 0.271 ± 0.224, respectively, and the ${}^{F}\beta_{sim}$ contributed 51.36%. During the research period, the ${}^{F}\beta_{sim}$ dominated the functional beta diversity of the Huanghe River estuary and adjacent seas, but its contribution showed a declining trend. The results of the two years of surveys were consistent with a normal distribution ($p > 0.05$, single-sample K–S test). The single-sample *t*-test did not show a significant difference between the functional beta diversity and its mean value between 2022 and 2023 ($p > 0.05$, two-tailed test). The paired-sample *t*-test showed a significant difference in functional beta diversity among sample pairs between 2022 and 2023 ($p < 0.001$, two-tailed test) (Table 2).

Table 2. Normality test and *t*-test results for functional beta diversity in 2022 and 2023.

Year	Normality Test		*t*-Test		
	Kolmogorov–Smirnov Z	P (Two-Tailed Test)	*t*-Value	d*f*	P (Two-Tailed Test)
2022	0.781	0.575	0.009	104	0.993
2023	0.895	0.400	−0.016	104	0.987
2022–2023	-	-	4.894	104	<0.001 ***

Asterisks denote a highly significant difference (*** $p < 0.001$).

The functional beta diversity of the fishes in the Huanghe River estuary and adjacent seas and the functional beta diversity components are shown in Figure 5. From 2022 to 2023, the ${}^{F}\beta_{sim}$ center of gravity significantly shifted toward lower values, although there was not a significant change in the ${}^{F}\beta_{sne}$ (Figure 5). This caused the similarity values ($1-{}^{F}\beta_{sor}$) of sample pairs to be pulled toward high values, which led to a reduction in the number of high-value sample pairs for the ${}^{F}\beta_{sor}$. The three-dimensional contour map of geographic distance distribution for sample pairs showed that, during the survey period, the geographic distance among sample pairs significantly increased as the values of ${}^{F}\beta_{sor}$ and ${}^{F}\beta_{sim}$ decreased. Therefore, there may be a negative correlation between the fish functional beta diversity and geographical distance in the Huanghe River estuary and adjacent seas.

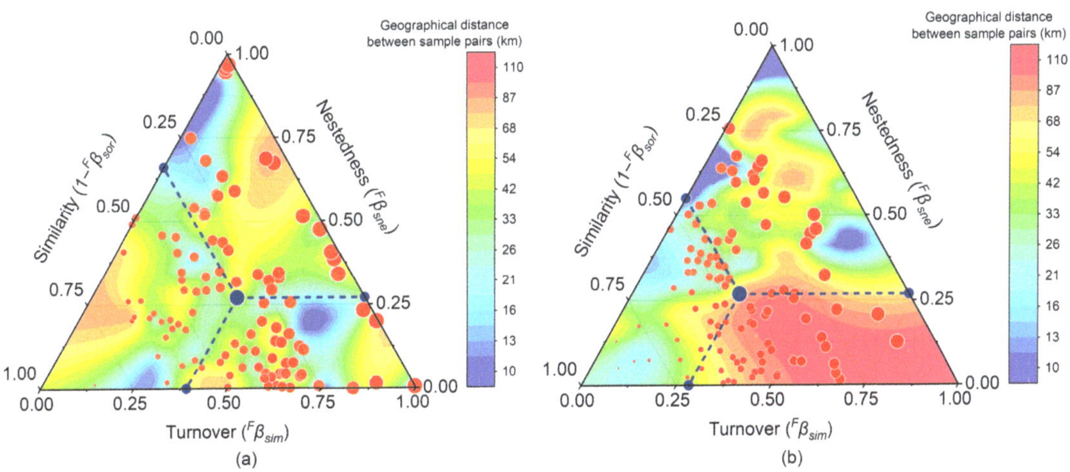

Figure 5. Functional beta diversity (${}^{F}\beta_{sor}$) and its turnover (${}^{F}\beta_{sim}$) and nestedness (${}^{F}\beta_{sne}$) components. (**a**) The 2022 values; (**b**) The 2023 values. Each red dot in the figure represents the data of one sample pair. The size of the red dot represents the ${}^{F}\beta_{sor}$ value of sample pair. The blue dots represent the mean values of the three components of functional beta diversity.

3.4. Relationship between Functional Diversity of Fishes and Environmental Factors in the Huanghe River Estuary and Adjacent Seas

Environmental factors were tested for correlation using a collinearity test, with correlation indicated by Spearman $\varrho^2 < 0.7$ (Appendix C Figure A2). Therefore, all environmental factors were incorporated into the MRM model, and geographic distance factors were also included in the analysis. The results indicated that the NH_4–N and TP levels were the principal environmental factors that influenced the *FRic* and *FEve* ($p < 0.05$). There was not a significant correlation between the geographical distance and functional alpha diversity indicators ($p > 0.05$). For the ${}^F\beta_{sor}$, only ammonia nitrogen showed a significant positive correlation ($p < 0.05$). The ${}^F\beta_{sim}$ exhibited a highly significant negative correlation with pH ($p < 0.01$) and a significant negative correlation with the geographic distance ($p < 0.05$). The ${}^F\beta_{sne}$ showed significant positive correlations with the TP and pH ($p < 0.05$). The correlation analysis between the *FRic* and ${}^F\beta_{sor}$ revealed an extremely significant negative correlation ($p < 0.001$) (Table 3).

Table 3. MRM results showing relationships of functional alpha and beta diversity indices and components with environmental factors.

Diversity Index	${}^F\beta_{sor}$		NH_4–N		TP		pH		Geographic Distance	
	Coefficient	*p*	Coefficient	*p*	Coefficient	*p*	Coefficient	*p*	Coefficient	*p*
FRic	−0.1157	0.0001 ***	−0.1550	0.0371 *	0.7977	0.0249 *	-	-	-	-
FEve	-	-	−0.0400	0.0161 *	-	-	-	-	-	-
FDiv	-	-	-	-	-	-	-	-	-	-
${}^F\beta_{sor}$	-	-	0.0317	0.0376 *	-	-	-	-	-	-
${}^F\beta_{sim}$	-	-	-	-	-	-	−0.0393	0.0066 **	−0.0510	0.046 *
${}^F\beta_{sne}$	-	-	-	-	0.1134	0.0318 *	0.0301	0.0379 *	-	-

Asterisks denote significant differences (* $p < 0.05$) and highly significant differences (** $p < 0.01$, *** $p < 0.001$).

The VPA results showed that NH_4–N and TP played more important roles than other variables in determining the functional alpha diversity of the fish communities in the Huanghe River estuary and adjacent seas. All environmental factors accounted for 54.44% of the functional alpha diversity, of which NH_4–N and TP each accounted for 26.22%. However, the geographical distance had minimal explanatory power on the distribution pattern of functional alpha diversity in the fish communities. In terms of the functional beta diversity, all environmental factors accounted for 44.31%, of which geographic distance variables accounted for 7.59%. However, the combined explanatory power of the alpha and beta diversity was relatively low. Of all the environmental factors, NH_4–N, TP, and pH contributed 32.90% to the functional beta diversity, whereas other environmental factors contributed 10.01% (Figure 6).

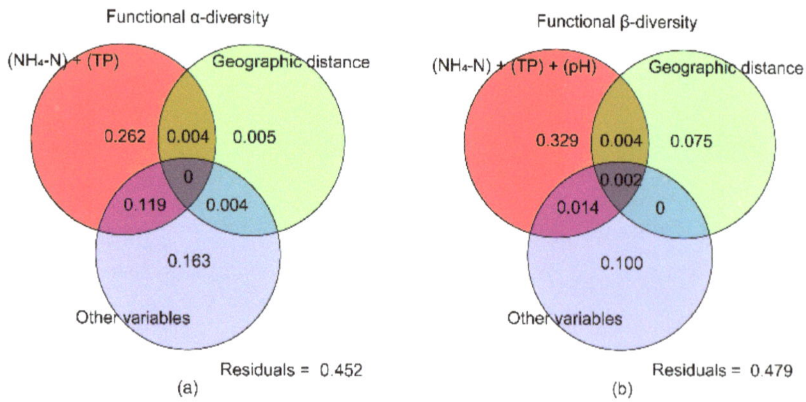

Figure 6. VPA of the functional diversity explained by environmental factors and geographic distance variables. (**a**) the functional alpha diversity; (**b**) the functional beta diversity.

4. Discussion

Numerous studies have been conducted on traditional fish species diversity, thus focusing on indices such as the Margalef richness index and Shannon–Wiener diversity index. These studies have primarily compared the species number and abundance within communities. This approach was referred to as taxonomic alpha diversity and has been used to quantify fish diversity. As research progressed, traits were introduced into the study of diversity. Studies have shown that differences in species traits can lead to differences in the β diversity in communities with the same α diversity [54]. Functional diversity research, which is based on "species–trait–abundance", can more effectively explain various ecological issues than traditional "species–abundance" research.

Factors that influence fish functional diversity include environmental gradients, geographical distance, and functional traits. The construction of a fish functional traits matrix is crucial to elucidating their impacts on fish communities, and functional traits must be selected that are associated with research objectives [38]. The morphological traits of fishes reflect the complex interplay between them and their environment. Existing research on fish functional traits has primarily relied on morphological characteristics and focused on freshwater fishes. For example, there are characteristics of motoric function that are reflected by ratios of body length to body height and ratios of minimum caudal peduncle height to maximum caudal fin height [55]. Additionally, there are measurements that reflect habitat, such as eye position, eye diameter, and head height [12]. Although morphological characteristics can effectively describe fish traits related to food acquisition, movement, and other functions, and they are the only functional traits that can be reasonably quantified [56], and they cannot fully reflect all the roles of fishes in the ecosystem. Therefore, some studies have also introduced other functional categories, such as life history strategy, habitat use, nutrient budget, and defense against predation [57,58]. There are few studies on the functional traits of marine fish [14–16]. In this study, in addition to selecting five morphological traits, such as mouth position, size, and body shape, 16 functional traits were also selected according to existing reports. These traits covered five characteristics of fishes, food acquisition, locomotion, reproduction, population dynamics, and ecological adaptation, which comprehensively reflect the functional diversity of fish communities in the Huanghe River estuary and adjacent seas.

Functional alpha diversity calculation revealed that the *FRic* was solely associated with the alteration in the "species–traits" matrix and the frequency of species occurrence within communities, and it was unaffected by species abundance. However, the influence of species abundance must be considered when calculating the *FEve*, *FDiv*, and *CWM*. In this study, 32 species were identified across 2022 and 2023 in the study area. Of these, 27 species were common in both years. However, the *FRic* was significantly higher in 2023 than 2022. This may be attributed to the significant increase in the frequency of various species across all sampling stations in 2023 compared to that in 2022. Moreover, variations in species occurrence between the two years may have altered the extreme traits within communities, although this was unlikely to be the primary factor. On the basis of these findings, we infer that the fish community in the studied area exhibited greater stability in 2023 compared to that in 2022.

The *CWM* indicated that there were significant changes in the composition of seven dominant traits, including feeding habits, trophic levels, and migration patterns, of the fish communities in the Huanghe River estuary and adjacent seas during 2022 and 2023. The survey results showed that the average abundance of *Chaeturichthys stigmatias* declined from 12,205 to 922 ind. h^{-1} over the two-year period, whereas the average abundance of *Acanthogobius ommaturus*, which is within the same family, dropped from 140 to 10 ind. h^{-1}. Over the two years, the influence of various previously dominant traits decreased, including benthivorous feeding habits, settlement migratory type, adhesive eggs, low growth coefficient, and low to medium resilience, and the abundance of Gobiidae also decreased. Conversely, the average abundance of *Konosirus punctatus* increased from 4 to 365 ind. h^{-1}, whereas that of *Setipinna taty* increased from 30 to 284 ind. h^{-1}. The dominance of herbivorous and planktivorous feeding habits, offshore migratory type, pelagic eggs, and high growth coefficients and resilience of the aforementioned two types were enhanced. Moreover, Engraulidae and Clupeidae primarily occupy the middle to upper water layers, whereas Gobiidae primarily occupy the bottom layer; this also contributed to the yearly variations in the habitat traits of the fish communities. A bottom trawl is unlikely to underestimate the amount of pelagic fishes because they tend to remain near the sea bottom during the day [59]. Therefore, we speculated that the significant fluctuation in fish abundance in the study area was a principal factor that influenced the dominant traits of the community. This fluctuation may be influenced by the annual average water temperature and fishing intensity.

Previous studies showed that, along a specific environmental gradient, community nestedness and turnover contribute to the spatial variation of beta diversity, which in turn depends, to a certain extent, on the attributes of functional groups. If environmental factors undergo a continuous gradual change, the fish community composition tends to be dominated by a nestedness structure [60]. However, if environmental factors undergo a sudden or abrupt shift, the fish community composition is dominated by turnover [61]. Studies by Oikonomou et al. [62] and Lansac-Tôha et al. [63] indicated that the beta diversity of fish communities primarily resulted from species turnover, with a negligible influence of nestedness components. This study revealed that the fish community functional beta diversity in the Huanghe River estuary and adjacent seas was dominated by turnover components, which is consistent with existing research findings. However, although the contribution of turnover was comparable to that of nestedness, it was significantly lower than what was observed in other sea areas [28,29] and exhibited an annual decline. These findings could differ because the scale of the research area was relatively small. The sampling stations were close to the shore (within a 20 m water depth). Habitat connectivity and fish adaptability to the environment contributed to their relatively strong dispersal ability in the research area. The functional homogeneity of fish communities was previously found to be relatively high in this area [28]. However, fishing pressure and environmental changes may have resulted in an increase in the proportion of shared functional traits among fish communities and a decrease in the diversity of functional traits.

Based on the correlation between functional diversity and environmental factors, it was observed that the functional alpha diversity of fish communities in the Huanghe River estuary and adjacent seas was negatively correlated with NH_4–N and positively correlated with TP. The functional beta diversity was positively correlated with the NH_4–N and TP gradients in sample pairs, which indicated that the functional beta diversity was influenced by a bottom-up control effect of the N/P ratio in the environment. Studies showed that increased nitrogen concentration and decreased phosphorus concentration in the Bohai Sea led to a decreased dominance of algae susceptible to phosphorus limitation and an increased dominance of algae susceptible to nitrogen limitation, which was not beneficial for the growth of high-trophic-level fishery resources [32,64]. We observed that the functional traits of high-trophic-level fish species in the study area were being gradually lost with a decrease in their species and abundance, which will weaken the top-down control effect of the population. This may be one of the reasons for the shift toward fish communities with lower trophic levels in the Huanghe River estuary and adjacent seas. This study further revealed that high pH gradients in the research area suppressed the turnover of functional traits in these fish communities while promoting the formation of nestedness patterns. The formation of a pH gradient may be closely related to the flow of the Huanghe River into the sea and the implementation of a water and sediment regulation project.

Apart from environmental factors, within this small-scale research area, an increase in the geographical distance between communities negatively impacted the turnover of the fish functional traits. The geographical distance between communities and the distance between communities and estuaries did not have an apparent impact on the functional alpha diversity. Zhang et al. [38] found that a vertical gradient distribution of river fishes had opposite beta and alpha diversity patterns, which is consistent with the findings of this study. These differences occur because alpha diversity describes the current status of the ecological space occupied by species within a community, whereas beta diversity is used to quantify the extent or differentiation of functional differences between communities. Additionally, different species exhibit different adaptability along one or several environmental gradients. This is reflected in differences in functional traits or abundance. Therefore, these two types of diversity may exhibit different spatial patterns along the same spatial gradient [65].

5. Conclusions

In this study we found that the *FRic*, *FEve*, and *FDiv* of the fish functional alpha diversity was 0.223, 0.461, and 0.805, respectively, and the functional beta diversity was 0.661 in the Huanghe River estuary. The fish abundance and turnover component were the main factors that affected the alpha and beta diversity. The ammonia nitrogen content, total phosphorus content, and pH were the main environmental factors that affected the fish functional diversity. The N/P ratio had a bottom-up control effect on the fish functional diversity. Within a small-scale research scope, high pH gradients and distant geographical distances can inhibit trait turnover in fish communities.

Author Contributions: S.L., Y.H. and H.M.: methodology, formal analysis and writing—original draft preparation. F.L., Y.L. and W.C.: supervision and project administration. X.Z. and X.W.: data curation. G.C. and T.W.: software and data analysis. All authors have read and agreed to the published version of the manuscript.

Funding: This research was funded by the Shandong Provincial Natural Science Foundation (General program, No. ZR2023MD039), the Science and Technology Innovation Program of the Laoshan Laboratory (No. LSKJ202203803), the Open Fund of Observation and Research Station of Bohai Strait Eco-Corridor (No. BH202301), and the Special Project of Yellow River Water Resources Protection Institute (No. KYY- KYZX-2022-02).

Institutional Review Board Statement: In this work, we used the data of the marine ecological monitoring project conducted by the authors for the Shandong Provincial Oceanic Bureau of China. We also used open databases (FishBase (online)) and literature sources. So, no additional approvals or permits are required for the studies.

Informed Consent Statement: Not applicable.

Data Availability Statement: The datasets that support the findings of this study are available from the corresponding author upon reasonable request.

Conflicts of Interest: The authors declare no conflicts of interest.

Appendix A

Table A1. List of fish species in the Huanghe River estuary and adjacent seas in 2022 and 2023 and their functional traits.

Family	Species	Feeding Habit	Mouth Position	Mouth Relative Size	Trophic Level	Habitat	Migration Type	Body Shape	Maximum Body Length (cm)	Body Length at First Sexual Maturity (cm)	Age at First Sexual Maturity (a)	Generation Time (a)	Ecological Type of Eggs	Growth Coefficient (k)	Vulnerability	Resilience	Thermo-phily
Apogonidae	*Apogon lineatus*	B	Front	CB	3.7	De	OM	2	9.00	6.8	0.8	1.0	PAE	0.50	12	Hr	WW
Callionymidae	*Callionymus beniteguri*	B	Front	S	3.3	De	ST	3	13.96	-	-	-	PE	0.52	13	Hr	T
Clupeidae	*Konosirus punctatus*	H	Front	CS	2.9	Pe	OM	2	32.00	18.1	2.9	3.2	PE	0.65	36	Hr	T
	Sardinella zunasi	P	Front	CS	3.4	Pe	OM	2	18.00	11.0	1.0	2.2	PE	0.50	22	Hr	T
Cottidae	*Trachidermus fasciatus*	B–N	Front	B	3.0	De	OM	10	17.00	10.9	1.0	-	DAE	-	10	Lr	CT
Cynoglossidae	*Cynoglossus joyeri*	B	Lower	S	4.3	De	CM	5	24.00	17.2	3.8	4.1	PE	0.20	40	Mr	T
	Cynoglossus semilaevis	B	Lower	S	3.7	De	CM	5	61.10	34.6	2.8	3.8	PE	0.26	44	Mr	T
Engraulidae	*Engraulis japonicus*	P	Lower	B	3.6	Pe	OM	2	18.00	10.0	1.0	4.0	PE	0.79	10	Hr	T
	Setipinna taty	P	Lower	B	3.6	Pe	OM	2	22.20	12.4	1.0	-	PE	0.57	17	Hr	WW
	Thrissa kammalensis	P	Lower	B	3.4	Pe	OM	2	18.00	8.9	2.4	3.3	PE	0.56	32	Mr	T
	Thrissa mystax	P	Lower	B	3.6	Pe	OM	2	19.00	13.0	1.0	-	PE	0.56	11	Hr	WW
Gobiidae	*Acanthogobius ommaturus*	B	Front	CB	3.4	De	ST	8	50.00	12.1	1.0	1.0	AE	0.20	50	Lr	T
	Amblychaeturichthys hexanema	B	Front	S	3.4	De	ST	4	17.40	11.4	1.9	2.5	AE	0.70	10	Mr	T
	Amoya pflaumi	B	Front	CS	3.1	De	ST	8	12.00	8.9	1.0	1.4	AE	0.44	10	Hr	WW
	Chaeturichthys stigmatias	B	Front	S	3.8	De	ST	4	28.20	11.8	2.7	2.9	AE	0.45	18	Mr	T
	Ctenotrypauchen chinensis	B	Front	S	3.8	De	ST	2	19.20	-	-	-	AE	-	13	Mr	T
	Myersina filifer	B	Front	CS	3.4	De	ST	2	13.20	-	-	-	AE	-	10	Hr	WW
	Odontamblyopus lacepedii	B	Front	CB	3.9	De	ST	4	33.40	20.3	3.7	4.2	AE	0.19	31	Mr	WW
	Tridentiger barbatus	B	Front	S	3.5	De	ST	8	10.40	7.3	1.5	2.0	AE	0.56	14	Hr	T
Hexagrammidae	*Hexagrammos otakii*	P–B	Front	S	3.8	De	OM	2	57.00	11.6	2.0	2.9	AE	0.36	34	Mr	CT
Mugilidae	*Liza haematocheila*	P	Lower	S	2.5	De	ST	11	80.00	48.0	2.5	8.1	PE	0.31	58	Mr	T
Paralichthyidae	*Paralichthys olivaceus*	N	Front	M	4.5	De	CM	5	103.00	40.0	2.0	7.3	PE	0.24	60	Mr	T
Pholididae	*Enedrias fangi*	P–B	Front	M	3.2	De	CM	9	16.80	11.2	1.1	1.5	O	0.61	10	Hr	CT
Platycephalidae	*Platycephalus indicus*	B–N	Front	M	3.6	De	OM	3	100.00	45.7	1.8	2.5	PE	0.30	35	Mr	WW
Pleuronectidae	*Kareius bicoloratus*	B	Front	S	3.7	De	CM	5	50.00	31.6	3.7	4.8	PE	0.18	46	Lr	CT
Sciaenidae	*Collichthys niveatus*	P–B	Front	CS	3.6	De	CM	2	17.00	8.8	1.0	4.0	PE	0.42	17	Hr	T
	Johnius belengerii	B	Lower	CS	3.3	De	OM	2	30.00	12.5	1.3	1.5	PE	0.53	20	Hr	WW
	Larimichthys polyactis	B–N	Front	M	3.7	De	OM	2	40.00	18.1	1.0	3.0	PE	0.45	27	Mr	T
	Pennahia argentata	B	Lower	M	4.1	De	OM	2	40.00	11.1	1.6	2.0	PE	0.42	34	Hr	T
Scombridae	*Scomberomorus niphonius*	N	Front	CB	4.8	Pe	DM	1	113.00	37.0	1.0	8.5	PE	0.53	34	Mr	T
Serranidae	*Lateolabrax maculatus*	N	Upper	M	4.7	De	CM	2	102.00	52.7	2.0	5.2	PE	0.42	52	Mr	T
Sillaginidae	*Sillago sihama*	B	Front	S	3.4	De	OM	6	31.00	16.7	1.4	1.5	PE	0.80	24	Hr	WW
Stromateidae	*Pampus argenteus*	P	Front	S	3.3	Pe	DM	2	60.00	25.3	1.3	1.5	PE	0.56	31	Mr	WW
Syngnathidae	*Syngnathus acus*	P	Front	S	3.3	De	ST	7	50.00	16.3	-	-	O	-	40	Mr	T
Tetraodontidae	*Takifugu niphobles*	P–B	Front	S	3.4	De	OM	6	20.00	11.0	1.0	-	DAE	0.31	44	Mr	CT
	Takifugu pseudommus	P–B	Front	S	3.4	De	OM	6	35.00	-	-	-	DAE	-	33	Mr	T
Trichiuridae	*Eupleurogrammus muticus*	B	Front	CB	4.1	De	OM	9	87.00	-	-	-	PE	-	48	Hr	WW

H: Herbivorous; P: Planktivorous; P–B: Planktivorous and benthivorous; B: Benthivorous; B–N: Benthivorous and nektivorous; N: Nektivorous; S: Small; CS: Comparatively small; M: Medium; CB: Comparatively big; B: Big; De: Demersal; Pe: Pelagic; ST: Settlement type; CM: Coastal migratory; OM: Offshore migratory; DM: Distantly migrating; (1) Fusiform; (2) Compressiform; (3) Depressiform; (4) Anguilliform; (5) Asymmetry; (6) Sub-cylinder; (7) Slightness; (8) Cylindrical anterior part and compressed posterior part; (9) Band shape; (10) Depressed anterior part and compressed posterior part; (11) Sub-cylindrical anterior part and compressed posterior part; PE: Pelagic eggs; AE: Adhesive eggs; DAE: Demersal adhesive eggs; PAE: Pelagic adhesive eggs; O: Ovoviviparous; Lr: Low resilience; Mr: Medium resilience; Hr: High resilience; CT: Cold temperate; T: Warm temperate; WW: Warm water.

Appendix B

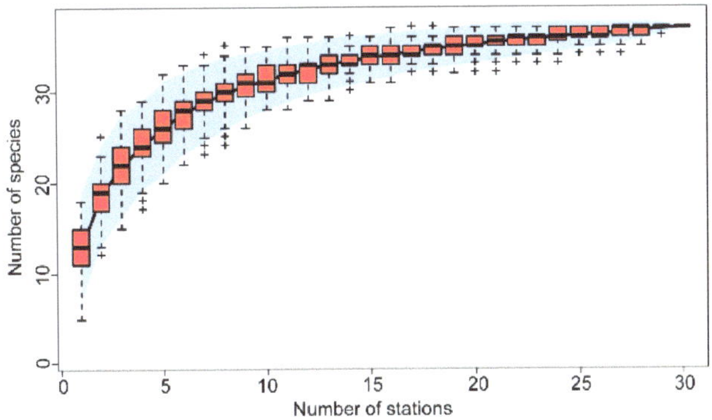

Figure A1. Species accumulation curves in the Huanghe River estuary and adjacent seas. The red box represents the number of species that may be obtained at the sampling station. The blue area represents the range of the number of species surveyed during continuous sampling.

Appendix C

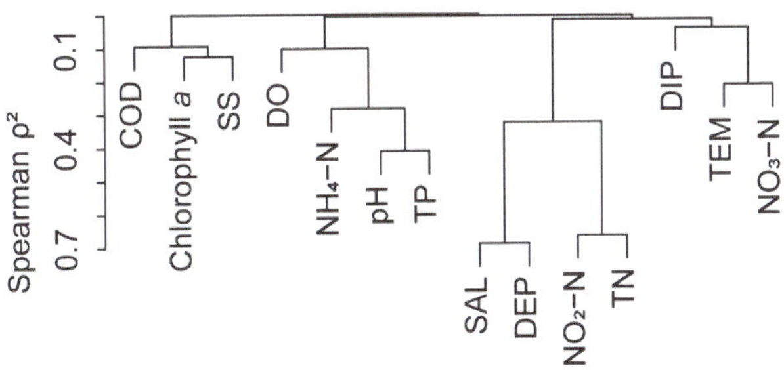

Figure A2. Correlation of environmental factors based on a collinearity test.

References

1. Niu, K.C.; Liu, Y.N.; Shen, Z.H.; He, F.L.; Fang, J.Y. Community assembly: The relative importance of neutral theory and niche theory. *Biodivers. Sci.* **2009**, *17*, 579–593. [CrossRef]
2. Chen, S.B.; Ouyang, Z.Y.; Xu, W.H.; Xiao, Y. A review of beta diversity studies. *Biodivers. Sci.* **2010**, *18*, 323–335. [CrossRef]
3. Violle, C.; Reich, P.B.; Pacala, S.W.; Enquist, B.J.; Kattge, J. The emergence and promise of functional biogeography. *Proc. Natl. Acad. Sci. USA* **2014**, *111*, 13690–13696. [CrossRef]
4. Whittaker, R.H. Vegetation of the Siskiyou Mountains, Oregon and California. *Ecol. Monogr.* **1960**, *20*, 279–338. [CrossRef]
5. Whittaker, R.H. Evolution and measurement of species diversity. *Taxon* **1972**, *21*, 213–251. [CrossRef]
6. Socolar, J.B.; Gilroy, J.J.; Kunin, W.E.; Edwards, D.P. How should beta-diversity inform biodiversity conservation? *Trends Ecol. Evol.* **2016**, *31*, 67–80. [CrossRef]
7. Si, X.F.; Zhao, Y.H.; Chen, C.W.; Ren, P.; Zeng, D.; Wu, L.B.; Ding, P. Beta-diversity partitioning: Methods, applications and perspectives. *Biodivers. Sci.* **2017**, *25*, 464–480. [CrossRef]
8. Laureto, L.M.O.; Cianciaruso, M.V.; Samia, D.S.M. Functional diversity: An overview of its history and applicability. *Nat. Conserv.* **2015**, *13*, 112–116. [CrossRef]
9. Swenson, N.G.; Erickson, D.L.; Mi, X.C.; Bourg, N.A.; Forero-Montaña, J.; Ge, X.J.; Howe, R.; Lake, J.K.; Liu, X.J.; Ma, K.P.; et al. Phylogenetic and functional alpha and beta diversity in temperate and tropical tree communities. *Ecology* **2012**, *93*, 112–125. [CrossRef]
10. Lavorel, S.; Grigulis, K.; Lamarque, P.; Colace, M.P.; Garden, D.; Girel, J.; Pellet, G.; Douzet, R. Using plant functional traits to understand the landscape distribution of multiple ecosystem services. *J. Ecol.* **2011**, *99*, 135–147. [CrossRef]
11. Higgins, S.N.; Zanden, M.J.V. What a difference a species makes: A metaanalysis of dreissenid mussel impacts on freshwater ecosystems. *Ecol. Monogr.* **2010**, *80*, 179–196. [CrossRef]

12. Villéger, S.; Grenouillet, G.; Brosse, S. Decomposing functional β-diversity reveals that low functional β-diversity is driven by low functional turnover in European fish assemblages. *Glob. Ecol. Biogeogr.* **2013**, *22*, 671–681. [CrossRef]
13. Mason, N.W.H.; Mouillot, D.; Lee, W.G.; Wilson, J.B.; Setl, H. Functional richness, functional evenness and functional divergence: The primary components of functional diversity. *Oikos* **2005**, *111*, 112–118.
14. Zhang, X.Z.; Wang, J.; Xu, B.D.; Zhang, C.L.; Xue, Y.; Ren, Y.P. Spatio-temporal variations of functional diversity of fish communities in Haizhou Bay. *Chin. J. Appl. Ecol.* **2019**, *30*, 3233–3244. [CrossRef] [PubMed]
15. He, Q.; Liu, S.D.; Tang, Y.L.; Dong, X.Q.; Zhao, W.; Feng, J.; Yu, M.J. Species and functional diversity of fish communities in artificial reef area of Pipa Island Sea, Shandong. *J. Fish. Sci. China* **2023**, *30*, 1479–1495.
16. Zhang, Y.X.; Shan, X.J.; Bian, X.D.; Wei, C.; Zhang, W.R.; Cui, P.D. Early life resources community structure and functional diversity to the Osteichthyes in the waters adjacent to the Changdao Islands. *Prog. Fish. Sci.* **2023**, *44*, 1–19. [CrossRef]
17. Swenson, N.G.; Anglada-Cordero, P.; Barone, J.A. Deterministic tropical tree community turnover: Evidence from patterns of functional beta diversity along an elevational gradient. *Proc. R. Soc. B Biol. Sci.* **2011**, *278*, 877–884. [CrossRef] [PubMed]
18. Si, X.F.; Baselga, A.; Leprieur, F.; Song, X.; Ding, P. Selective extinction drives taxonomic and functional alpha and beta diversities in island bird assemblages. *J. Anim. Ecol.* **2016**, *85*, 409–418. [CrossRef]
19. Harrison, S.; Ross, S.J.; Lawton, J.H. Beta diversity on geographic gradients in Britain. *J. Anim. Ecol.* **1992**, *61*, 151–158. [CrossRef]
20. Williams, P.H. Mapping variations in the strength and breadth of biogeographic transition zones using species turnover. *Proc. R. Soc. B Biol. Sci.* **1996**, *263*, 579–588. [CrossRef]
21. Lennon, J.J.; Koleff, P.; Greenwood, J.J.D.; Gaston, K.J. The geographical structure of British bird distributions: Diversity, spatial turnover and scale. *J. Anim. Ecol.* **2001**, *70*, 966–979. [CrossRef]
22. Greve, M.; Gremmen, N.J.M.; Gaston, K.J.; Chown, S.L. Nestedness of Southern Ocean island biotas: Ecological perspectives on a biogeographical conundrum. *J. Biogeogr.* **2005**, *32*, 155–168. [CrossRef]
23. Wang, Y.P.; Bao, Y.X.; Yu, M.J.; Xu, G.F.; Ding, P. Nestedness for different reasons: The distributions of birds, lizards and small mammals on Islands of an inundated lake. *Divers. Distrib.* **2010**, *16*, 862–873. [CrossRef]
24. Baselga, A. Partitioning the turnover and nestedness components of beta diversity. *Glob. Ecol. Biogeogr.* **2010**, *19*, 134–143. [CrossRef]
25. Baselga, A. The relationship between species replacement, dissimilarity derived from nestedness, and nestedness. *Glob. Ecol. Biogeogr.* **2012**, *21*, 1223–1232. [CrossRef]
26. Podani, J.; Ricotta, C.; Schmera, D. A general framework for analyzing beta diversity, nestedness and related community-level phenomena based on abundance data. *Ecol. Complex.* **2013**, *15*, 52–61. [CrossRef]
27. Li, X.T.; Xu, B.D.; Xue, Y.; Ren, Y.P.; Zhang, C.L. Variation in the β diversity of fish species in Haizhou Bay. *J. Fish. Sci. China* **2021**, *28*, 451–459.
28. Li, X.T.; Xu, B.D.; Xue, Y.; Ren, Y.P.; Zhang, C.L. *β* diversity and its components of the fish community in the Haizhou Bay during autumn and the relationships with environmental factors. *Haiyang Xuebao* **2022**, *44*, 46–56.
29. Zhang, X.M.; Su, H.X.; Li, S.W.; Yang, Y.Y.; Wang, T.T.; Xu, B.Q.; Li, F.; Wang, X.X. The Driving Factors of *β*-diversity and its Components Spatial Variation of Fish Community in Coastal Area of Yantai and Weihai. *Oceanol. Et Limnol. Sin.* **2023**, *54*, 1125–1133.
30. Wang, J.Y.; Li, H.Y. Temporal and spatial differentiation characteristics of economy in the country Yellow River Basin and its influencing factors. *Yellow River* **2022**, *44*, 1–6.
31. Zhao, J.J.; Liu, Y.; Zhu, Y.K.; Qin, S.L.; Wang, Y.H.; Miao, C.H. Spatiotemporal differentiation and influencing factors of the coupling and coordinated development of new urbanization and ecological environment in the Yellow River Basin. *Resour. Sci.* **2020**, *42*, 159–171. [CrossRef]
32. Zhang, J.J.; Li, F.; Lv, Q.M.; Wang, Y.B.; Yu, J.B.; Gao, Y.J.; Ren, Z.H.; Zhang, X.M.; Lv, Z.B. Impact of the water–sediment regulation scheme on the phytoplankton community in the Yellow River estuary. *J. Clean. Prod.* **2021**, *294*, 40–57. [CrossRef]
33. Fan, J.J. Alteration of Hydrologic Regime and Its Driving Factors of the Yellow River Basin. Master Thesis, Chinese Academy of Sciences, Beijing, China, 2023.
34. Zhang, M.L.; Yu, G.L.; Wang, F.; Li, B.; Han, H.Z.; Qi, Z.H.; Wang, T.T. Terrestrial dissolved organic carbon consumption by heterotrophic bacterioplankton in the huanghe river estuary during water and sediment regulation. *J. Oceanol. Limnol.* **2019**, *37*, 1062–1070. [CrossRef]
35. Mu, B.; Cui, T.W.; Qin, P.; Gong, J.L.; Xiao, Y.F.; Zheng, R.E. Remote sensing retrieval and temporal-spatial distribution characteristics of particulate organic carbon concentration in seawater near Yellow River estuary. *Acta Opt. Sin.* **2017**, *37*, 33–43. [CrossRef]
36. Cui, B.S.; Hua, Y.Y.; Wang, C.F.; Liao, X.L.; Tan, X.J.; Tao, W.D. Estimation of ecological water requirements based on habitat response to water level in huanghe river delta, China. *Chin. Geogr. Sci.* **2010**, *20*, 318–329. [CrossRef]
37. Gu, Y.Z.; Xu, C.L.; Zhang, Z.H.; Bi, N.S.; Zhao, L.L.; Liu, Y.F.; Zhou, B. Response of fresh water from Yellow River to marine ecological regulation. *Yellow River* **2019**, *41*, 68–75.
38. Zhang, D. Longitudinal Pattern and Formation Mechanism in α and β Diversity of Taxonomic and Functional of Stream Fish Assemblages in the Xin'an River. Ph.D. Thesis, Anhui Normal University, Wuhu, China, 2018.
39. FishBase. Available online: https://fishbase.se/search.php (accessed on 15 February 2024).

40. Zhang, J.X.; Wang, J.; Niu, M.X.; Zuo, T.; Chang, W.; Chen, R.S. Biological characteristics of Liza haematocheila in the shallow coastal waters of the Yellow River estuary. *Prog. Fish. Sci.* **2023**, *44*, 20–29. [CrossRef]
41. Niu, M.X.; Zuo, T.; Wang, J.; Chen, R.S.; Zhang, J.X. Egg and larval distribution of Liza haematocheila and their relationship with environmental factors in the coastal waters of the Yellow River Estuary. *J. Fish. Sci. China* **2022**, *29*, 377–387.
42. Bi, X.X. Studies on the Morphology and Genetics of *Trachidermus fasciatus* Populations. Ph.D. Thesis, Ocean University of China, Qingdao, China, 2018.
43. Fan, H.Y.; Ji, Y.P.; Zhang, S.H.; Yuan, C.T.; Gao, T.X. Research of Fishery Biology of the Neritic Fish Synechogobius ommaturus in the Area of the Huanghe Delta. *Period. Ocean Univ. China* **2005**, *35*, 733–736. [CrossRef]
44. Zhang, L.C. Study on the Community Structure and Fishery Biological Characteristics of Gobioidei in Laizhou Bay. Master's Thesis, Shanghai Ocean University, Shanghai, China, 2019.
45. Mu, X.X.; Zhang, C.; Zhang, C.L.; Xu, B.D.; Xue, Y.; Tian, Y.J.; Ren, Y.P. The fisheries biology of the spawning stock of Scomberomorus niphonius in the Bohai and Yellow Seas. *J. Fish. Sci. China* **2018**, *25*, 1308–71316. [CrossRef]
46. Qu, J.Y.; Yang, G.M.M.; Fang, Z.; Chen, X.J. A Review of Research Advancement on Fisheries Biology of Japanese Mackerel *Scomberomorus niphonius*. *Fish. Sci.* **2021**, *40*, 643–650. [CrossRef]
47. Xu, S.Y. Study on Morphology and Genetics of *Setipinna tenuifilis*. Master Dissertation, Ocean University of China, Qingdao, China, 2014.
48. Han, Q.P.; Wu, Q.; Shan, X.J.; Jin, X.S.; Su, C.C. Stock assessment of Setipinna taty in Shandong inshore waters based on lengthbased model and ensemble model. *J. Fish. China* **2024**, 1–11. Available online: https://link.cnki.net/urlid/31.1283.s.20240115.1908.004 (accessed on 16 January 2024).
49. Villeger, S.; Mason, N.W.H.; Mouillot, D. New multidimensional functional diversity indices for a multifaceted framework in functional ecology. *Ecology* **2008**, *89*, 2290–2301. [CrossRef]
50. Cornwell, W.K.; Schwilk, D.W.; Ackerly, D.D. A trait-based test for habitat filtering: Convex hull volume. *Ecology* **2006**, *87*, 1465–11471. [CrossRef]
51. Shuai, F.M.; Li, X.H.; Chen, F.C.; Li, Y.F.; Yang, J.P.; Li, J.; Wu, Z. Functional diversity of freshwater fishes and methods of measurement. *Acta Ecol. Sin.* **2017**, *37*, 5228–5237.
52. Dong, J.Y.; Sun, X.; Zhan, Q.P.; Zhang, Y.Y.; Zhang, X.M. Patterns and drivers of beta diversity of subtidal macrobenthos community on the eastern coast of Laizhou Bay. *Biodivers. Sci.* **2022**, *30*, 21388. [CrossRef]
53. Dong, J.Y.; Zhao, L.L.; Sun, X.; Hu, C.Y.; Wang, Y.H.; Li, W.T.; Zhang, P.D.; Zhang, X.M. Response of macrobenthic communities to heavy metal pollution in Laoshan Bay, China: A trait-based method. *Mar. Pollut. Bull.* **2021**, *167*, 112292. [CrossRef]
54. Li, S.W.; Zhang, X.M.; Su, H.X.; Chen, S.; Chen, W.; Li, F.; Wang, T.T.; Sun, Z.N.; Weng, X.N. Taxonomic diversity of the macrobenthos and its relationship with environmental factors in the Huanghe River (Yellow River) estuary and adjacent seas, China. *J. Sea Res.* **2023**, *194*, 102402. [CrossRef]
55. Poff, N.L.; Allan, J.D. Functional organization of stream fish assemblages in relation to hydrological variability. *Ecology* **1995**, *76*, 606–627. [CrossRef]
56. Langerhans, R.B.; Gifford, M.E.; Joseph, E.O. Ecological speciation in Gambusia fishes. *Evolution* **2007**, *61*, 2056–2074.
57. Pease, A.A.; González-Díaz, A.A.; Rodiles-Hernández, R.; Winemiller, K.O. Functional diversity and trait–environment relationships of stream fish assemblages in a large tropical catchment. *Freshw. Biol.* **2012**, *57*, 1060–1075. [CrossRef]
58. Villeger, S.; Brosse, S.; Mouchet, M.; Mouillot, D.; Vanni, M.J. Functional ecology of fish: Current approaches and future challenges. *Aquat. Sci.* **2017**, *79*, 783–801. [CrossRef]
59. Yin, M.C. *Fish Ecology*; China Agriculture Press: Beijing, China, 1995; pp. 54–96.
60. Taylor, C.M.; Warren, M.L. Dynamics in species composition of stream fish assemblage: Environmental variability and nested subsets. *Ecology* **2001**, *82*, 2320–2330. [CrossRef]
61. McGarvey, D.J.; Hughes, R.M. Longitudinal zonation of Pacific Northwest (USA) fish assemblages and the species–discharge relationship. *Copeia* **2008**, *2008*, 311–321.
62. Oikonomou, A.; Stefanidis, K. α- and β-diversity patterns of macrophytes and freshwater fishes are driven by different factors and processes in lakes of the unexplored Southern Balkan biodiversity hotspot. *Water* **2020**, *12*, 1984. [CrossRef]
63. Lansac-Tôha, F.M.; Heino, J.; Quirino, B.A.; Moresco, G.A.; Peláez, O.; Meira, B.R.; Rodrigues, L.C.; Jati, S.; Lansac-Tha, F.A.; Velho, L.F.M. Differently dispersing organism groups show contrasting beta diversity patterns in a dammed subtropical river basin. *Sci. Total Environ.* **2019**, *691*, 1271–1281. [CrossRef]
64. Xu, S.S. Decline Mechanisms of Fishery Resources in the Bohai Sea under Anthropogenic Activities. Ph.D. Thesis, The Institute of Oceanology, Chinese Academy of Sciences, Qingdao, China, 2011.
65. Tylianakis, J.M.; Klein, A.M.; Tscharntke, T. Spatiotemporal variation in the diversity of Hymenoptera across a tropical habitat gradient. *Ecology* **2008**, *86*, 3296–3302.

fishes

MDPI

Article

Assessing Prospects of Integrating Asian Carp Polyculture in Europe: A Nature-Based Solution under Climate Change?

Oksana Nekrasova [1,2,3,*], Mihails Pupins [2], Volodymyr Tytar [3], Leonid Fedorenko [4], Oleksandr Potrokhov [4], Arturs Škute [2], Andris Čeirāns [2], Kathrin Theissinger [5,6] and Jean-Yves Georges [1]

1 CNRS, IPHC UMR 7178, Université de Strasbourg, 67037 Strasbourg, France; jean-yves.georges@iphc.cnrs.fr
2 Institute of Life Sciences and Technologies, Daugavpils University, LV5400 Daugavpils, Latvia; mihails.pupins@gmail.com (M.P.); arturs.skute@du.lv (A.Š.); cuskisa@gmail.com (A.Č.)
3 I. I. Schmalhausen Institute of Zoology, National Academy of Sciences of Ukraine, 01030 Kyiv, Ukraine; vtytar@gmail.com
4 Institute of Hydrobiology, National Academy of Sciences of Ukraine, 03057 Kyiv, Ukraine; fedorenko20071979@gmail.com (L.F.); potrokhov@nas.gov.ua (O.P.)
5 Institute for Insect Biotechnology, Justus Liebig University Giessen, Heinrich-Buff-Ring 35, 35392 Giessen, Germany; kathrin.theissinger@agrar.uni-giessen.de
6 Fraunhofer Institute for Molecular Biology and Applied Ecology, Ohlebergsweg 12, 35392 Giessen, Germany
* Correspondence: oneks22@gmail.com; Tel.: +380-672171970

Abstract: Aquaculture's role in addressing food security has grown, with a spotlight on Asian carp species. Polyculture, i.e. cultivating multiple fish species in a single system, is being increasingly adopted for its resource efficiency and economic benefits. This practice for Asian fish for food and ornamental purposes is gaining traction in Europe despite their invasive potential. Rising temperatures due to climate change offer an opportunity for thermophilic Asian carps (*Hypophthalmichthys molitrix* and *Aristichthys nobilis*). Using GIS modeling (Maxent), we identified the possible settlement of Asian carp in Northern Europe amidst climate change. We analyzed carp global distribution centers, assessed the potential carp spread in Europe, and evaluated their potential suitability for polyculture systems. By 2050, *H. molitrix* may extend its range to 58–62° N latitude, with a potential 1.7-fold habitat increase, while *A. nobilis*, which are more heat-tolerant, may move north to 52–58° N latitude, with a 1.3-fold potential increase. Despite the slight ecological differences in their native habitats, niche modeling indicates that these carp can occupy similar niches in Europe (proven statistically). The eventuality of using Asian species for polyculture in Europe presents both opportunities and challenges in the face of a changing climate as long as invasion risks are prevented. Envisaging such polyculture, yet very carefully for the protection of ecosystems, can help food security.

Keywords: aquaculture; invasion; climate change; thresholds; species distribution modeling; Europe

Key Contribution: This manuscript presents novel insights into the feasibility of integrating Asian carp polyculture in Europe as a nature-based solution to the effects of climate change. Through comprehensive GIS modeling and analysis, the study elucidates the potential expansion of Asian carp populations in European waters, highlighting both opportunities and challenges for sustainable aquaculture practices in a changing climate, as long as invasion risks are appropriately prevented.

check for updates

Citation: Nekrasova, O.; Pupins, M.; Tytar, V.; Fedorenko, L.; Potrokhov, O.; Škute, A.; Čeirāns, A.; Theissinger, K.; Georges, J.-Y. Assessing Prospects of Integrating Asian Carp Polyculture in Europe: A Nature-Based Solution under Climate Change? *Fishes* **2024**, *9*, 148. https://doi.org/10.3390/fishes9040148

Academic Editors: Heng Zhang, Ying Xue, Yunzhi Yan and Luoliang Xu

Received: 2 March 2024
Revised: 15 April 2024
Accepted: 19 April 2024
Published: 22 April 2024

1. Introduction

The utilization of aquaculture to address food security concerns has gained significant attraction, with a particular focus on Asian carp species, renowned for their long-standing history in Chinese aquaculture dating back to 475 BC [1]. Polyculture, a practice involving the cultivation of multiple fish species in a single system, has emerged as an intriguing and widely adopted approach [2,3]. By fostering complementary interactions among species at various levels (trophic, ecological, genetic, etc.), polyculture promotes the efficient utilization of natural resources and enhances economic returns [4]. As a result, the polyculture of Asian fish, for both food and ornamental purposes, has recently gained popularity in Europe despite the associated risks of biological invasion [5,6]. Cyprinids constitute approximately 80% of the total aquaculture production in freshwater across Central and Eastern Europe [7]. The projected rise in temperatures due to climate change [7–12] presents an opportunity to envisage these thermophilic Asian fish species in European fish farms, as long as invasion risks are prevented.

Two particularly intriguing candidates are the silver carp *Hypophthalmichthys molitrix* (Valenciennes, 1844) and the bighead carp *Aristichthys* (=*Hypophthalmichthys*) *nobilis* (Richardson, 1846), both belonging to the Cyprinidae family and capable of hybridizing in polyculture conditions [13]. The silver carp (*H. molitrix*) is native to most major Pacific drainages of East Asia, ranging from the Amur River in China to the Xi Jiang River in Vietnam. It has a wide distribution, extending up to 55° N and even reported from 58° N in warm waters near an electricity power plant [14]. Silver carps (*H. molitrix*) prefer brackish freshwater environments and inhabit benthopelagic and potamodromous habitats up to 20 m deep. Their optimal temperature range is between 6 and 30 °C, with spawning occurring at temperatures above 15 °C (typically 18–26 °C) in conditions with a high current (0.5–1.7 m/s), turbidity, and oxygen concentration. The silver carp is often mistaken for its more heat-tolerant relative, the bighead carp (*A. nobilis*), which prefers temperatures ranging from 1 °C to 38 °C. The bighead carp's optimal conditions are typically found at 48° N, but it can also thrive in higher latitudes through aquaculture, utilizing warm water from the cooling systems of power stations, nuclear power plants, and thermal power plants. The bighead carp is found in both European and Asian rivers, including the Syrdarya, Amudarya, Karakum Canal, Terek, Kuban, Dniester, and Prut. It favors deep, turbid waters with temperatures above 18 °C (usually 22–30 °C), strong currents, and high oxygen concentrations. Spawning in bighead carp ceases when water levels drop too low and resumes when water levels rise. Both species have the potential to expand their range northward to 55° N and beyond in Canada [15].

Asian carps were introduced to Europe in the mid-20th century [16–18] and subsequently spread to North America, where they have established themselves in the Mississippi and Illinois River basins, the Great Lakes, and other waterways [19]. Asian carps have been introduced into European waters for phytoplankton control in eutrophic environments, as these large and voracious filter feeders consume up to 40% of their body weight per day of phytoplankton. In Ukraine, the introduction of Asian carps has been associated with positive effects, including the active control of summer phytoplankton blooms, recycling of detritus, and cleansing of vegetation from reclamation channels [18]. Although Asian carps have significant invasive potential, they also constitute a potential source of food. The intentional introduction of Asian carps in many European countries has led to cases of self-reproduction in the wild, such as in the Danube and Tisza River basins, an irrigation canal network in Northern Italy [17–22].

However, the voracious appetite of Asian carps has led to concerns about their ecological impacts, as they can potentially displace native fish species. For instance, Asian carps have been found to account for 97% of the total fish biomass in a Missouri National Wildlife Refuge (MICRA 2002) [23]. Opinions on the ecological impacts of Asian carps vary. Some argue that their voracious consumption of phytoplankton can positively control water blooms, while others express concerns about their potential to reduce aquatic vegetation, negatively affecting native aquatic herpetofauna, such as the European pond turtle (*Emys orbicularis* (Linnaeus, 1758)) and amphibians (e.g., newts) [24]. A comprehensive assessment of their ecological impacts is still lacking [12,23,25].

The literature emphasizes the need for GIS modeling and ecological niche models for studying and managing the processes of fish species invasions. However, the literature only mentions either local models for other regions (such as North America, Africa, etc.) or the use of other fish species for modeling, which highlights the uniqueness of our work, especially in assessing the ecological niches of these fish species—*H. molitrix* and the bighead carp *A. nobilis* [26–30]. The application of GIS modeling not only allows for accurate prediction of the potential spread of invasive species in new ecosystems but also effectively plans measures for their control and the prevention of negative impacts on local biodiversity and economy.

The feasibility of raising Asian carp species in more northerly European regions remains a topic of debate, particularly in light of the potential freshwater shortages and general warming expected to accompany climate change (IPCC, 2021) [31,32]. Our study aims to address this issue by investigating three key areas:

1. Identifying global carp distribution centers: Utilizing species distribution modeling techniques, we seek to pinpoint and compare the primary centers of carp distribution worldwide.
2. Analyzing carp spread in Europe: We delve into the likelihood and prospects for the expansion of Asian carp populations in Europe, particularly in northeastern regions, and explore the factors influencing this spread.
3. Assessing polyculture potential: We examine the viability of incorporating Asian carp species into polyculture systems by comparing their natural biotopes (ecological niches, niche clustering) to those found in their new distribution centers in Europe and America.

The results of our study will contribute to a more comprehensive understanding of the potential for Asian carp aquaculture in Northern Europe, shedding light on the opportunities and challenges associated with this practice in the context of climate change and biological invasions.

Conducted against the backdrop of climate change, this research underscores the pressing need for sustainable and adaptive measures in conservation biology. With a specific focus on Asian carps, regarded as a pivotal species for polyculture in Europe, this study seeks to address challenges posed by invasive species through environmentally friendly strategies.

2. Materials and Methods

The methodology employed leverages GIS models, offering a spatial approach to comprehensively evaluate the feasibility and impact of Asian carp polyculture in Europe. This assessment incorporates crucial considerations of environmental, geographical, and climatic factors, providing a holistic perspective for sustainable conservation solutions.

2.1. Occurrence Data Collection

An extensive literature review was conducted, and databases were compiled to include point records for both *H. molitrix* and *A. nobilis*. These records encompassed original and literary sources, UKRBIN data, and GBIF.org data (2022) [33,34]. All records were verified to eliminate duplicates. To address sampling bias, the nearest neighbor distance method (implemented using the 'ntbox' package in R) [35] was employed to thin the data. Occurrence points located within 0.1 units (km) of each other were removed to mitigate errors arising from spatial autocorrelation. Consequently, the number of points was substantially reduced to 6552: 982 for *H. molitrix* (5896 records from GBIF) and 692 for *A. nobilis* (2290 records from GBIF).

2.2. Environmental Data

For building the species distribution models (SDMs), we used bioclimatic variables from the CliMond dataset (https://www.climond.org/, (accessed on 27 December 2020), A1B) [9,10,36]. We used 35 bioclimatic variables for the time periods around 1975 (1970–2000) and 2050 (2041–2060) for GIS modeling. The following bioclimatic variables (CliMond) were employed (resolution of 2.5 arc-minutes, i.e., ~5 km). Of the 35 bioclimatic variables, the highly correlated (>0.7) predictors were removed using the 'virtualspecies' package in R, resulting in a selection of 16 bioclimatic variables for both 1975 (1970–2000) and 2050

(2041–2060). The selected bioclimatic variables (CliMond) were (Table S1): bio01, annual mean temperature (°C); bio02, mean diurnal temperature range (mean (period max–min)) (°C); bio03, isothermality (Bio02÷Bio07); bio04, temperature seasonality (C of V); bio06, min temperature of coldest week (°C); bio07, temperature annual range (Bio05-Bio06) (°C); bio08, mean temperature of wettest quarter (°C); bio10, mean temperature of warmest quarter (°C); bio11, mean temperature of coldest quarter (°C); bio12, annual precipitation (mm); bio14, precipitation of driest week (mm); bio15, precipitation seasonality (C of V); bio17, precipitation of driest quarter (mm); bio25, radiation of driest quarter (W m^{-2}); bio30, lowest weekly moisture index; bio34, mean moisture index of warmest quarter.

We also used 13 environmental variables—12 variables from the near-global environmental information for freshwater ecosystems at a 1 km resolution (EarthEnv NGEI, characterizing land cover, http://www.earthenv.org/, accessed on 20 January 2023) [37] and 1 the Global Aridity Index_ET0 (Global-AI_ET0). Aridity is usually expressed as a generalized function of precipitation, temperature, and reference evapo-transpiration (ET0). The Aridity Index (https://csidotinfo.wordpress.com/, accessed on 20 January 2023) [38] can be used to quantify the precipitation availability over atmospheric water demand (Table S2). The variables were as follows: (1) evergreen/deciduous needleleaf trees; (2) evergreen broadleaf trees; (3) deciduous broadleaf trees; (4) mixed/other trees; (5) shrubs; (6) herbaceous vegetation; (7) cultivated and managed vegetation; (8) regularly flooded shrub/herbaceous vegetation; (9) urban/built-up; (10) snow/ice; (11) barren lands/sparse vegetation; (12) open water; (13) Global Aridity Index.

Ecological niche and species distribution modeling (SDM) techniques have been applied to investigate the potential home range of invasive species in new environments. The Maxent algorithm (version 3.4.4) [39,40], which was run with 25 replicates, was employed using the default settings. Unlike other distributional modeling approaches, Maxent only requires presence and background data.

2.3. Model Building

Ecological niche and species distribution modeling (SDM) techniques have been extensively applied to explore the potential home range of invasive species in new environments. In this study, we employed the Maxent algorithm (version 3.4.4), developed by Phillips et al. (2005) and Peterson et al. (2008) [39,40], to model the potential distribution of Asian carps in Europe. Maxent is a robust and widely used SDM algorithm that can effectively model species distributions using presence-only data, unlike other distributional modeling techniques that require both presence and absence data. We conducted modeling at the scale of the entire distribution area, utilizing distribution points from both the natural range and recorded introductions on other continents. This approach allowed us to incorporate the full extent of the species' potential ecological niche and identify areas that may be suitable for their establishment and spread. This revision clarifies the purpose of using the Maxent algorithm and emphasizes its ability to utilize presence-only data, which is particularly relevant for invasive species management. It also highlights the importance of considering distribution points from both the natural range and recorded introductions to capture the full extent of the species' potential distribution.

Evaluation metrics were used to measure the performance of the SDMs and included the partial ROC [9,40] and the confusion matrix. The performance of the Maxent model is usually evaluated by the threshold-independent receiver operating characteristic (ROC) approach, where the calculated area under the ROC curve (AUC) is considered as a measure of prediction success [41]. The ROC curve is a graphical method that represents the relationship between the false-positive fraction (one minus the specificity) and the sensitivity for a range of thresholds. It has a range of 0–1, with a value greater than 0.5 indicating a better-than-random performance event [42]. A rough classification guide is the traditional academic point system: very good (0.8–0.9) and excellent (0.9–1.0). Models were evaluated using the true skill statistic (TSS) [43]. TSS values range from −1 to +1, with −1 corresponding to systematically wrong predictions, and +1 to systematically correct predictions;

TSS values <0.4 are considered poor, and >0.4 as useful and good. GIS-modeling was accomplished using SagaGis, QGIS. Statistical processing of the obtained data was carried out in PAST (principal components analysis (PCA)) [44].

3. Results

As a result of the modeling, the potentially most suitable territories for the distribution of both species of Asian carps were identified, including their native range—Asia. According to the available data and GIS modeling, three main centers could be distinguished where the most favorable current conditions were identified for carps (Figure 1, Binomial tests [35]): (1) southeast of North America, (2) Europe (quite a wide range), and (3) Southeast Asia and the Far East. From the perspective of time, by 2050, three additional areas were identified: (4) north and east of South America, (5) northern tropical Africa and Indian area, and (6) southeast Australia.

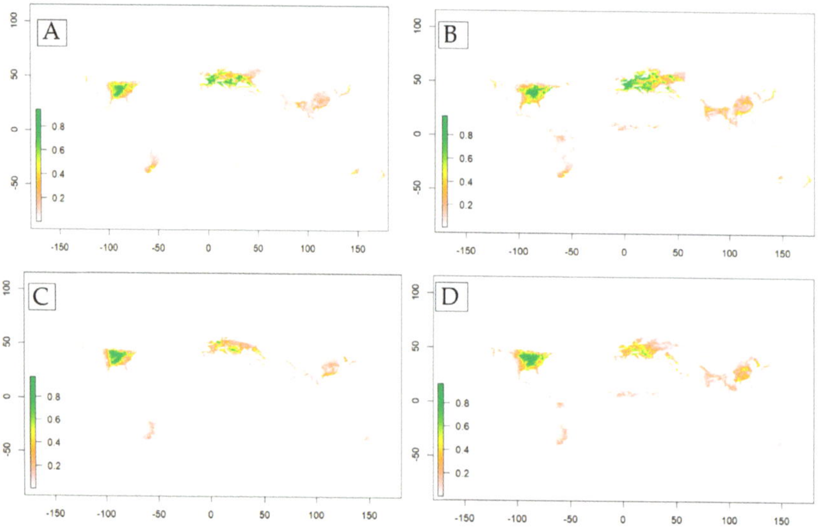

Figure 1. Results of the analysis of SDM binomial tests (green color > 0.5, CliMond dataset [35]). *H. molitrix:* (**A**) current; (**B**) 2050 (2041–2060). *A. nobilis*: (**C**) current; (**D**) 2050 (2041–2060).

In terms of discrimination accuracy, the Maxent model (SDMs, CliMond) showed excellent performance: AUC > 0.9. For both *H. molitrix* and *A. nobilis*, useful TSS values were obtained: TSS = 0.636 and TSS = 0.572, respectively.

The distribution of two carp species was influenced by temperature factors, particularly the "annual mean temperature" (bio1), which contributed 22–24% to the model and is most likely due to the dependence of juvenile reproduction and growth on sufficiently warm waters (Table 1).

For the more thermophilic *A. nobilis*, the favorable range of the "annual mean temperature" (bio1) was 7 °C to 27 °C, and the optimal range was from 9 °C to 17 °C, with the response curve peaking around 10 °C (Figure 2). However, these species have specific requirements under certain conditions, particularly during breeding. This is evident from the influence on the distribution of these fish species of other important factors besides temperature, such as "precipitation of driest quarter" (bio17), "isothermality" (bio3), and "lowest weekly moisture index" (bio30) (Table 1).

Table 1. Results of modeling the spread of Asian carps (CliMond and EarthEnv NGEI, Maxent).

Variable	Contribution Percent	Permutation Importance
	H. molitrix	
	CliMond (Table S1)	
bio1—Annual mean temperature (°C)	22.0	26.6
bio17—Precipitation of driest quarter (mm)	19.2	1.2
bio3—Isothermality	14.1	6.3
	EarthEnv NGEI (Table S2)	
lc_avg_09—Urban/built-up	34.4	10.0
lc_avg_04—Mixed/other trees	12.1	18.1
	A. nobilis	
	CliMond (Table S1)	
bio1—Annual mean temperature (°C)	24.0	44.0
bio3—Isothermality	18.3	8.8
bio30—Lowest weekly moisture index	9.5	0.4
	EarthEnv NGEI (Table S2)	
lc_avg_09—Urban/built-up	29.1	11.8
lc_avg_07—Cultivated and managed vegetation	16.4	14.2

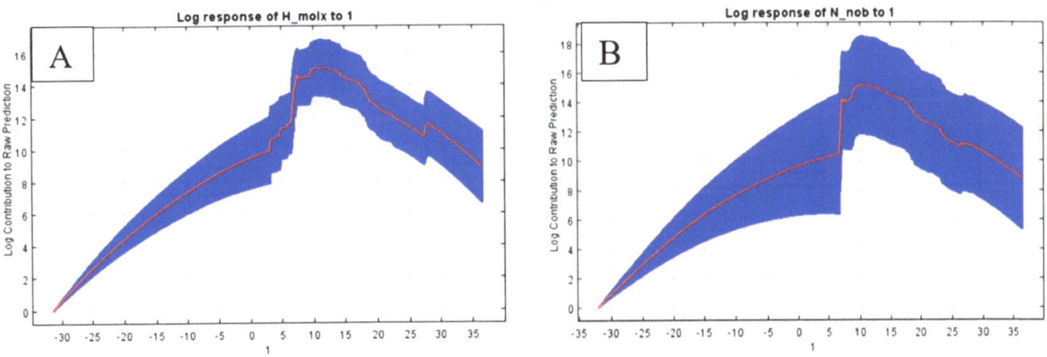

Figure 2. Response curves generated by Maxent of Asian carps for the variable "annual mean temperature" (°C, bio1): *H. molitrix* (**A**); *A. nobilis* (**B**) (Table 1).

For *H. molitrix*, the favorable range of the "annual mean temperature" (bio1) was between 3 °C and 27 °C, with an optimal range between 8 °C and 17 °C (Figures 2 and 3).

GIS modeling revealed that Asian carp are anticipated to shift their range northward in Europe in response to climate change [12,24]. By 2050, projections suggest that *H. molitrix* may extend its range, with a potential 1.7-fold increase in habitat, while *A. nobilis*, being more heat-tolerant, may migrate northward, with a potential 1.3-fold increase (Figure 3).

The results of niche clustering, employing techniques such as k-means clustering, were visualized to illustrate variations in the global climatic conditions across continents for two species of carps. This approach categorizes ecological niches by integrating geographic and environmental variables, offering insights into habitat distribution patterns and ecological diversity (Figure 4).

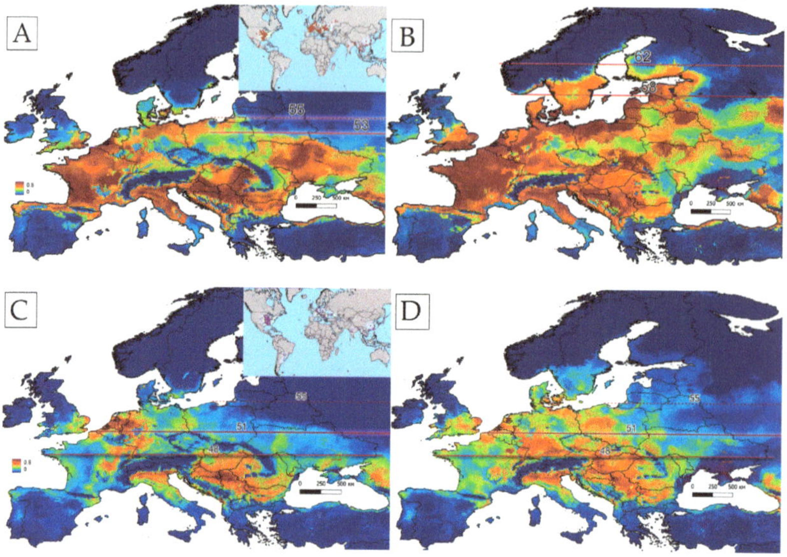

Figure 3. Potential (probabilistic) model of Asian carp expansion built in the Maxent program in Europe based on CliMond dataset (red color > 0.5). *H. molitrix*: (**A**) current (dotted line marks—55° N, red line marks—53° N); (**B**) CliMond dataset 2050 (thin line—62° N, thick red line—58° N). *A. nobilis* (dotted line—55° N; thin line—51° N, thick red line—48° N): (**C**) current; (**D**) CliMond dataset 2050 (Table S1).

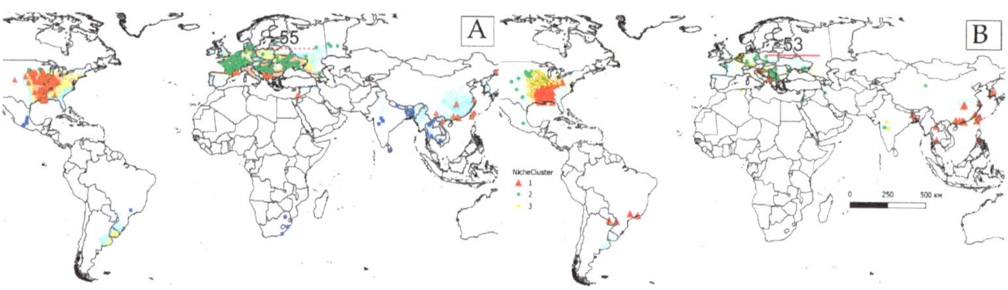

Figure 4. Niche clustering (geographic space, current CliMond) for (**A**) *H. molitrix*, 55°N; (**B**) *A. nobilis*, 53° N. Differently colored marks represent different biotopes where the fish species were registered according to the data collected and which have the combination of different climatic factors. Red triangle (1)—close-to-home region; green circle (2)—westernmost finds; and orange circles (3)—intermediate conditions.

4. Discussion

The silver carp *H. molitrix* and bighead carp *A. nobilis* are two species of Asian carps that have become invasive in many parts of the world, including Europe. These fish can have significant negative impacts on native ecosystems. Understanding the factors that affect their distribution is important for managing their spread. Because these species are actively used in aquaculture, particularly in fisheries, it is crucial to investigate the influence of other climatic factors on their distribution in the northern portion of their range.

Our results show that the temperature, namely the annual mean temperature, is the main factor affecting the distribution of Asian carps in Europe. Given the critical role of temperature factors in shaping the distribution of these Asian fish species, our analysis suggests that the potential suitable habitat for *H. molitrix* in Europe currently lies within latitudes between 53 and 55° N. However, by 2050, projections indicate a northward shift in its range, extending up to 58–62° N (Figure 3). This expansion will encompass Latvia and areas further north of Norway, Sweden, and Finland. Potentially suitable habitats for the more thermophilic species *A. nobilis* are primarily located in the Danube, Dniester, and Seversky Donets basins, spanning up to 48–51° N. Forecast projections suggest that by 2050, this species could potentially expand its range northward to 52–55° N, reaching up to 58° N around the Baltic Sea (Figure 3). These findings corroborate former studies reporting that most thermophilic freshwater animals are likely to experience range shifts in the future [12,24,45]. According to former forecasts, self-reproduction of silver carps will likely occur in the Netherlands by 2050 [30]. While temperature factors play a significant role in determining the distribution of Asian carps, these fish species have demonstrated a remarkable ability to adapt to colder climates. In northern regions, they are more likely to survive in artificially created water systems, such as ponds, reservoirs, and canals. These environments provide a more stable temperature regime compared to natural water bodies, enabling the fish to persist even in colder conditions. Adapting to colder temperatures also requires the ability to migrate to wintering areas where there is sufficient oxygen. As water temperatures drop below 10 °C, the behavior of fish changes. They form schools and prepare for wintering by moving to deeper parts of water bodies, where oxygen levels are typically higher. In the context of aquaculture, maintaining adequate oxygen levels in stagnant water bodies during the winter months is crucial for the successful wintering of the fish population. This may necessitate additional aeration measures, such as the use of air pumps or bubblers. Overall, the adaptability of Asian carps to colder environments poses a significant challenge for managing their spread. Understanding their survival strategies and implementing effective control measures are essential for mitigating the negative impacts of these invasive species.

As species cultured in warm waters discharged from power plants, Asian carps have demonstrated an ability to thrive in more northern regions, as documented in open databases (Fishbase.org, FAO.org). This adaptation is partly attributed to their ability to utilize warm water discharges from power plants. Additionally, the widespread distribution of these thermophilic herbivorous Asian fish is facilitated by the stocking of captive-bred fry. For these species to successfully reproduce in the wild, specific conditions are required. Water temperatures between 22 °C and 28 °C, along with suitable water flow and changes in water levels, can stimulate natural fish reproduction and favorably impact the development of benthopelagic fish eggs. While fry growth can occur in stagnant water bodies, these conditions are unsuitable for reaching maturity and triggering reproduction.

Utilizing niche clustering analysis, such as clustering algorithms like K-means clustering [46], our study has yielded detailed insights into the ecological preferences of both Asian carp species (Figure 4). This approach, known as Hutchison's duality, enables researchers to identify distinct clusters or groups of similar ecological conditions within a geographic area. Notably, *H. molitrix* exhibits a broader distribution across various ecological niches compared to second species. Moreover, this analysis has illuminated distinctions in bioclimatic parameters within the native habitats of these Asian alien fish species, as illustrated in Figure 4. Remarkably, regions with analogous conditions have been identified in the southern sectors of North America and in limited areas of South America and Europe, as indicated by red triangles in Figure 4. These findings underscore the remarkable adaptability of Asian carps and their potential to thrive across a broader range of climatic conditions, including those conventionally considered inhospitable to their survival.

Despite subtle ecological disparities from their native environments, niche modeling provides compelling evidence suggesting that these carp species can effectively occupy similar ecological niches within Europe. This conclusion is bolstered by rigorous statistical analysis, underscoring the adaptability of Asian carps to diverse environmental conditions and their potential to establish sustainable populations in European waters.

To strengthen the justification for calibrating Asian carp models for our purpose using occurrences from both the native and the invaded range, and to detect niche shift, we employed principal component analysis (PCA) [47]. This technique allowed us to compare the position of occurrences in these ranges in the bioclimatic space of the variables (i.e., predictors) with the greatest permutation importance (listed in Table 1). The PCA of the predictors revealed two significant axes (PC1 and PC2) of climatic variation, which explain around 60% of the total variance. Plotting the occurrences in the dimensions of the first two PCs revealed the environmental space occupied by the introduced population of Asian carps in Europe and America (round labels). This occupied space appears to be partially overlapped with that of the Asian and similar (including native) occurrences (red triangle), indicating a niche shift (Figure 5).

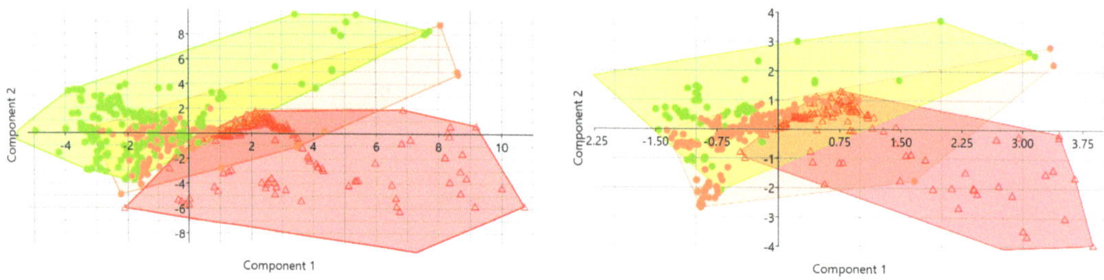

(A) **(B)**

Figure 5. Triplot showing the bioclimatic spaces of the three groups of occurrence records (niche clustering, Figure 4) of Asian carps (*H. molitrix* (**A**); *A. nobilis* (**B**)), representing the suitability for breeding, including native (red triangle) and invaded (green and orange circles) ranges (marked in Figure 4).

Despite their shared thermophilic nature, the two Asian carp species exhibit subtle ecological distinctions, indicating a degree of specialization under natural conditions. Compared to the modeling results presented in the literature, where conditions in the native range of carps are compared with those in North America, they also emphasize the unique conditions and advantages specifically in the native range [26–28]. However, when introduced to new environments, these fish demonstrate remarkable adaptability, a characteristic common among many invasive fish species. This revision emphasizes the ecological differences between the two carp species and their ability to adapt to new environments, highlighting the potential for further range expansion and colonization. The presence of specific abiotic conditions, such as non-freezing waters with sufficient depth and adequately mobile water with a high oxygen content, plays a critical role in the distribution of Asian carps.

To delineate the key determinants of potential suitable habitats for fish in Northern Europe [32], we employed GIS modeling and 13 environmental variables [37,38]—comprehensive ecological data for freshwater ecosystems, focusing on Latvia as a case study. These data encompass spatially continuous and freshwater-specific environmental variables, including river system characteristics, land cover features, and anthropogenic impacts. Our study reveals that the acclimatization and distribution of *H. molitrix* and *A. nobilis* in Europe are strongly linked to factors such as "urban/built-up areas (9)", which may be due to the use of thermophilic waters; "mixed/other trees (4)" and "cultivated and managed vegetation (7)", which may be due to the use of artificial fish ponds (Figure 6, Table 1). Under the current conditions, *H. molitrix* appears to be the more likely successful species between the two Asian carps examined, given its enhanced tolerance to extreme temperatures and anthropogenic factors, especially in Northern Europe (Figure 6).

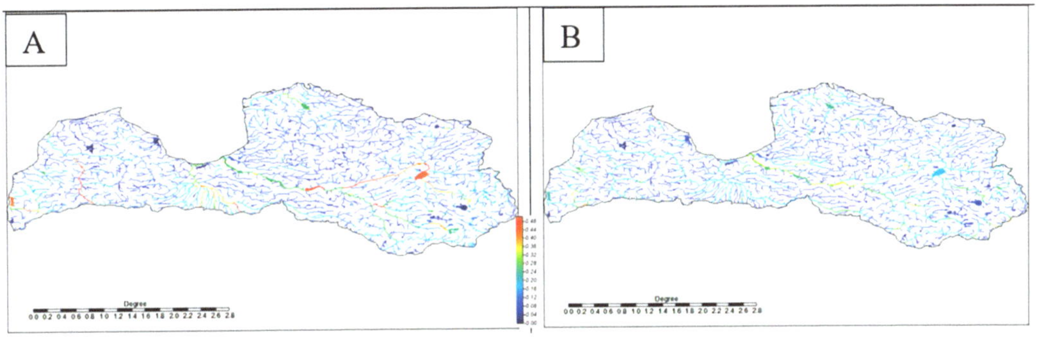

Figure 6. Potential (probabilistic) model of Asian carps (*H. molitrix* (**A**); *A. nobilis* (**B**)) in Northern Europe using the example of Latvia. The expansion was built in the Maxent program based on a dataset of 13 environmental variables (red color > 0.5; EarthEnv NGEI, Table S2).

Thus, our findings demonstrate that under the current climatic conditions in Northern Europe, Asian carps rely on anthropogenic factors and support, but in the future, more favorable conditions may emerge for these thermophilic fish. Additionally, given the hybridization of these species, such adaptation of invasive fish holds evolutionary significance under new conditions. However, questions remain regarding the ecological safety of native amphibian and fish species and ecosystems as a whole, necessitating ongoing monitoring and control.

5. Conclusions

Our research shows that the distribution of Asian carps in Europe is likely to shift northward in response to climate change. This necessitates the development of proactive pan-European management plans to curb the future expansion of these exotic invasive species. We all but call for the introduction of invasive alien species in intact European ecosystems for aquaculture or any other reasons. This is why understanding which geographical areas are potentially suitable for these fish species and their potential spread is crucial. Our models indicate that by 2050, waters, including those of Northern Europe, will provide suitable conditions for Asian carps. Interestingly, despite exhibiting slight ecological niche differences in their native range, these two species can occupy the same biotopes and thrive in polyculture, as demonstrated by our models. This not only proves the advantage of polycultures for their economic and food value but also shows the evolutionary potential of such aquacultures, because these species, by hybridizing, provide new opportunities for offspring to survive in new extreme conditions where native species may not. However, the northward range shift of Asian carps poses a significant challenge for European ecosystems, and effective management strategies are essential to prevent the negative impacts of these invasive species. Understanding the main key points can help develop sustainable management plans at the international level [47]. To harness the potential benefits of polyculture while minimizing ecological disruptions, several key measures are recommended:

- Thorough impact assessments: Comprehensive evaluations of the potential ecological consequences of Asian carp species occurrence in Northern European ecosystems are essential (including the introduction of health-threatening infections and parasites).
- Monitoring and management strategies: Effective monitoring and management strategies are critical for mitigating any negative impacts on native fauna and ensuring the sustainable development of aquaculture practices.
- Collaborative approach: Collaboration among scientists, policymakers, and stakeholders is paramount to developing sustainable aquaculture practices that balance food production with the preservation of native biodiversity.

By understanding the distribution patterns and ecological requirements of Asian carps, we can develop targeted interventions to control their spread while protecting native biodiversity.

Supplementary Materials: The following supporting information can be downloaded at: https://www.mdpi.com/article/10.3390/fishes9040148/s1, Table S1: Bioclimatic variables (35) from the CliMond dataset, [36], (https://www.climond.org/), Table S2: Environmental variables (13) from the near-global environmental information for freshwater ecosystems (EarthEnv NGEI, 12 variables characterizing land cover, [37], http://www.earthenv.org/) and the Global Aridity Index_ET0 (Global-AI_ET0, [38]).

Author Contributions: Conceptualization, O.N., M.P., A.Š. and J.-Y.G.; methodology, O.N., M.P. and V.T.; software, O.N. and V.T.; validation, O.N.; formal analysis, O.N. and V.T.; investigation, O.N., M.P., L.F. and O.P.; resources, O.N., M.P., A.Š. and J.-Y.G.; data curation, O.N., M.P., A.Š. and J.-Y.G.; writing—original draft preparation, O.N.; writing—review and editing, O.N., M.P., V.T., A.Č., L.F., O.P., K.T. and J.-Y.G.; visualization, O.N.; supervision, O.N., M.P., V.T., K.T. and J.-Y.G.; project administration, O.N., M.P., A.Š. and J.-Y.G. All authors have read and agreed to the published version of the manuscript.

Funding: We acknowledge the Emys-R project (https://emysr.cnrs.fr, accessed on 7 January 2022) under the Biodiversa+ and Water JPI joint call for research projects, under the BiodivRestore ERA-NET Cofund (GA N°101003777), with the EU and the funding organizations Agence Nationale de la Recherche (ANR, France, grant ANR-21-BIRE-0005), Bundesministerium für Bildung und Forschung (BMBF, Germany, grant 16LW015), the State Education Development Agency (VIAA, Latvia, grant ES RTD/2022/2), and the National Science Center (NSC, Poland, grant 2021/03/Y/NZ8/00101). We are also thankful for the support from the Collège de France and Agence Nationale de la Recherche (ANR) through the PAUSE ANR Ukraine program (Nekrasova O., grant ANR-23-PAUK-0074), and the project "Ecological and socioeconomic thresholds as a basis for defining adaptive management triggers in Latvian pond aquaculture" (lzp-2021/1-0247), project 16-00-F02201-000002 for the use of the mobile complex of scientific laboratories for research purposes.

Institutional Review Board Statement: Ethics Committee or Institutional Review Board approval are not required for this manuscript. In this work, we used the data of national monitoring programs conducted by the authors for the Nature Conservation Agency of France, Latvia, and Ukraine. We also used open databases (GBIF (online)) and literature (links to these data are in the list of references). According to Latvian and Ukraine legislation, no additional approvals or permits are required for the studies. According to Latvian legislation, only certified experts (who are certified for working with particular species/groups) can work with the animals.

Informed Consent Statement: Informed consent was obtained from all subjects involved in the study.

Data Availability Statement: The data were derived from public domain resources. Environmental data: https://www.climond.org/, http://www.earthenv.org/, and https://csidotinfo.wordpress.com/. Occurrence data collection: *Aristichthys nobilis* (Richardson, 1846); *Hypophthalmichthys nobilis;* GBIF.org (accessed on 7 January 2022); GBIF Occurrence Download: https://doi.org/10.15468/dl.rbcrky (accessed on 7 January 2022); *Hypophthalmichthys molitrix* (Valenciennes, 1844); GBIF.org (accessed on 7 January 2022); GBIF Occurrence Download: https://doi.org/10.15468/dl.46yqnd (accessed on 7 January 2022). The data that support the findings of this study are available in the Supplementary Materials of this article.

Conflicts of Interest: The authors declare no conflicts of interest.

References

1. Liao, I.C.; Chao, N.H. Aquaculture and food crisis: Opportunities and constraints. *Asia Pac. J. Clin. Nutr.* **2009**, *18*, 564–569. [PubMed]
2. Neori, A.; Shpige, M.; Guttman, L.; Israel, A. Development of polyculture and integrated multi-trophic aquaculture (IMTA) in Israel: A review. *Isr. J. Aquac.-Bamidgeh* **2017**, *69*, 1–19. [CrossRef]
3. Rakocy, J.E. Aquaponics—Integrating fish and plant culture. In *Aquaculture Production Systems;* John Wiley & Sons, Inc.: Hoboken, NJ, USA, 2012; pp. 344–386. [CrossRef]
4. Billard, R.; Berni, P. Trends in cyprinid polyculture. *Cybium* **2004**, *28*, 255–261.

5. Jeney, Z.; Jian, Z. Use and exchange of aquatic resources relevant for food and aquaculture: Common carp (*Cyprinus carpio* L.). *Rev. Aquac.* **2009**, *1*, 163–173.
6. Woynarovich, A.; Bueno, P.B.; Altan, O.; Jeney, Z.; Reantaso, M.; Xinhua, Y.; Van Anrooy, R. *Better management practices for carp production in Central and Eastern Europe, the Caucasus and Central Asia*; Food and Agriculture Organization of the United Nations: Rome, Italy, 2011.
7. Szücs, I.; Stündl, L.; Váradi, L. Carp farming in Central and Eastern Europe and a case study in multifunctional aquaculture. In *Species and System Selection for Sustainable Aquaculture*; Springer: Berlin/Heidelberg, Germany, 2007; pp. 389–414.
8. Kuybida, V.V.; Nekrasova, O.D.; Kutsokon, Y.K.; Lopatynska, V.V. Summer fish kills in the Kaniv Reservoir. *Hydrobiol. J.* **2019**, *55*, 103–106. [CrossRef]
9. Nekrasova, O.D.; Tytar, V.M.; Kuybida, V.V. *GIS Modeling of Climate Change Vulnerability of Amphibians and Reptiles in Ukraine*; NAS of Ukraine, Shmalgausen Institute of Zoology NAS: Kyiv, Ukraine, 2019; pp. 1–204. ISBN 978-966-02-8956-7. (In Ukrainian)
10. Nekrasova, O.; Tytar, V.; Pupins, M.; Čeirāns, A.; Marushchak, O.; Skute, A. GIS Modeling Study of the Distribution of Viviparous Invasive Alien Fish Species in Eastern Europe in Terms of Global Climate Change, as Exemplified by *Poecilia reticulata* Peters, 1859 and *Gambusia holbrooki* Girarg, 1859. *Diversity* **2021**, *13*, 385. [CrossRef]
11. Tytar, V.; Nekrasova, O.; Pupins, M.; Čeirāns, A.; Skute, A. Modelling the range expansion of pumpkinseed *Lepomis gibbosus* across Europe, with a special focus on Ukraine and Latvia. *North-West. J. Zool.* **2022**, *18*, 143–150.
12. Georges, J.-Y.; Pupins, M.; Nekrasova, O.; Marushchak, O.; Tytar, V.; Skute, A.; Ceirans, A.; Petrov, I.; Garkajs, A.; Theissinger, K. Distribution trends of invasive Asian carps (*Hypophthalmichthys molitrix* and *Aristichthys nobilis*) in Europe in the context of climate change. In Proceedings of the 36th Congress of the International Society of Limnology, Berlin, Germany, 7–10 August 2022; p. 335.
13. Wang, J.; Lamer, J.T.; Gaughan, S.; Wachholtz, M.; Wang, C.; Lu, G. Transcriptomic comparison of invasive bigheaded carps (*Hypophthalmichthys nobilis* and *Hypophthalmichthys molitrix*) and their hybrids. *Ecol. Evol.* **2016**, *6*, 8452–8459. [CrossRef]
14. Karpova, E.P.; Boltachev, A.R. *Fish of the Inner Water Reservoirs in the Crimean Peninsula*; Business-Inform: Simferopol, Crimea, 2012; p. 200.
15. Ayles, G.B.; Campbell, K.; Gillis, D.; Saunders, L.; Scott, K.J.; Tallman, R.; Traverse, N. Technical assessment of the status, health and sustainable harvest levels of the lake Winnipeg fisheries resource. In *Lake Winnipeg Quota Review Task Force Report*, 2nd ed.; Conover, G., Simmonds, R., Whalen, M., Eds.; Asian Carp Working Group, Aquatic Nuisance Species Task Force: Washington, DC, USA, 2011; Volume 3, pp. 154–196.
16. Zambrano, L.; Martínez-Meyer, E.; Menezes, N.; Peterson, A.T. Invasive potential of common carp (*Cyprinus carpio*) and Nile tilapia (*Oreochromis niloticus*) in American freshwater systems. *Can. J. Fish. Aquat. Sci.* **2006**, *63*, 1903–1910. [CrossRef]
17. Milardi, M.; Chapman, D.; Lanzoni, M.; Long, J.M.; Castaldelli, G. First evidence of bighead carp wild recruitment in Western Europe, and its relation to hydrology and temperature. *PLoS ONE* **2017**, *12*, e0189517. [CrossRef]
18. Jawdhari, A.; Mihăilescu, D.F.; Fendrihan, S.; Jujea, V.; Stoilov-Linu, V.; Negrea, B.M. Silver Carp (*Hypophthalmichthys molitrix*) (Asian Silver Carp) Presence in Danube Delta and Romania—A Review with Data on Natural Reproduction. *Life* **2022**, *12*, 1582. [CrossRef]
19. Conover, G.; Simmonds, R.; Whalen, M. Management and control plan for bighead, black, grass, and silver carps in the United States. In *Asian Carp Working Group, Aquatic Nuisance Species Task Force, Eds.*; Asian Carp Working Group, Aquatic Nuisance Species Task Force: Washington, DC, USA, 2007; Volume 3, pp. 154–196.
20. Novitskiy, R.A. Scale, Direction and Consequences of Alien Fish Species Invasions in the Dnieper Reservoirs: Avt. Ph.D. Thesis, Dnipro State Agrarian and Economic University, Kyiv, Ukraine, 2019; p. 41.
21. Ciolac, A. Migration of fishes in Romanian Danube River. *Appl. Ecol. Environ. Res.* **2004**, *2*, 143–163. [CrossRef]
22. Molnár, T.; Lehoczky, I.; Edviné Meleg, E.; Boros, G.; Specziár, A.; Mozsár, A.; Vitál, Z.; Józsa, V.; Allele, W.; Urbányi, B.; et al. Comparison of the Genetic Structure of Invasive Bigheaded Carp (*Hypophthalmichthys* spp.) Populations in Central-European Lacustrine and Riverine Habitats. *Animals* **2021**, *11*, 2018. [CrossRef] [PubMed]
23. Robinson, D.C.E.; Knowler, D.; Kyobe, D.; de la Cueva Bueno, P. *Preliminary Damage Estimates for Selected Invasive Fauna in B.C.*; Report Prepared for Ecosystems Branch; B.C. Ministry of Environment, Victoria, B.C. by ESSA Technologies Ltd.: Vancouver, BC, USA, 2013; p. 62.
24. Pupins, M.; Nekrasova, O.; Marushchak, O.; Tytar, V.; Theissinger, K.; Čeirāns, A.; Skute, A.; Georges, J.-Y. Potential Threat of an Invasive Fish Species for Two Native Newts Inhabiting Wetlands of Europe Vulnerable to Climate Change. *Diversity* **2023**, *15*, 201. [CrossRef]
25. Soes, D.M.; Leuven, R.S.E.W.; Matthews, J.; Broeckx, P.B.; Haenen, O.L.M.; Engelsma, M.Y. A risk analysis of bigheaded carp (Hypophthalmichthys sp.) in the Netherlands. 2011. 118p. Available online: https://repository.ubn.ru.nl/handle/2066/123477 (accessed on 20 April 2024).
26. DeVaney, S.C.; McNyset, K.M.; Williams, J.B.; Peterson, A.T.; Wiley, E.O. A tale of four "carp": Invasion potential and ecological niche modeling. *PLoS ONE* **2009**, *4*, e5451. [CrossRef] [PubMed]
27. Chen, P.; Wiley, E.O.; Mcnyset, K.M. Ecological niche modeling as a predictive tool: Silver and bighead carps in North America. *Biol. Invasions* **2007**, *9*, 43–51. [CrossRef]
28. Kulhanek, S.A.; Leung, B.; Ricciardi, A. Using ecological niche models to predict the abundance and impact of invasive species: Application to the common carp. *Ecol. Appl.* **2011**, *21*, 203–213. [CrossRef] [PubMed]

29. Lübcker, N.; Zengeya, T.A.; Dabrowski, J.; Robertson, M.P. Predicting the potential distribution of invasive silver carp *Hypophthalmichthys molitrix* in South Africa. *Afr. J. Aquat. Sci.* **2014**, *39*, 157–165. [CrossRef]
30. Smith, A. Bigheaded carps (*Hypophthalmichthys molitrix* and *H. nobilis)* An Annotated Bibliography on Literature Composed from 1970 to 2014. Publication 2890. Mississippi State University Extension, 2015. 43 p. Available online: https://extension.msstate.edu/sites/default/files/publications/publications/p2890.pdf (accessed on 4 September 2023).
31. Masson-Delmotte, V.; Zhai, P.; Pirani, A.; Connors, S.L.; Péan, C.; Berger, S.; Caud, N.; Chen, Y.; Goldfarb, L.; Gomis, M.I.; et al. (Eds.) *IPCC, 2021: Climate Change 2021: The Physical Science Basis. Contribution of Working Group I to the Sixth Assessment Report of the Intergovernmental Panel on Climate Change*; Cambridge University Press: Cambridge, UK, 2021; p. 35.
32. *Historical Aquaculture in Northern Europe. Historical Aquaculture in Northern Europe, Eds. Madeleine Bonow, Håkan Olsén and Ingvar Svanberg*; Elanders: Stockholm, Sweden, 2016; p. 204.
33. *Aristichthys nobilis* (Richardson, 1846) (=*Hypophthalmichthys nobilis*) GBIF.org. Available online: https://www.gbif.org/occurrence/download/0097149-210914110416597 (accessed on 7 January 2022).
34. *Hypophthalmichthys molitrix* (Valenciennes, 1844) GBIF.org. Available online: https://www.gbif.org/occurrence/download/0097129-210914110416597 (accessed on 7 January 2022).
35. Osorio-Olvera, L.; Lira-Noriega, A.; Soberón, J.; Peterson, A.T.; Falconi, M.; Contreras-Díaz, R.G. ntbox: An r package with graphical user interface for modelling and evaluating multidimensional ecological niches. *Methods Ecol. Evol.* **2020**, *11*, 1199–1206. [CrossRef]
36. Kriticos, D.J.; Jarošik, V.; Ota, N. Extending the suite of Bioclim variables: A proposed registry system and case study using principal components analysis. *Methods Ecol. Evol.* **2014**, *5*, 956–960. [CrossRef]
37. Domisch, S.; Amatulli, G.; Jetz, W. Near-global freshwater-specific environmental variables for biodiversity analyses in 1 km resolution. *Sci. Data* **2015**, *2*, 150073. [CrossRef]
38. Trabucco, A.; Zomer, R. Global Aridity Index and Potential Evapotranspiration (ET0) Climate Database v2. Figshare. Fileset. Available online: https://figshare.com/articles/dataset/Global_Aridity_Index_and_Potential_Evapotranspiration_ET0_Climate_Database_v2/7504448 (accessed on 4 September 2023).
39. Phillips, S.J. A brief tutorial on Maxent. *ATT Res.* **2005**, *190*, 231–259.
40. Peterson, A.T.; Papes, M.; Soberón, J. Rethinking receiver operating characteristic analysis applications in ecological niche modeling. *Ecol. Model.* **2008**, *213*, 63–72. [CrossRef]
41. Anderson, R.P.; Lew, D.; Peterson, A.T. Evaluating predictive models of species' distributions: Criteria for selecting optimal models. *Ecol. Model.* **2003**, *162*, 211–232. [CrossRef]
42. Fielding, A.H.; Bell, J.F. A review of methods for the assessment of prediction errors in conservation presence/absence models. *Environ. Conserv.* **1997**, *24*, 38–49. [CrossRef]
43. Allouche, O.; Tsoar, A.; Kadmon, R. Assessing the accuracy of species distribution models: Prevalence, Kappa and the True Skill Statistic (TSS). *J. Appl. Ecol.* **2006**, *43*, 1223–1232. [CrossRef]
44. Hammer, Ø.; Harper, D.A.T.; Ryan, P.D. PAST: Paleontological Statistics Software Package for Education and Data Analysis. *Palaeontol. Electron.* **2001**, *4*, 9. Available online: http://palaeo-electronica.org/2001_1/past/issue1_01.htm (accessed on 29 February 2024).
45. Rubenstein, M.A.; Weiskopf, S.R.; Bertrand, R.; Carter, S.L.; Comte, L.; Eaton, M.J.; Thompson, L.M. Climate change and the global redistribution of biodiversity: Substantial variation in empirical support for expected range shifts. *Environ. Evid.* **2023**, *12*, 1–21. [CrossRef]
46. Colwell, R.K.; Thiago, F. Rangel. Hutchinson's duality: The once and future niche. *Proc. Natl. Acad. Sci. USA* **2009**, *106*, 19651–19658. [CrossRef]
47. Broennimann, O.; Treier, U.A.; Müller-Schärer, H.; Thiuller, W.; Peterson, A.T.; Giusan, A. Evidence of climatic niche shift during biological invasion. *Ecol. Lett.* **2007**, *10*, 701–709. [CrossRef]

fishes

MDPI

Article

Distribution Characteristics and Driving Factors of *Collichthys lucidus* Species in Offshore Waters of Zhejiang Province, China

Wendan Xuan [1,2], Hongliang Zhang [1,2,3,4], Haobo Zhang [5], Tian Wu [1,2], Yongdong Zhou [1,2,3,4] and Wenbin Zhu [1,2,3,4,*]

1 Zhejiang Marine Fisheries Research Institute, Zhoushan 316021, China; xwdalways@163.com (W.X.); hidalgo310@163.com (H.Z.); zyd511@126.com (Y.Z.)
2 Marine and Fishery Institute, Zhejiang Ocean University, Zhoushan 316021, China
3 Scientific Observing and Experimental Station of Fishery Resources for Key Fishing Grounds, Ministry of Agriculture and Rural Affairs of the People's Republic of China, Zhoushan 316021, China
4 Key Laboratory of Sustainable Utilization of Technology Research for Fishery Resources of Zhejiang Province, Zhoushan 316021, China
5 State Key Laboratory of Water Resources Engineering and Management, Wuhan University, Wuhan 430072, China; z1393679095@163.com
* Correspondence: wenbinzhu@zjou.edu.cn

Abstract: *Collichthys lucidus* is a small fish found in offshore waters that is economically important for China. It is imperative to understand its distribution characteristics and driving factors. Based on survey data of trawl fishery resources offshore of Zhejiang province, China, in spring (April) and autumn (November) from 2018 to 2022, the spatial and temporal distributions of *C. lucidus* in this area were analyzed. The random forest (RF) model was used to determine the important marine factors affecting the distribution of *C. lucidus*. The relationship between the distributions of the important variables was analyzed. The results showed that *C. lucidus* was mainly distributed in coastal waters. The tail density of the species exhibited obvious seasonal variation and was significantly greater in autumn than in spring. The most important factor affecting the distribution of this species in spring and autumn was water depth. The bottom temperature, bottom salinity and dissolved oxygen concentration were also important influencing factors. The importance of these factors differed among the different seasons, while the chlorophyll a concentration and pH had no significant effect on the species distribution. This study revealed the distribution pattern of *C. lucidus* in offshore waters of Zhejiang Province and the influence of important marine factors on its distribution. This study can enrich the survey data on *C. lucidus* and provide basic data for its scientific management and protection.

Keywords: Zhejiang offshore; *Collichthys lucidus*; distribution characteristics; impact factors; random forest model

Key Contribution: This study not only enriches the resource change survey data in the coastal and offshore areas but also helps to understand the response of *Collichthys lucidus* to environmental factors. This study provides a theoretical basis for the scientific management and effective utilization of *C. lucidus* species.

Citation: Xuan, W.; Zhang, H.; Zhang, H.; Wu, T.; Zhou, Y.; Zhu, W. Distribution Characteristics and Driving Factors of *Collichthys lucidus* Species in Offshore Waters of Zhejiang Province, China. *Fishes* **2024**, *9*, 83. https://doi.org/10.3390/fishes9030083

Academic Editors: Heng Zhang, Ying Xue, Yunzhi Yan and Luoliang Xu

Received: 3 January 2024
Revised: 20 February 2024
Accepted: 21 February 2024
Published: 23 February 2024

1. Introduction

China's Zhejiang offshore region is located in the north-central part of the East China Sea and is an important habitat for a variety of marine organisms [1]. Due to the interactions among monsoons, the Kuroshio current and multiple water masses, such as the Yangtze River estuary diluted water, the sea area has high water temperature and a large salinity gradient. Adequate food and nutrient availability provide good breeding, feeding and overwintering conditions for fish, forming a species-rich ecosystem, which is an important area for producing offshore fishery resources in China [2]. However, due to the frequent interference of global climate change and human activities, the offshore marine environment in Zhejiang is constantly changing, as indicated by changes in temperature and frequent hypoxia in coastal waters [3]. Changes in environmental conditions have had profound impacts on the distribution of many fish and other marine organisms. Some fish may migrate to new areas to find more suitable living conditions [4,5]. This will lead to changes in fishing catches and species in the coastal waters of Zhejiang, with a series of economic impacts on fisheries and related industries.

Collichthys lucidus (Spinyhead Croaker) belongs to Perciformes, Sciaenidae and *Collichthys*, and is a warm-water bottom fish [6]. It is mainly distributed in the Yellow Sea, Bohai Sea, East China Sea and South China Sea in the western and central Pacific Oceans. It is the main economic fish in the coastal areas of China, Japan, North Korea, the Philippines, Indonesia and other regions [7]. The fish is rich in nutrients and has broad market prospects. It is an important fresh seafood for coastal residents and an important part of estuarine ecosystems and has good development value [8]. In offshore waters of Zhejiang Province, *C. lucidus* is not only used as bait for traditional fish such as *Larimichthys crocea*, *Larimichthys polyactis* and *Trichiurus lepturus* but is also a predator of the larvae and eggs of these economic fish. It is an important part of the marine food chain [9]. Its presence is conducive to maintaining the stability of marine ecosystems, and it plays a connecting role in marine ecosystems. In addition, *C. lucidus* exhibits a rapid growth rate, early sexual maturity, strong fecundity and no long-distance migratory habits. With the decline in traditional main economic fish (enemy fish), *C. lucidus* has been fully reproduced and grown, and the number of species have increased significantly. It has gradually become a dominant species in the northern part of Hangzhou Bay, Zhongjieshan Islands Marine Reserve and other sea areas [10,11]. The spatial activity and distribution of *C. lucidus* are affected by many marine factors. Understanding the spatial distribution characteristics of *C. lucidus* and determining the key index factors are highly important for guiding the fishing production and habitat prediction of *C. lucidus*.

The interaction effect of ecosystems and various environmental and biological factors on the offshore distribution of *C. lucidus* in the Zhejiang Province makes it difficult for the traditional generalized additive model (GAM) to reveal the dominant factors affecting the distribution of *C. lucidus*. The random forest (RF) model is a machine learning model based on classification and regression trees. By constructing a certain number of decision trees and combining the generated decision trees according to the corresponding criteria, an additional layer of randomness is added to reduce the influence of outliers on the prediction results. Therefore, it can be used to simulate multiple nonlinear relationships [12]. Liu [13] used the RF model and the GAM model to analyze the relationship between the catch per unit fishing effort of *Euphausia superba* and environmental factors, and reported that the RF model had a better fitting effect than the GAM model. Wu [14] used the RF model to analyze the influence of the marine environment on the change in spawning ground of *Sepiella japonica* in offshore waters of Zhejiang Province, and reported that water depth, sea surface salinity, dissolved oxygen and chlorophyll a were important driving factors. The application of a random forest model to analyze the relationships between *C. lucidus* and marine factors in offshore waters of Zhejiang Island allowed us to calculate the influence of each variable on the resources of *C. lucidus*, and it was convenient to identify the dominant factors affecting the spatial distribution of *C. lucidus*.

Several studies have shown that on the coast of China, *C. lucidus* can be divided into southern and northern populations bounded by the sea area near Zhejiang [15]. As the boundary between the two populations, the importance of Zhejiang offshore waters is self-evident. At present, the research data on the changes in *C. lucidus* species in Zhejiang Province are relatively old, the research area is small and the marine factors affecting these changes have not been fully recognized [16]. In addition, *C. lucidus* is a short-lived fish [17]. The spatial and temporal distributions of short-lived species are particularly sensitive to changes in the marine environment. Based on survey data of bottom trawl fishery resources in Zhejiang offshore waters from 2018 to 2022, this study used a random forest model to explore the spatial distribution characteristics of *C. lucidus*. The objective of this study was to explore the interannual variation in the spatial distribution of *C. lucidus* and its driving factors by using fishery survey information from this species in Zhejiang offshore waters for five consecutive years and in close real-time. The aim was to enrich the survey data of *C. lucidus* in coastal and offshore areas, and the scientific management and effective utilization of this species were promoted.

2. Materials and Methods

2.1. Data Source

The fisheries data were derived from the bottom trawl survey of fishery resources in offshore areas of Zhejiang Province, China. The survey times were April (spring) and November (autumn) from 2018 to 2022. The survey range was 120°30′–123°45′ E, 27°00′–30°45′ N, and included northern Zhejiang (A), central Zhejiang (B) and southern Zhejiang (C). A total of 120 stations were set up (Figure 1). The fishing vessel was named "Zhepu Fishery 43019". The length of the vessel was 30.4 m, the width was 6.0 m and the depth was 2.7 m. The main engine power was 202 kW. The engine model of the vessel was "Weichai WP4.1Q95E41" (Weichai Power Co., Ltd., Weifang, China). A single bottom trawling method was used. According to the National "Specifications for oceanographic survey -Marine Biological Survey" (GB/T 12763.6-2007) [18], the mesh of the cod-end net was set to 25 mm, the width of the net mouth was 9.9 m and the circumference of the net mouth was 50 m. The designed average towing speed at each station was 3 kn, and the trawling time was 30 min. The samples were collected, processed and analyzed according to the National "Specifications for oceanographic survey -Marine Biological Survey" (GB/T 12763.6-2007) and "The specification for marine monitoring" (GB 17378.3-2007) [19].

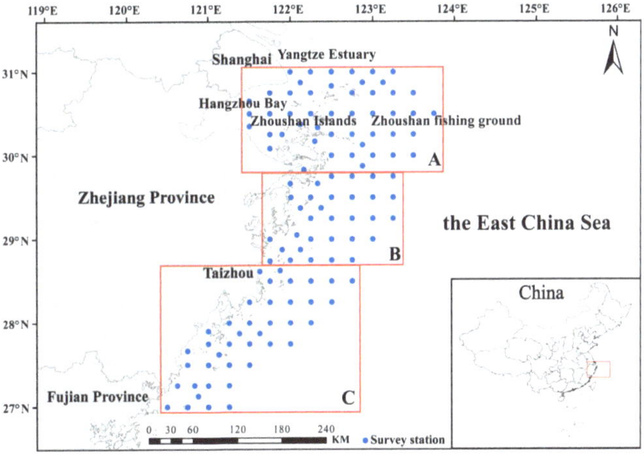

Figure 1. Map of the bottom trawl survey area in Zhejiang Province and adjacent waters of China. A: Northern Zhejiang, B: central Zhejiang and C: southern Zhejiang. The blue dots represent the survey stations.

Bottom temperature, bottom salinity, bottom pH, chlorophyll a concentration and dissolved oxygen are known to directly influence the spawning and distribution of *C. lucidus*. Different water depths lead to varying changes in environmental factors such as temperature and light in the ocean, consequently impacting the ecological habits and distribution range of marine organisms [20,21]. Therefore, the variables selected in this study included the physical factor water depth (depth) and environmental factors such as bottom temperature (BST), bottom salinity (BSS), bottom dissolved oxygen (BDO), bottom chlorophyll a (BChla) and bottom pH (BpH). The above data were measured synchronously by a multifunction water quality instrument SeaBird SBE 37-SM MicroCAT (Nanjing Dongguo Marine Technology Co., Ltd., Nanjing, China).

2.2. Data Processing

This study investigated the changes in resource density of *C. lucidus* based on tail density. The formula for calculating the tail density of each station is as follows:

$$D = \frac{C}{q \times A} \quad (1)$$

D is the mantissa density (ind/km^2). C is the number of tails in the sampling area per hour (ind). q is the net capture rate, and referring to the research of Shi [22] and Yang [23], q is set to 0.5. A is the area of the net sweeping the sea per hour, which is obtained by the product of the ship speed and the sweeping width.

All of the samples were grouped according to the sampling time, and all of the stations were divided into April and November. Using R 4.0.3 software, based on the generalized linear model, the Poisson regression method was selected to evaluate the significant differences in the tail density of *C. lucidus* in different months.

The variance inflation factor (VIF) represents the degree of correlation between variables. The larger the VIF is, the more serious the collinearity. In general, if the VIF > 3, there is strong collinearity between variables. This should be left out before modeling [24]. Using SPSS 24.0 software, the factors with VIF less than three were selected by a multicollinearity test for subsequent analysis.

2.3. Construction and Analysis of the Random Forest Model

The random forest (RF) model is a machine learning model based on classification and regression trees [25]. It combines bagging and random feature selection to add an extra layer of randomness to bootstrap and reduce the impact of outliers on prediction results. The model has strong generalizability and is widely used in species distribution models [26]. In this study, an RF model was used to analyze the distribution and marine variables of *C. lucidus*. A total of 75% of the data were randomly selected from the samples as the training set for modeling, and the remaining 25% of the data were verified. Fivefold cross-validation was used to evaluate the prediction performance and accuracy of the model [27], and the calculation was repeated 100 times. According to the results of cross-validation, the number of decision trees (n_{tree}) was set to 1000, and the number of features (mtry) was 2.

In the fitting process, to reduce the variable scale, the natural logarithm transformation of the density value Y of *C. lucidus* at each station was carried out. The ln (Y + 1) was obtained as the response variable. The spatial factors longitude (Lon) and latitude (Lat) and physical factor depth were selected; the environmental factors included BST, BSS, BDO, BChla, and BpH. A total of eight factors were considered explanatory variables. These factors were subsequently added to the model for fitting. According to the root mean squared error (*RMSE*) and mean absolute error (*MAE*), the difference between the predicted value and the actual value of the model was tested. The coefficient of determination R^2 was used to judge the fitting degree and prediction performance of the model.

The calculation formula is as follows:

$$RMSE = \sqrt{\frac{\sum_{i=1}^{N}(y_i - \hat{y}_i)^2}{N}} \quad (2)$$

$$MAE = \sqrt{\frac{\sum_{i=1}^{N}|y_i - \hat{y}_i|}{N}} \quad (3)$$

$$R^2 = 1 - \frac{SSE}{SST} \quad (4)$$

where y_i represents the true value, $\hat{y}_i$ represents the predicted value, N represents the number of true values, SSE represents the sum of squared errors and SST represents the sum of squared total deviations. Smaller $RMSE$ and MAE values and greater R^2 values indicate higher reference values of the model [28].

The calculation formula of the RF model for the importance ranking of explanatory variables is as follows:

$$V_i = \frac{1}{n_{tree}} \sum v \in S_{Xi}\, G\,(X_i,\, v) \tag{5}$$

where V_i represents the explanatory rate of the independent variable X_i to the model, S_{Xi} represents the set of nodes split by X_i under the condition of n_{tree}, and $G\,(X_i, v)$ represents the Gini information gain of X_i at the split node v.

R 4.0.3 software was used for model construction, model prediction and verification of the above steps.

2.4. Results Visualization

ArcGIS 10.1 software was used to visualize sampling sites and the distributions of *C. lucidus*. In order to better analyze the nonlinear relationship between tail density changes and explanatory variables, each explanatory variable was used as the *X*-axis, and the tail density was used as the *Y*-axis. Based on the "ggplot2" package in the R 4.0.3 software, locally weighted regression (LWR) was used for nonlinear fitting.

3. Results and Analysis

3.1. Changes in the Density of C. lucidus

Due to weather and other reasons, the actual number of trawl sites per year is slightly different, as detailed in Table 1. Overall, the occurrence rate of *C. lucidus* in spring (April) was lower than that in autumn (November). The maximum biomass density and tail density were 220.42 kg/km^2 and 7.43 × 10^3 ind/km^2, respectively. The maximum biomass density and individual density in autumn were 611.10 kg/km^2 and 25.31 × 10^3 ind/km^2, respectively, which were significantly greater than those in spring. By calculating the ratio of biomass density to tail density of *C. lucidus* at each station, the relative weight of *C. lucidus* individuals in spring and autumn was obtained. The results showed that the relative weight of individuals was greater in the spring.

Table 1. Statistics on occurrence stations and resource density of *C. lucidus* in Zhejiang coastal waters from 2018 to 2022.

Seasons		Stations		Biomass (kg/km^2)		Tail Density (×10^3 ind/km^2)		Average Individual Relative Weight (g)
		Investigation Stations	Occurrence Stations	Maximum Value	Average Value	Maximum Value	Average Value	
Spring	2018	98	56	121.30	16.10	6.89	0.60	26.83
	2019	108	54	97.47	9.02	4.14	0.34	26.16
	2020	79	39	220.42	14.53	5.07	0.43	34.06
	2021	111	55	114.78	7.58	3.84	0.29	25.93
	2022	114	65	148.68	14.47	7.43	0.68	21.27
Autumn	2018	90	61	354.89	40.23	21.76	2.13	18.93
	2019	94	53	269.30	27.27	11.03	1.21	22.48
	2020	99	52	184.02	15.78	9.92	0.71	22.31
	2021	110	61	611.09	35.79	25.31	1.59	22.49
	2022	112	80	314.96	34.24	14.82	1.79	19.17

According to the annual changes, the tail density in spring and autumn showed a trend of decreasing first and then increasing (Figure 2). The tail density of *C. lucidus* in spring changed little among the different years and was in a stable state. The lowest average tail density, 0.29×10^3 ind/km^2, occurred in 2021. The highest value, 0.68×10^3 ind/km^2, occurred in 2022. The average tail density in autumn fluctuated greatly from year to year. The lowest value, 0.71×10^3 ind/km^2, occurred in 2020. The highest value, 2.13×10^3 ind/km^2, occurred in 2018. In addition, the tail density of *C. lucidus* was greater in autumn than in spring. The tail density index in autumn is 3.12 times that in spring. Poisson regression results showed that there was a significant difference in tail density between spring and autumn in 2020 ($p < 0.05$). Except from 2020, there were extremely significant differences in other years ($p < 0.01$).

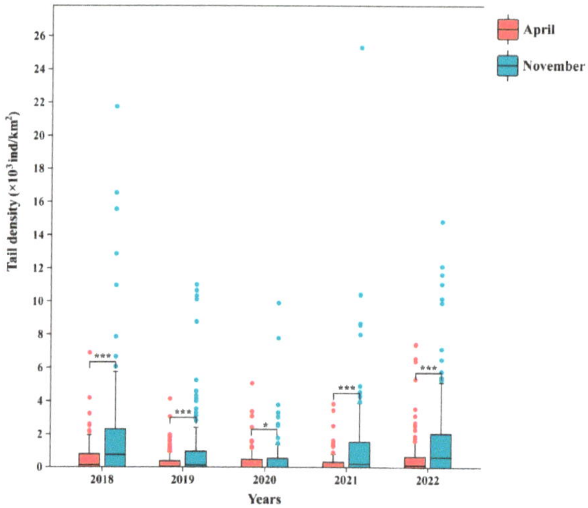

Figure 2. Changes in tail density of *C. lucidus* species in spring and autumn. Red represents April, and green represents November. "*" and "***" represent the significant difference and extremely significant difference of Poisson regression, respectively.

3.2. Screening of Environmental Variables

Based on the VIF, a multicollinearity test was carried out on the six variables of depth, BST, BSS, BChla, BDO and BpH. The VIFs were all less than 3 (Table 2). Therefore, there is no multicollinearity problem between variables, which can be retained and added to the RF model for analysis.

Table 2. Multicollinearity test of variables for the distribution of *C. lucidus*.

Parameter	Month	Explanatory Variables					
		Depth	BST	BSS	BChla	BDO	BpH
VIF	April	2.787	2.188	2.410	1.113	2.792	1.152
	November	2.441	1.805	1.984	1.214	1.769	1.057

3.3. Model Performance Assessment

The fitting results of the RF model for the explanatory variables reveal that the R^2 values in spring and autumn are 0.873 and 0.912, respectively, and the *RMSE* values are 0.180 and 0.263, respectively, indicating that the fitting effect of the model is good. The accuracy and stability of the model were tested by fivefold cross-validation. The R^2 values in spring and autumn are 0.251 ± 0.054 and 0.395 ± 0.072, respectively. The standard deviation is small, indicating that the stability of the model is good. The model interpretation rates in spring and autumn are 23.26% and 34.88%, respectively (Table 3).

Table 3. Results of model fitting and fivefold cross-validation.

Month	Model Fitting			Fivefold Cross-Validation			Explanation (%)
	RMSE	R^2	*MAE*	*RMSE*	R^2	*MAE*	
April	0.180	0.873	0.113	0.355 ± 0.060	0.251 ± 0.054	0.234 ± 0.028	23.26
November	0.263	0.912	0.179	0.554 ± 0.045	0.395 ± 0.072	0.390 ± 0.024	34.88

3.4. Importance Ranking of Explanatory Variables

The results of the relative importance of the explanatory variables showed that the longitude and latitude of spatial factors had significant impacts on the distribution of *C. lucidus* species in spring and autumn. Among the marine variables, the four variables with the most influence in spring were depth, BSS, BST and BDO. The four most influential variables in autumn were depth, BDO, BSS and BST. BChla and BpH had little effect on the distribution of C. *lucidus* in spring or autumn, so no detailed analysis was carried out later (Figure 3).

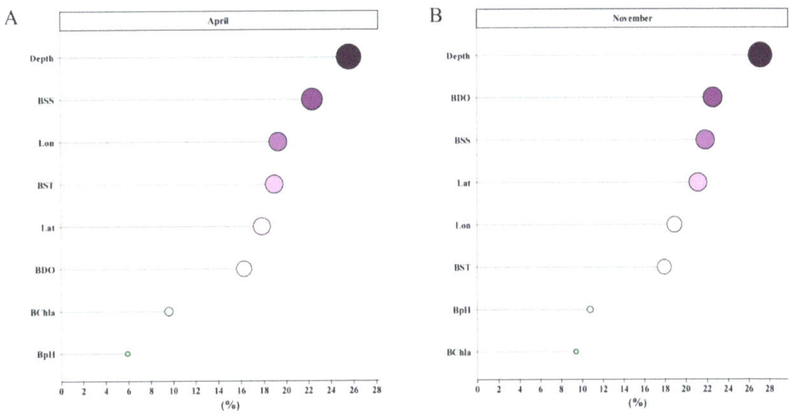

Figure 3. The relative importance of explanatory variables for determining the distribution of *C. lucidus*. (**A**) Importance ranking of explanatory variables in April. (**B**) Importance ranking of explanatory variables in November. The *X*-axis represents the importance score of each explanatory variable, and the *Y*-axis represents different explanatory variables. The larger the shape of the point, the deeper the color, and the more important the variable.

3.5. The Relationships between the Distributions of C. lucidus Species and Important Explanatory Variables

According to Figure 4, the trends of tail density in spring and autumn with different explanatory variables are nonlinear. For longitude, the optimum range in spring is between 120.5°~121° E and 122°~122° E. The change trend in autumn is small, and the resource quantity is higher within 123° E. For latitude, the change trend in spring is not obvious, and the peak value of resource is concentrated near 31° N. In autumn, the resource quantity decreases first and then increases with the increase in latitude, and the peak value is concentrated between 27°~28° N and 30.5°~31° N. For marine variables, the amount of *C. lucidus* in spring and autumn showed a trend of increasing first and then decreasing with the increase in depth, BSS, BST and BDO concentrations. In spring, the peak resources were mainly in the depth range of 4~30 m, the BSS range was 27~34, the BST range was 13~16 °C and the BDO range was 5~9 mg/L. In autumn, the peak resources were mainly in the depth range of 4~45 m, the BSS range was 13~34, the BST range was 17~21 °C and the BDO range was 6~9 mg/L.

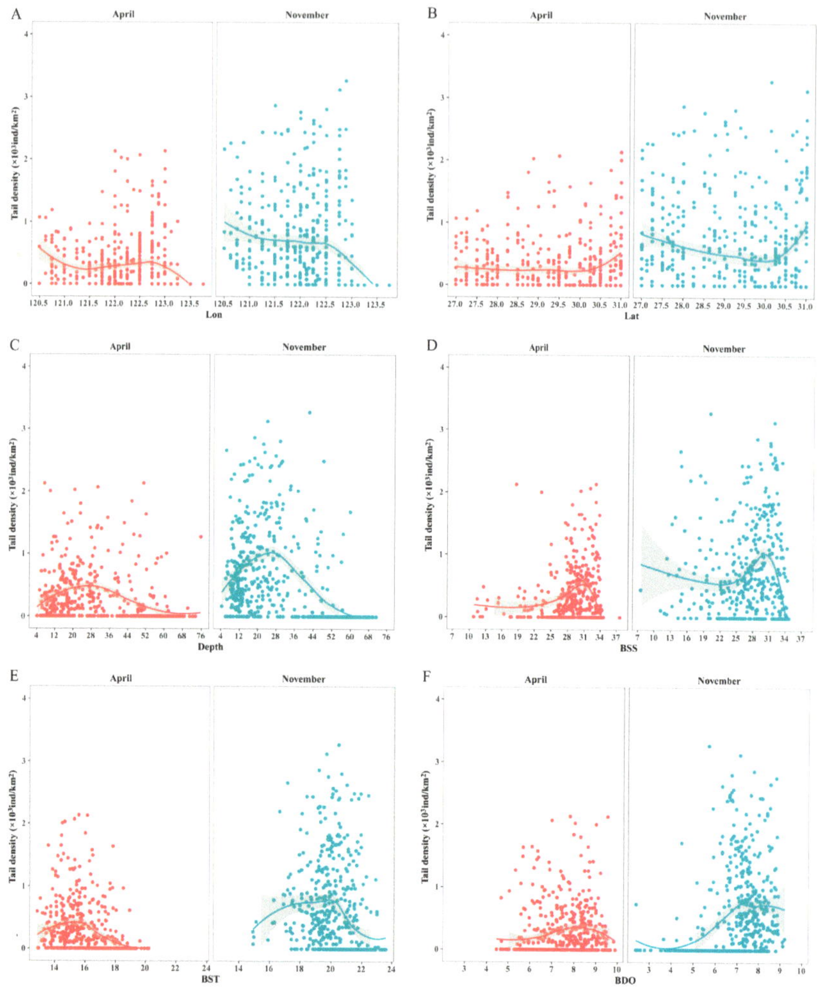

Figure 4. The relationships between the distributions of *C. lucidus* species in spring and autumn and important environmental variables according to the RF model. (**A**) Lon; (**B**) lat; (**C**) depth; (**D**) BSS; (**E**) BST; (**F**) BDO. The *X*-axis is the range of each variable, and the *Y*-axis is the corresponding tail density. The curve represents the nonlinear fitting curve, and the shadow represents the confidence interval.

3.6. Spatial and Temporal Distributions of *C. lucidus* Species

In this study, the most important explanatory variable affecting the spatial and temporal distributions of *C. lucidus* was water depth. Therefore, the water depth data were superimposed to construct a map of the distribution of *C. lucidus* species in April and November from 2018 to 2022 (Figure 5) to determine the distribution of *C. lucidus* in Zhejiang offshore waters, China. The distribution of the fish species is consistent with the distribution trend of the water depth contour, mainly in the northeast–southwest direction. In spring, *C. lucidus* was mainly distributed in the northern and central parts of Zhejiang Province and was distributed offshore and adjacent to the open seas. It was mainly distributed within 31 m of the water depth contour. The high-abundance areas were concentrated near the Yangtze estuary and Zhoushan fishing ground. In autumn, the distribution area of *C. lucidus* species expanded to southern Zhejiang and was distributed in the north, middle and south. It was mainly distributed within 53 m of the water depth contour. The high-abundance areas were mainly in the coastal waters.

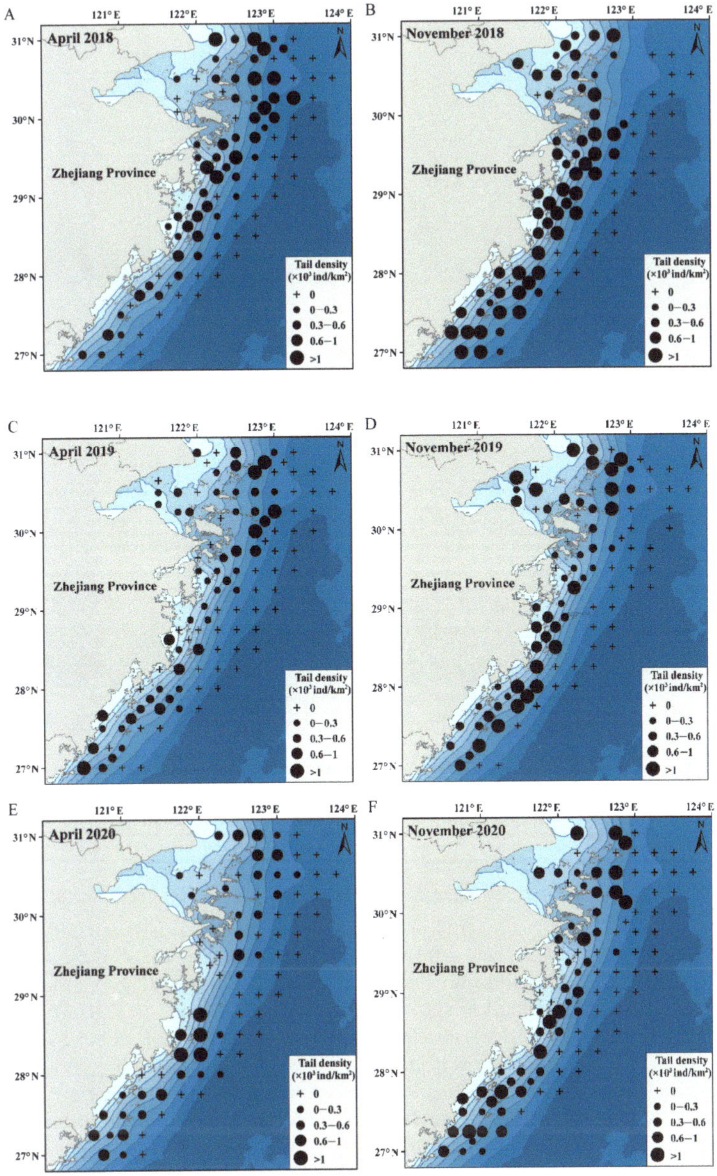

Figure 5. *Cont.*

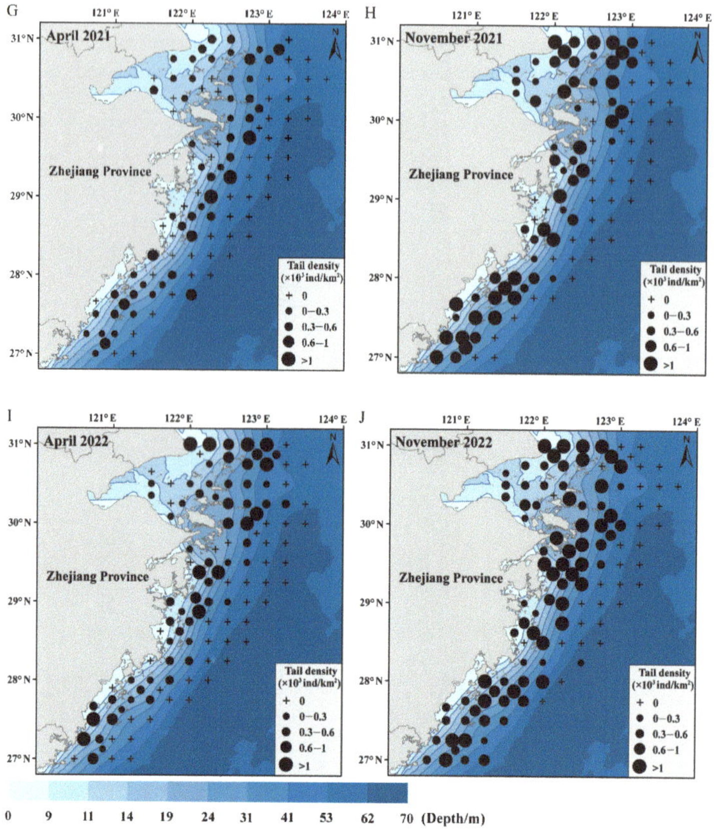

Figure 5. The distribution of *C. lucidus* in spring and autumn in Zhejiang coastal waters from 2018 to 2022. (**A**) April 2018; (**B**) November 2018; (**C**) April 2019; (**D**) November 2019; (**E**) April 2020; (**F**) November 2020; (**G**) April 2021; (**H**) November 2021; (**I**) April 2022; (**J**) November 2022.

4. Discussion

4.1. Spatiotemporal Changes in the Distribution of C. lucidus

In terms of time, the distribution of *C. lucidus* in offshore Zhejiang changed significantly between different months and different years. The tail density of *C. lucidus* in April was significantly lower than that in November (Figures 2 and 5). The distribution change was related to spawning, feeding and migration [29]. There are two spawning periods for *C. lucidus* in offshore areas of the Zhejiang Province. The spawning period in spring and summer is from late April to July. The spawning period in autumn is from September to October [30]. In spring, with increasing water temperature, the population of *C. lucidus* began to migrate from the deep-water area to the offshore area for feeding to prepare for life activities such as spawning and reproduction. It is speculated that the survey object of this study, which was conducted in April, was mainly spawning parents. The tail density increased significantly in November. At this time, many spawning parents had completed their breeding activities. After breeding in that year, the larvae could be captured by bottom trawling in November, which increased the tail density. It is speculated that the survey objects in November included spawning parents and supplementary groups. This also explained why the relative weight of the individuals in November was lower than that in April (Table 1). In terms of interannual variation, the tail density of *C. lucidus* first decreased and then decreased from 2018 to 2022. It was relatively low in 2019 and 2020 (Figure 2). The optimum bottom temperature of *C. lucidus* in the study area was between 17 and

21 °C in autumn, and the density was negatively correlated with the water temperature (Figure 4). According to relevant reports from the China Meteorological Administration, due to global warming, 2019 was the warmest year in which the ocean temperature was recorded [31]. The synchronous environmental data of this survey showed that the average bottom temperature in the autumn of 2019 and 2020 reached 21 °C. It is speculated that higher temperatures may affect the reproductive development of *C. lucidus*. In addition, the sudden outbreak of coronavirus (COVID-19) in 2019 has had a significant impact on global marine fishing [32]. During the coronavirus infection period, due to transportation restrictions and declining consumer demand, the amount of *C. lucidus* fishing was reduced. It is known that fishing is a key factor influencing fish resources, particularly in the short term [33], which may be the reason for the low density of *C. lucidus* in 2019–2020. However, the decline in industrial and daily activities during the coronavirus has also played a positive role in the recovery of marine species to a certain extent [34]. In addition to the gradual recovery of fishing activities after coronavirus, the resources have gradually increased since 2021.

Spatially, *C. lucidus* species were mainly distributed in the northern and central parts of Zhejiang in spring and in the northern, central and southern parts of Zhejiang offshore regions in autumn. *C. lucidus* is a low-grade carnivorous fish of which plankton are the main food. The food composition includes *Acetes chinensis*, *Myctophum pterotum* and *Calanus sinicus*. The frequency and abundance of most food species were greater in autumn than in spring. These species are distributed mainly along offshore coasts and near islands and reefs [35]. Studies have reported that the abundance of fish species is closely related to the abundance of bait, and the abundant food resources in the sea area are an important reason for fish aggregation [36,37]. In autumn, most of the spawning parents began to feed, and the supplementary group began to fatten. The abundant food in the coastal waters caused *C. lucidus* to gather there. In the same year, the supplementary group of juvenile fish bred in shallow waters moved offshore to southern Zhejiang so that the species were distributed in the north, middle and south of the offshore region in autumn. In addition, the difference in the spatial distribution of *C. lucidus* may be related to climate. The northern and central parts of Zhejiang Province are temperate sea areas, and the southern part of Zhejiang belongs to the subtropical sea area. Due to the influence of the Taiwan warm current, the Kuroshio warm current and its branches, the water temperature in the southern part of the country is higher [38]. These regional differences in climate and environmental conditions may lead to differences in the distributions of *C. lucidus* in the north and south. Many studies have shown that the genetic diversity of *C. lucidus* is high. In China's coastal areas, the population is clearly divided into northern and southern populations near Taizhou, Zhejiang Province. There are obvious differences in gene exchange and habitat utilization patterns [39,40]. It is speculated that most of the *C. lucidus* in northern and central Zhejiang in this study were northern, whereas most of them in southern Zhejiang were southern. This phenomenon of north–south differentiation may also explain the spatial difference in the distribution.

4.2. Effects of Marine Variables on the Distribution of C. lucidus *Species*

The distribution of *C. lucidus* is not only affected by the reproductive habits of parents but also closely related to the marine environment. The results of relative importance showed that the most important factor affecting the distribution of *C. lucidus* in spring and autumn was water depth (Figure 3). Water depth can not only reflect changes in pressure and light intensity but also lead to changes in physical and chemical factors such as water temperature and salinity [41]. The model fitting results and the distribution map (Figures 4 and 5) showed that the tail density of *C. lucidus* was the highest near a water depth of 28~30 m in spring. A high tail density in autumn occurred in the shallow water area near a water depth of 40 m. This occurred because April was the main spawning month for *C. lucidus*, while November was the supplementary population month and a small number of parents spawned in autumn. The suitable habitat depths of fish individuals at different developmental stages are different [42].

In addition to water depth, environmental factors such as bottom salinity, bottom temperature and dissolved oxygen also have important influences on the distribution of *C. lucidus*. Salinity can affect the survival rate of fish eggs and juveniles through osmotic pressure, which plays an important role in the growth and development of fish [43]. The results of the RF model showed that the suitable salinity range for *C. lucidus* in spring was between 27 and 34. The suitable salinity range in autumn was wide, ranging between 13 and 34 (Figure 4). Many studies have shown the characteristics of *C. lucidus* tolerance to a wide range of salinities [44], but Liu [45] reported that the buoyancy of fish eggs and the development of juveniles are closely related to the salinity of the habitat. A high-salinity water environment is often needed to stimulate parent spawning, reproduction and early juvenile growth. Compared with November, bottom salinity had a greater impact on the distribution of *C. lucidus* in April. In April, most *C. lucidus* plans began or were ready to enter the life stage of spawning and reproduction, which may be the main reason why *C. lucidus* often appears in high-salinity waters.

Seawater temperature can not only affect the reproduction and development of fish, but also change the distribution and quantity of bait, which indirectly affects the migration and distribution of fish [46]. This study revealed that the suitable temperatures for *C. lucidus* in offshore waters of Zhejiang Province in spring and autumn were 13~16 °C and 17~21 °C, respectively (Figure 4). Compared with the suitable temperature range for *C. lucidus* in the Pearl River estuary of China [20], these values were slightly lower. This difference may be related to the large scale of water temperature changes and the difference in investigation times between different sea areas. Moreover, studies have confirmed that the *C. lucidus* strains in the offshore region of Zhejiang and the Pearl River estuary belong to different populations [37], and there may be some differences in their adaptability to seawater temperature. In addition, the bottom water temperature had a greater impact on the distribution of *C. lucidus* species in spring. This difference may be related to the combined action of multiple water masses and ocean currents in offshore regions of Zhejiang. In spring, the coastal current in the Zhejiang offshore region gradually increased, and the offshore region exhibited low-temperature characteristics, while the fish were mainly spawning parents. Changes in water temperature strongly influence the gonadal development of *C. lucidus*, so this species is more sensitive to the changes in water temperature. In autumn, affected by the Taiwan warm current and the Kuroshio warm current, the water temperature and salinity in offshore regions of Zhejiang could be maintained at their levels. This approach can provide the necessary environmental conditions for the growth of juvenile fish, thereby reducing the impact of water temperature changes on the distribution of fish species.

Dissolved oxygen is one of the most basic parameters of oceanography. Its concentration difference affects the classification and size of plankton communities, which in turn affects a series of ecological activities in fish [47]. Usually, most fish prefer to live in habitats with high concentrations of dissolved oxygen [48]. In recent years, due to human activities, a large amount of nutrients and industrial wastewater have been transported to estuaries and coastal waters. This leads to the aggravation of eutrophication in offshore waters and a decrease in the dissolved oxygen concentration [49]. In the offshore region of Zhejiang, especially in the Yangtze River estuary and adjacent waters, hypoxia is more common [50], which affects fishery production, community structure and the ecological stability of the sea area. In the present study, the suitable dissolved oxygen concentration range for *C. lucidus* was 5~9 mg/L. The highest tail density in spring was nearly 7 mg/L, and that in autumn was nearly 6 mg/L. It is speculated that *C. lucidus* is a kind of fish that can adapt to low-oxygen environments. *C. lucidus* has a large gill area, abundant hemoglobin, and high water content in their muscle [51]. These physiological characteristics can increase the absorption and transport efficiency of oxygen. The same volume of fish can dissolve more oxygen, thereby reducing the demand for dissolved oxygen. Some low-oxygen-tolerant organisms can use low-oxygen waters to evade natural enemies [52]. In shallow offshore areas with low dissolved oxygen concentrations, some predators and competitors of *C. lucidus* may be excluded, which is also the reason why *C. lucidus* density is greater under lower dissolved oxygen conditions.

5. Conclusions

Surveying revealed that the distribution of *C. lucidus* in offshore regions of Zhejiang, China, varied greatly during different seasons. In spring, it is mainly distributed in the northern and central Zhejiang Sea areas. In autumn, the species is distributed in offshore waters in the north, middle and south of Zhejiang. In this study, the random forest model was used to screen the important driving factors influencing the distribution of *C. lucidus* species, and the relationships between them were fitted and analyzed. Compared with the traditional linear regression model, the RF model has the characteristics of ensemble learning. By constructing many different regression trees to improve the accuracy of the results, the generalization ability of the model is strengthened, and the fitting effect is good. However, the RF model may also exhibit overfitting [53]. Understanding the relationship between species distribution characteristics and the environment is the premise of rational utilization and scientific management. This study revealed that different variables have different effects on the distribution of *C. lucidus* in spring and autumn, suggesting that the species have different levels of adaptability to habitat environments at different developmental stages. Water depth, bottom salinity, bottom temperature and bottom dissolved oxygen have important effects in spring and autumn. Due to the reasons of data acquisition, this study analyzed the influence of only water depth, bottom temperature, bottom salinity, bottom dissolved oxygen, bottom chlorophyll a and bottom pH on the distribution of *C. lucidus*. However, the distribution of food organisms, ocean currents and other factors are also important factors affecting the spatial and temporal distributions of fish. With the continuous improvement in data acquisition, important environmental and biological factors will be introduced into the model to further improve the fitting and prediction accuracy of the model.

Author Contributions: W.X., methodology, formal analysis and writing—original draft preparation. H.Z. (Hongliang Zhang), data curation. H.Z. (Haobo Zhang) and T.W., software analysis. Y.Z., supervision and project administration. W.Z., writing—review and editing, visualization, supervision and project administration. All authors have read and agreed to the published version of the manuscript.

Funding: This study was funded by the Key Technology and System Exploration of Quota Fishing, Ministry of Agriculture and Rural Affairs, Fishery Management Fund Project [Grant number: 36, 2017], and the Zhejiang Fishery Resources Survey Special Project [HYS-CZ-202314].

Institutional Review Board Statement: Not applicable.

Data Availability Statement: The datasets that support the findings of this study are available from the corresponding author upon reasonable request.

Acknowledgments: We thank the staff of Zhejiang Marine Fisheries Research Institute for their help and support in our experiment.

Conflicts of Interest: All authors declare no conflicts of interest.

References

1. Zhao, S.L.; Xu, H.X.; Zhong, J.S.; Chen, J. *Zhejiang Marine Fishes*; Zhejiang Science and Technology Press: Hangzhou, China, 2016; pp. 86–92.
2. Zhu, Y.D.; Zhang, C.L.; Cheng, Q.T. *Fishes of East China Sea*; Science Press: Beijing, China, 1963; pp. 241–243.
3. Zhou, F.; Qian, Z.Y.; Liu, A.Q.; Ma, X.; Ni, X.B.; Zeng, D.Y. Recent progress on the studies of the physical mechanisms of hypoxia off the Changjiang (Yangtze River) Estuary. *J. Mar. Sci.* **2021**, *39*, 22–38. [CrossRef]
4. Gao, F.; Chen, X.J.; Guan, W.J.; Li, G. A New model to forecast fishing ground of *Scomber japonicus* in the Yellow Sea and East China Sea. *Acta Oceanol. Sin.* **2016**, *35*, 74–81. [CrossRef]
5. Hu, W.J.; Du, J.G.; Su, S.K.; Tan, H.J.; Yang, W.; Ding, L.K.; Dong, P.; Yu, W.W.; Zheng, X.Q.; Chen, B. Effects of climate change in the seas of China: Predicted changes in the distribution of fish species and diversity. *Ecol. Indic.* **2022**, *134*, 108489. [CrossRef]

6. Ji, Q.; Xie, Z.L.; Gan, W.; Wang, L.M.; Song, W. Identification and Characterization of PIWI-Interacting RNAs in Spinyhead Croakers (*Collichthys lucidus*) by Small RNA Sequencing. *Fishes* **2022**, *7*, 297. [CrossRef]
7. Song, W.; Gan, W.; Xie, Z.L.; Chen, J.; Wang, L.M. Small RNA sequencing reveals sex-related miRNAs in *Collichthys lucidus*. *Front. Genet.* **2022**, *13*, 955645. [CrossRef]
8. Ou, Y.J.; Liao, R.; Li, J.E.; Gou, X.W. Studies on the spawning period and growth of *Collichthys lucidus* in estuary of Pearl River based on otolith daily annulus. *J. Oceanogr. Taiwan Strait.* **2012**, *31*, 85–88. [CrossRef]
9. Zhang, H.Y.; Lin, L.S. Spatial heterogeneity of *Trichiurus japonicus* and small-scale fish in East China Sea and their spatial relationships. *Chin. J. Appl. Ecol.* **2005**, *16*, 708–711. [CrossRef]
10. Wang, M.; Hong, B.; Zhang, Y.P.; Sun, Z.Z. Spring and summer fish community structure in northern Hangzhou bay. *J. Hydrol.* **2016**, *37*, 75–81. [CrossRef]
11. Zhang, H.L.; Wang, Y.; Liang, J.; He, Z.T.; Zhou, Y.D. Seasonal Variations of the Biological Characteristics and Abundance Density of *Collichthys lucidus* in Zhongjieshan Islands Marine Protected Area. *J. Zhejiang Ocean Univ. Nat. Sci.* **2015**, *34*, 407–410.
12. DeLuca, N.M.; Mullikin, A.; Brumm, P.; Rappold, A.G.; Hubal, E.C. Using Geospatial Data and Random Forest To Predict PFAS Contamination in Fish Tissue in the Columbia River Basin, United States. *Environ. Sci. Technol.* **2023**, *57*, 14024–14035. [CrossRef] [PubMed]
13. Liu, J.C.; Jia, M.X.; Feng, W.D.; Liu, C.D.; Huang, L.Y. Spatial-temporal distribution of Antarctic Krill (*Euphausia superba*)resource and its association with environment factors revealed with RF and GAM Models. *J. Ocean Univ. China* **2021**, *51*, 20–29. [CrossRef]
14. Wu, T.; Liang, J.; Zhou, Y.D.; Xuan, W.D.; Fang, G.J.; Zhang, Y.Z.; Chen, F. The factors driving the spatial variation in the selection of spawning grounds for *Sepiella japonica* in offshore Zhejiang Province, China. *Fishes* **2024**, *9*, 20. [CrossRef]
15. Zhang, S.; Li, M.; Yan, S.; Kong, X.l.; Zhu, J.F.; Xu, S.N.; Chen, Z.Z. Population genetic structure analysis of big head croaker (*Collichthys lucidus*) based on mitochondrial cytochrome b gene sequences. *J. Fish. Sci. China* **2021**, *28*, 90–99. [CrossRef]
16. Wu, C.W.; Wang, W.H. Distribution biology and resource changes of *Collichthys lucidus* in Zhejiang coastal waters. *Marine Fish.* **1991**, *13*, 6–10.
17. Wu, Z.X.; Wu, C.W.; Wang, W.H.; Hu, N.Q. Growth law of *Collichthys lucidus* in Zhejiang offshore research on group composition. *Fish. Sci. Technol. Inf.* **1990**, *6*, 170–174. [CrossRef]
18. *GB/T 12763.6-2007*; Specifications for Oceanographic Survey-Part 6: Marine Biological Survey. General Administration of Quality Supervision, Inspection and Quarantine of the People's Republic of China: Beijing, China, 2007.
19. *GB/T 17378.3-2007*; The Specification for Marine Monitoring-Part 3: Sample Collection, Storage and Transportation. General Administration of Quality Supervision, Inspection and Quarantine of the People's Republic of China: Beijing, China, 2007.
20. Xiong, P.L.; Xu, S.N.; Chen, Z.Z.; Zhang, S.; Jiang, P.W.; Fan, J.T. Spatiotemporal distribution of *Collichthy lucidus* in the Pearl River Estuary and its relationship with environmental factors. *Mar. Sci.* **2022**, *46*, 79–87. [CrossRef]
21. Zou, M.X.; Chen, Y.G.; Song, X.J.; Li, S.F.; Zhong, J.S. Distribution and drift trend of *Collichthys lucidus* larvae and juveniles in the coastal waters of the southern Yellow Sea. *J. Fish. China* **2022**, *46*, 557–568. [CrossRef]
22. Shi, Y.R.; Chao, M.; Quan, W.M.; Tang, F.H.; Shen, X.Q.; Yuan, Q.; Huang, H.J. Spatial variation in fish community of Yangtze River estuary in spring. *J. Fish. China* **2011**, *18*, 1141–1151. [CrossRef]
23. Yang, C.H.; Li, X.D.; Yang, Q.; Wang, Y.B. Application of machine learning methods for estimating the biomass of economically important crabs in the Zhoushan fishery. *Mar. Sci.* **2023**, *47*, 61–70. [CrossRef]
24. Gómez, S.R.; Sánchez, R.A.; García, G.C.; Pérez, J.G. The VIF and MSE in Raise Regression. *Mathematics* **2020**, *8*, 605. [CrossRef]
25. Flores, A.; Wiff, R.; Donovan, C.; Gálvez, P. Applying machine learning to predict reproductive condition in fish. *Ecol. Inform.* **2024**, *80*, 102481. [CrossRef]
26. Rahman, L.F.; Marufuzzaman, M.; Alam, L.; Bari, M.A.; Sumaila, U.R.; Sidek, L.M. Application of Machine Learning to Investigate the Impact of Climatic Variables on Marine FishLandings. *Natl. Acad. Sci. Lett.* **2022**, *45*, 245–248. [CrossRef]
27. Kang, H.; Jeon, D.J.; Kim, D.; Jung, K. Estimation of fish assessment index based on ensemble artificial neural network for aquatic ecosystem in South Korea. *Ecol. Indic.* **2022**, *136*, 108708. [CrossRef]
28. Xu, B.D.; Zhang, C.L.; Xue, Y. Optimization of sampling effort for a fishery-independent survey with multiple goals. *Environ. Monit. Assess.* **2015**, *187*, 252. [CrossRef]
29. Hitomi, O.; Satoshi, S.; Daisuke, A. Modeling the growth, transport, and feeding migration of age-0 Pacific saury *Cololabis saira*. *Fish Sci.* **2022**, *88*, 131–147.
30. Wang, M.; Xu, K.D.; Liang, J. Preliminary analysis of biological characteristics of reproductive stocks of *Collichthys lucidus* in Northern Hangzhou Bay. *J. Ocean Univ. Shanghai* **2018**, *27*, 781–788. [CrossRef]
31. Wang, Y.L.; Wang, J.; Zhou, Y.D.; Xu, K.D.; Jiang, R.J.; Li, Z.H.; Zhu, H.C.; Wang, J.; Cui, G.C. Spatial and temporal distribution characteristics of *Larimichthys polyactis* in Zhoushan fishing ground and the adjacent waters based on two-stage GAM. *Fish. J. Fish. China* **2022**, *29*, 633–641. [CrossRef]
32. Papcunová, V.; Gregáňová, R.H. Global Impacts of COVID-19 on the Financing of Local Self-Governments: Evidence from Slovak municipalities. *SHS Web Conf.* **2021**, *92*, 01038. [CrossRef]
33. Jackson, J.B.; Kirby, M.X.; Berger, W.H.; Bjorndal, K.A.; Botsford, L.W.; Bourque, B.J.; Bradbury, R.; Cooke, R.G.; Erlandson, J.M.; Estes, J.A.; et al. Historical overfishing and the recent collapse of coastal ecosystems. *Science* **2001**, *293*, 629–637. [CrossRef]
34. Al Shehhi, M.R.; Samad, Y.A. Effects of the Covid-19 pandemic on the oceans. *Remote Sens. Lett.* **2021**, *12*, 325–334. [CrossRef]

35. He, Z.T.; Zhang, Y.Z.; Xue, L.J.; Jin, H.W.; Zhou, Y.D. Seasonal and ontogenetic diet composition variation of *Collichthys lucidus* in inshore waters in the north of East China Sea. *Mar. Fish.* **2012**, *34*, 270–276. [CrossRef]
36. Pothoven, A.S. The influence of ontogeny and prey abundance on feeding ecology of age-0 Lake Whitefish (*Coregonus clupeaformis*) in southeastern Lake Michigan. *Ecol. Freshw. Fish* **2020**, *29*, 103–111. [CrossRef]
37. Sun, X.H.; Song, M.P.; Li, Z.G.; Song, Y.; Yuan, X.N.; Dong, B.; Zhang, L.; Zhu, L.X.; Liang, Z.L. Selective feeding of the mullet larvae *Liza haematocheila* during ontogeny in Laizhou Bay, Bohai Sea, China: The importance of small copepods in mesozooplankton as prey. *Front. Mar. Sci.* **2023**, *10*, 1147886. [CrossRef]
38. Cai, R.S.; Chen, J.L.; Huang, R.H. The response of marine environment in the offshore area of China and its adjacent ocean to recent global climate change. *Chin. J. Atmos. Sci.* **2006**, *30*, 1019–1033.
39. Xiao, J.Z.; Zou, Y.; Xiao, S.J.; Chen, J.N.; Wang, Z.Y.; Wang, Y.L.; Jie, X.M.; Cai, Y.M. Development of a PCR-based genetic sex identification method in spinyhead croaker (*Collichthys lucidus*). *Aquac. Rep.* **2020**, *522*, 735130. [CrossRef]
40. Song, N.; Ma, G.Q.; Zhang, X.M.; Gao, T.X.; Sun, D.R. Genetic structure and historical demography of *Collichthys lucidus* inferred from mtDNA sequence analysis. *Environ. Biol. Fish.* **2014**, *97*, 69–77. [CrossRef]
41. Francis, M.P.; Hurst, R.J.; McArdle, B.H.; Bagley, N.W.; Anderson, O.F. New Zealand demersal fish assemblages. *Environ. Biol. Fish.* **2002**, *65*, 215–234. [CrossRef]
42. Jiang, R.J.; Sun, H.Q.; Li, X.F.; Zhou, Y.D.; Chen, F.; Xu, K.D.; Li, P.F.; Zhang, H.L. Habitat suitability evaluation of *Harpadon nehereus* in nearshore of Zhejiang province, China. *Front. Mar. Sci.* **2022**, *9*, 961735. [CrossRef]
43. Bhat, M.; Nayak, V.N.; Chandran, M.D.S.; Ramachandra, T.V. Fish distribution dynamics in the Aghanashini estuary of Uttara Kannada, west coast of India. *Curr. Sci. India* **2014**, *106*, 1739–1744. [CrossRef]
44. Zhang, S.; Jiang, Y.E.; Li, M.; Zhu, J.F.; Xu, S.N.; Chen, Z.Z. Life history of spinyhead croaker *Collichthys lucidus* (Sciaenidae) differentiated among populations from Chinese coastal waters. *Aquat. Biol.* **2022**, *31*, 65–76. [CrossRef]
45. Liu, H.B.; Jiang, T.; Huang, H.H.; Shen, X.Q.; Zhu, J.B.; Yang, J. Estuarine dependency in *Collichthys lucidus* of the Yangtze River Estuary as revealed by the environmental signature of otolith strontium and calcium. *Environ. Biol. Fish.* **2015**, *98*, 165–172. [CrossRef]
46. Lewin, W.C.; Mehner, T.; Ritterbusch, D.; Brämick, U. The influence of anthropogenic shoreline changes on the littoral abundance of fish species in German lowland lakes varying in depth as determined by boosted regression trees. *Hydrobiologia* **2014**, *724*, 293–306. [CrossRef]
47. Rabalais, N.N.; Diaz, R.J.; Levin, L.A.; Turner, R.E.; Gilbert, D.; Zhang, J. Dynamics and distribution of natural and human-caused hypoxia. *Biogeosciences* **2010**, *7*, 585–619. [CrossRef]
48. Maes, J.; Stevens, M.; Breine, J. Modelling the migration opportunities of diadromous fish species along a gradient of dissolved oxygen concentration in a European tidal watershed. *Estuar. Coast. Shelf Sci.* **2007**, *75*, 151–162. [CrossRef]
49. Rabalais, N.N.; Cai, W.J.; Carstensen, J.; Conley, D.J.; Fry, B.; Hu, X.P.; Quiñones-Rivera, Z.; Rosenberg, R.; Slomp, C.P.; Turner, R.E.; et al. Eutrophication-Driven deoxygenation in the coastal ocean. *Oceanography* **2014**, *27*, 172–183. [CrossRef]
50. Wang, H.J.; Dai, M.H.; Liu, J.W.; Kao, S.J.; Zhang, C.; Cai, W.J.; Wang, G.Z.; Qian, W.; Zhao, M.X.; Sun, Z.Y. Eutrophication-Driven hypoxia in the East China Sea off the Changjiang Estuary. *Environ. Sci. Technol.* **2016**, *50*, 2255–2263. [CrossRef]
51. Zhang, Y. Analysis and evaluation of nutritional component in muscle of *Collichthys lucidus*. *Sci. Fish Farming* **2022**, *43*, 75–76. [CrossRef]
52. Sato, K.N.; Levin, L.A.; Schiff, K. Habitat compression and expansion of sea urchins in response to changing climate conditions on the California continental shelf and slope (1994–2013). *Deep Sea Res. Part II* **2017**, *137*, 377–389. [CrossRef]
53. Culter, D.R.; Edwards, T.C.; Beard, K.H.; Hess, K.T. Random forests for classification in ecology. *Ecology* **2007**, *88*, 2783–2792. [CrossRef]

fishes

MDPI

Article

The Factors Driving the Spatial Variation in the Selection of Spawning Grounds for *Sepiella japonica* in Offshore Zhejiang Province, China

Tian Wu [1,2], Jun Liang [1,2,3,*], Yongdong Zhou [1,2,3], Wendan Xuan [1,2], Guangjie Fang [1,3], Yazhou Zhang [1,3] and Feng Chen [1,3]

1 Zhejiang Marine Fisheries Research Institute, Zhoushan 316021, China; wutian530@126.com (T.W.); zyd511@126.com (Y.Z.); xwdalways@163.com (W.X.); guangjiefang@yeah.net (G.F.); 13868237659@163.com (Y.Z.); cf0421223@163.com (F.C.)

2 Marine and Fishery Institute, Zhejiang Ocean University, Zhoushan 316021, China

3 Key Laboratory of Sustainable Utilization Research for Fishery Resources of Zhejiang Province, Zhoushan 316021, China

* Correspondence: jliang@zjou.edu.cn

Abstract: Due to overfishing and marine pollution, the status of fishery resources and the ecological environment in the sea areas have considerably changed. The proliferation and release of fishery resources have become the main means of human intervention to maintain and conserve the resource populations. The annual output of *Sepiella japonica* (*S. japonica*), once one of the four major seafood species in the East China Sea, has now recovered to a level of 4000 t. However, do the traditional spawning grounds of *S. japonica* still exist? Have the spawning grounds shifted? These scientific questions are worthy of attention. Based on the survey data of fishery resources and environment at 120 stations in coastal Zhejiang waters from 2015 to 2022, the spatial and temporal distribution and resource center of gravity of *S. japonica* in springtime were analyzed. Random forest (RF) was used to explain the importance of six environmental variables, including water depth, sea surface temperature, sea surface salinity, chlorophyll a, pH, and dissolved oxygen. The generalized additive model (GAM) nonparametric smoothing function was used to analyze the relationship between environmental factors and the distribution of *S. japonica* inhabiting the offshore areas of Zhejiang province, and the effects of environmental factors on spawning habitat selection of *S. japonica* were revealed. This study found that there was a significant interannual variation in *S. japonica* resources, with an overall increasing trend in the resource. The spawning grounds were mainly distributed in the Jiushan Islands Marine Reserve, the Dachen Islands Marine Reserve, and their nearby sea areas. The resource peaked at latitude 28.3° N. Additionally, the most important variables affecting the distribution of *S. japonica* were depth of water, followed by sea surface salinity, pH, dissolved oxygen, sea surface temperature, and chlorophyll a. *S. japonica* mainly inhabited sea areas with a depth of 15~25 m and a sea surface salinity of 26~32. When the pH ranged from 7.6 to 8.3, dissolved oxygen ranged from 6 to 9 mg/L, sea surface temperature ranged from 14 to 17 °C, and chlorophyll a ranged from 2.5 to 5 μg/L, *S. japonica* was more likely to be present. This study provides insights into the spatial distribution of *S. japonica* in offshore Zhejiang province, offering a reference for the rational utilization and scientific protection of this resource.

Keywords: *Sepiella japonica*; spatiotemporal distribution characteristics; resource center; marine environment; random forest; generalized additive model

Key Contribution: This study analyzed the spatial and temporal distributions and center changes of *Sepiella japonica* and the effects of environmental factors on its distribution based on the 2015 to 2022 spring (April) fishery resources survey data in offshore Zhejiang province, China. The findings of this study offer significant insights for the protection and restoration of *Sepiella japonica* spawning grounds, resource conservation, and management.

Citation: Wu, T.; Liang, J.; Zhou, Y.; Xuan, W.; Fang, G.; Zhang, Y.; Chen, F. The Factors Driving the Spatial Variation in the Selection of Spawning Grounds for *Sepiella japonica* in Offshore Zhejiang Province, China. *Fishes* **2024**, *9*, 20. https://doi.org/10.3390/fishes9010020

Academic Editors: Heng Zhang, Ying Xue, Yunzhi Yan and Luoliang Xu

Received: 29 November 2023
Revised: 25 December 2023
Accepted: 29 December 2023
Published: 31 December 2023

1. Introduction

Due to factors such as high-intensity fishing activities and poor fishery management, global catch species have gradually shifted from long-lived, high-trophic, bottom-feeding fish species to short-lived, low-trophic pelagic, invertebrate species [1]. Cephalopods are among the most promising fisheries species globally due to their short lifecycles, rapid growth, and adaptability to climate change, leading to a continuous increase in their production over the past 60 years. Future research on cephalopod and related industries is expected to remain a long-term topic of interest. Currently, China is one of the leading countries in cuttlefish fishing. In recent years, changes in marine habitats and fishery structures resulting from human activities and climate change have led to a decline in the cuttlefish species resources and changes in their spatial distribution, which have affected the structure and function of marine ecosystems [2–4].

Sepiella japonica (*S. japonica*) is the main economic species of cuttlefish in China, Japan, and South Korea. It is distributed along the coast of Russia in the northwestern Pacific Ocean, from the Kanto area of Honshu Island in Japan to South Korea, the East China Sea and the South China Sea, mainly in the offshore area of Zhejiang, the sea area in eastern Fujian and Guangdong in China, and the sea area west of Japan [5,6]. *S. japonica* is the most populous and productive cuttlefish along the coast of China, and it was once named one of the four major seafoods in the East China Sea, together with *Larimichthys crocea*, *Larimichthys polyactis*, and *Trichiurus japonicus* [7]. In 1959, the highest annual production of *S. japonica* in the East China Sea District exceeded 70,000 t. The main distribution areas were in the Zhejiang's offshore area, the seas of eastern Fujian, and the Zhejiang's offshore area, the latter of which had the largest production, accounting for about 80% of the total production of this species of cuttlefish in the offshore areas of China [5,6,8]. From the mid-1970s to the early 1980s, with the diversification of operating tools and methods (cuttlefish cages, trawls, paired nets, custom-made tension nets, etc., with motor-sailed fishing vessel trawling gradually becoming the main operation method for catching cuttlefish), the number of fishing vessels increased sharply, and the fishing effort continuously increased, with a total fishing effort that was more than double that of the 1960s. In 1993, the development of single-trawler operation in Zhejiang province caused significant changes in the resource status of cuttlefish. These changes affected not only the quantity and spatial distribution of resources but also the replacement of catch species. Among them, the dominant species of cuttlefishes have undergone wide changes compared with those before the 1990s. *Sepia esculenta* and *Sepia kobiensis* have replaced *S. japonica* as the most abundant species, while it has been difficult for *S. japonica* to form a fishing season [5,9].

All organisms living in specific marine environmental conditions are closely tied to their living environment, and the importance of different environmental factors on biological effects significantly differ. Changes in fish resources and the distribution of fishing grounds are often related to multiple factors, such as time, space, and the marine environment [10]. *S. japonica* is an offshore migratory species, with parents that reproduce in groups near the coast from April to May. The reproductive period is a crucial stage in the life cycle of *S. japonica*; recruitment is primarily determined by the number of fertilized eggs and the survival rate of offspring [11]. *S. japonica* makes some choices regarding the spawning grounds; therefore, this resource is susceptible to the influence of the natural environment and human factors [5]. This study analyzed the spatiotemporal distribution characteristics and resource centers of *S. japonica* during the reproductive period using resource survey data from the Zhejiang's offshore area between 2015 and 2022. This study investigated the effect of environmental elements on their distribution using random forest (RF) and a generalized additive model (GAM), combined with environmental data. This study aims to establish a scientific foundation for the conservation and management of *S. japonica* resources, serving as a reference for the sustainable use of offshore cephalopod resources.

3. Results and Analysis

3.1. Distribution Characteristics of *S. japonica*

3.1.1. The Interannual Distribution Characteristics of *S. japonica*

The annual variation in *S. japonica* resources was obvious, and the overall resources fluctuated and increased. Among them, the relative resource abundance fluctuated greatly from 2015 to 2019, and the resources were relatively stable from 2020 to 2022. From 2015 to 2022, the spatial distribution pattern of *S. japonica* was roughly the same, which was concentrated in the central and south-central islands and reefs in the study area (Figure 2).

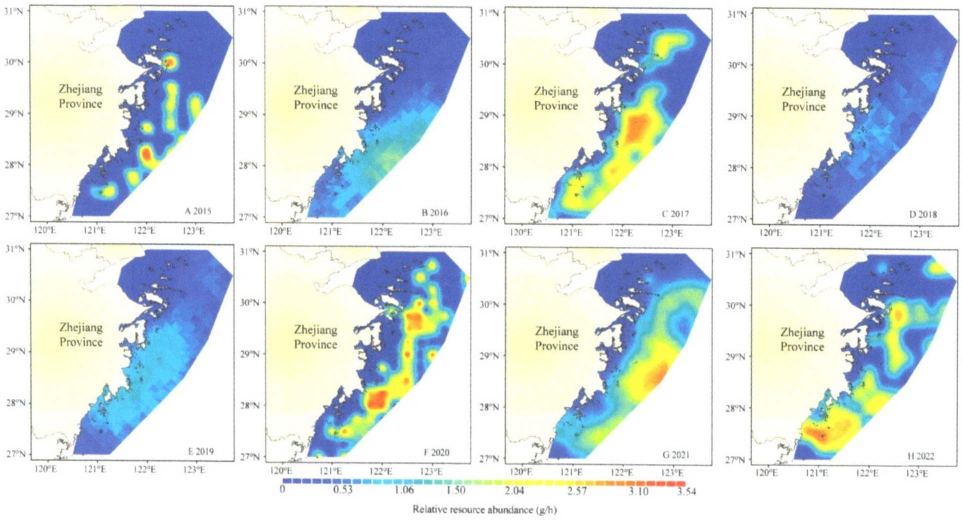

Figure 2. Spatial distribution of *S. japonica* from 2015 to 2022. ((**A**): 2015; (**B**): 2016; (**C**): 2017; (**D**): 2018; (**E**): 2019; (**F**): 2020; (**G**): 2021; (**H**): 2022).

3.1.2. Resource Center of *S. japonica*

From 2015 to 2022, the distribution range of the resource center of *S. japonica* was 122°00′ E–122°30′ E and 28°20′ N–29°00′ N, mainly located near the waters of the Jiushan Islands protected spawning ground area and Dachen Islands protected spawning ground area in the central part of the Zhejiang's offshore area. From the perspective of the spatial change trajectory of the resource center: in 2015–2016, the resource center of *S. japonica* shifted to the southwest; in 2017–2018, the resource center shifted to the northeast; the change pattern in 2019 was consistent with that in 2016; and in 2020–2022, the resource center shifted in the southwest offshore direction (Figure 3).

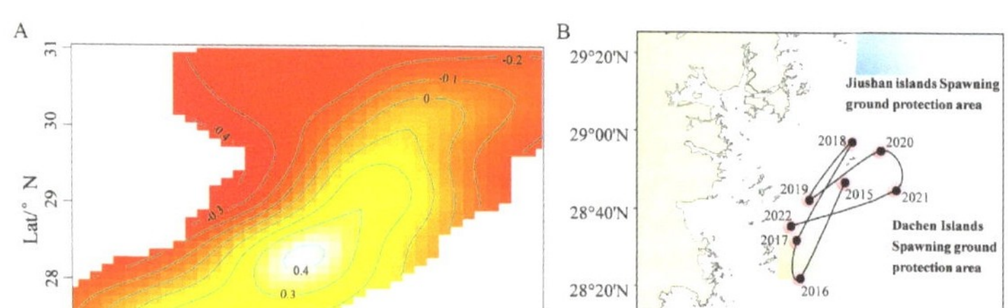

Figure 3. Distribution of *S. japonica* resource center: (**A**): 2015–2021 spatial distribution of *S. japonica* resource center; (**B**): distribution and change trajectory of resource center of gravity each year.

3.2. *FR Analysis*

3.2.1. Multicollinearity Test

The variance inflation factor (VIF) is used to measure the correlation between explanatory variables. A higher VIF value indicates a stronger correlation, and collinearity becomes more severe. A VIF value greater than 10 generally indicates a strong collinearity problem. After conducting tests, it was found that the VIF values of each explanatory variable in this study were less than 10 (Table 2), indicating the absence of any serious collinearity problem. Therefore, the model could be expanded without any concern.

Table 2. Multicollinearity test of explanatory variables.

Parameter	Explanatory Variable					
	Depth	SST	SSS	Chl-a	pH	DO
VIF	2.664	1.258	1.806	1.148	1.102	2.776

3.2.2. Importance of Explanatory Variables

The importance of the environmental factors regarding the distribution of *S. japonica* was ranked using the RF output (Figure 4). MDG represents the heterogeneous effects of the observed values of the nodes on the classification trees and environmental factors. Its value represents and is proportional to the importance of the explanatory variables. It is therefore an important indicator for determining the importance of each factor. The environmental factors affecting the distribution of *S. japonica* distribution were ranked as Depth, SSS, pH, DO, SST, and Chl-a from first to last. The model's accuracy, precision, recall, and F1 score were calculated using the data from the confusion matrix (Table 3). The mean accuracy value of 0.74 indicated a high proportion of correct model tests, and the mean precision value of 0.78 was also noteworthy. The mean recall of 0.90 showed that the model could correctly identify most positive samples. The results indicated that the model's accuracy improved when testing for the presence of *S. japonica*. The mean F1 score of 0.83 indicated that the model achieved a good balance between accuracy and recall (Table 3). The random forest model performed well in this test.

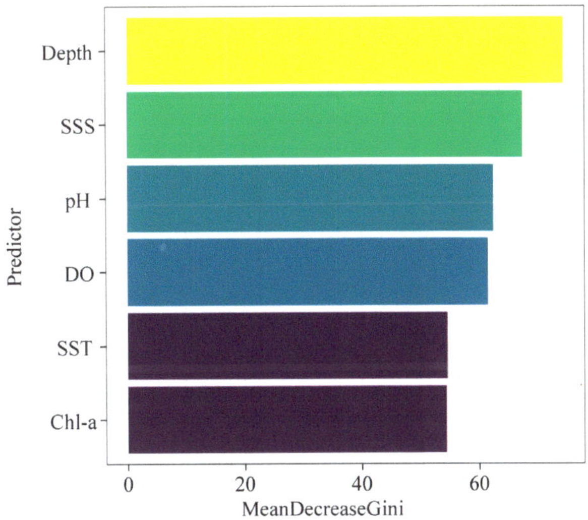

Figure 4. Environmental factor importance sorting results of RF model.

Table 3. RF model's performance parameters.

	Accuracy (95% CI)	Precision	Recall	F1 Score
RF 1 (2015)	0.88 (0.80, 0.93)	0.91	0.95	0.93
RF 2 (2016)	0.79 (0.71, 0.86)	0.79	0.99	0.88
RF 3 (2017)	0.76 (0.67, 0.83)	0.79	0.86	0.82
RF 4 (2018)	0.85 (0.77, 0.91)	0.88	0.95	0.92
RF 5 (2019)	0.63 (0.54, 0.72)	0.78	0.70	0.74
RF 6 (2020)	0.73 (0.64, 0.80)	0.74	0.95	0.83
RF 7 (2021)	0.65 (0.56, 0.73)	0.65	0.96	0.78
RF 8 (2022)	0.65 (0.56, 0.73)	0.67	0.87	0.76

3.3. GAM Analysis

Figure 5 displays the results of the GAM analysis of six explanatory variables: Depth, SSS, pH, DO, SST, and Chl-a. The species were mainly found at water depths ranging from 15 to 60 m, being concentrated at depths of 15~25 m. The optimal sea surface salinity ranged from 26 to 32, while the optimal pH range was 7.6–8.3. The optimal DO range was 6~9 mg/L, and the optimal sea surface temperature range was 14~17 °C. The optimal range of Chl-a was 2.5~5 μg/L. The results of the test on *S. japonica* and environmental variables (Table 4) indicate that Depth, SST, DO, and pH were significantly correlated with the distribution of *S. japonica* ($p < 0.05$). Specifically, the chi-squared values of pH and DO for *S. japonica* were 18.325 and 23.665, respectively ($p < 0.01$). The chi-squared value of the depth of the distribution of *S. japonica* was 39.384 ($p < 0.001$). SSS and Chl-a were not significantly correlated with the distribution of *S. japonica* ($p > 0.05$).

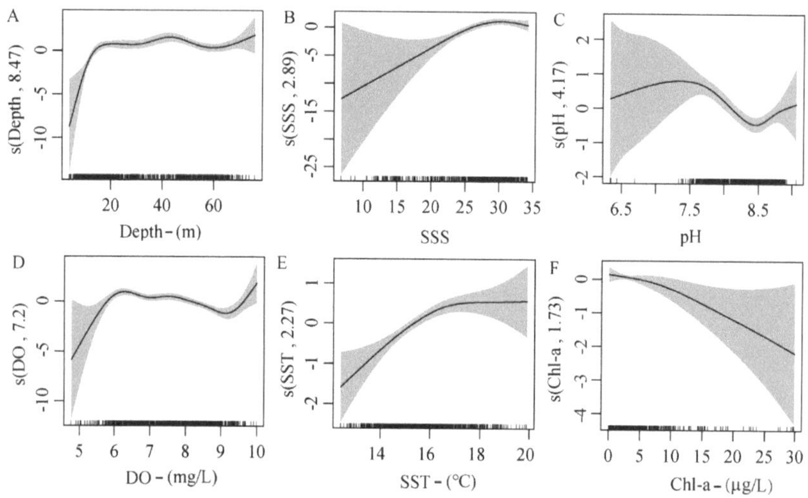

Figure 5. GAM-derived Effects of environmental factors on the *S. japonica*, based on the model constructed using: (**A**) Depth, (**B**) SSS, (**C**) pH, (**D**) DO, (**E**) SST, and (**F**) Chl-a. The plot includes a spline function (s) and the corresponding edf value. The shaded area represents the point-by-point standard deviation of the fit function, which indicates the upper and lower bounds of the 95% confidence interval.

Table 4. GAM verification results.

Variable	edf	Ref.df	Chi.sq	p-Value
Depth	8.764	8.969	39.384	<0.001 ***
SSS	3.457	4.273	10.586	0.05691
pH	3.020	3.764	18.325	0.00139 **
DO	5.584	6.732	23.665	0.00117 **
SST	1.001	1.002	17.239	<0.001 ***
Chl-a	1.730	2.170	2.065	0.42100

*** significant correlation at the 0.001 level; ** significant correlation at the 0.01 level.

4. Discussion

4.1. Spatial–Temporal Distribution Characteristics of S. japonica

The characteristics of the habitat and distribution of *S. japonica* are complex and influenced by various factors. In spring, the northeasterly wind strengthens and the offshore water temperature rises. *S. japonica* migrates from the open sea to the offshore area for reproduction, showing a northwestward trend [5]. The distribution of *S. japonica* in the offshore area of Zhejiang from 2015 to 2022 followed the same trend as the water depth contour and sea surface salinity contour, mainly in the northeast–southwest direction. It was primarily distributed in sea area with water depth contours of 14~56 m and salinity of 26~34, and the fish resources showed an overall increasing trend (Figure 2; Figure 6). Since 2013, Zhejiang province has taken fishery restoration and revitalization actions. The implementation of the summer fishing moratorium system, the ecological restoration measures and the release tens of billions of aquatic species seedlings have had a decided effect on the replenishment of the *S. japonica* resources in the offshore area of Zhejiang.

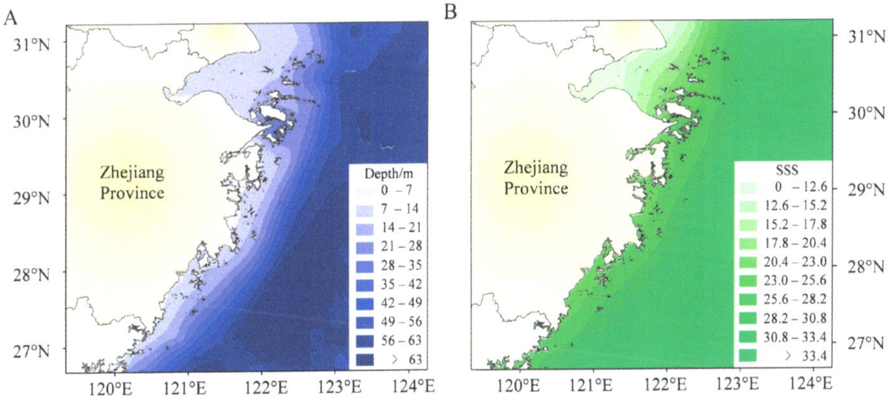

Figure 6. Sea depth and sea surface salinity contours in Zhejiang. (**A**): Mean bathymetry contours from 2015 to 2022; (**B**): mean surface salinity of seawater contour from 2015 to 2022.

For marine life, global climate change is mainly reflected in changes in sea surface temperatures. The results showed that the mean sea surface temperature in the spring of 2015 to 2022 was increasing. According to the results of the analysis of the trend, the relative resource abundance of *S. japonica* increased with slight increases in sea surface temperature. As the surface water temperature continued to rise, the amount of the resource showed a downward trend (Figure 7). A small increase in temperature was found to have some stimulating effect on the resources of *Octopus ocellatus* by Cui et al. [19]. However, if the temperature rises too high, it can have an adverse effect on *O. ocellatus* resources, indicating that cephalopods are sensitive to environmental changes. The ocean warming model driven by climate in China is parallel to and is exceeding the trend of the rising global average temperature. Different species or different life stages of a species have different heat tolerance limits. Although warmer water temperatures may be beneficial to the growth and reproduction of some species, cephalopod responses to climate change can be extremely complex [20]. Changes in ocean temperature may lead to earlier life cycle events such as cephalopod migration, maturation, and reproduction. At the same time, changes in ocean temperature may affect population flow and abundance by changing the type and amount of the food eaten by cephalopods and the characteristics of their spawning habitats [21–23]. As an annual cephalopod, climate change may have profound effects on the species' biological, physiological, and behavioral changes, which can further impact its distribution.

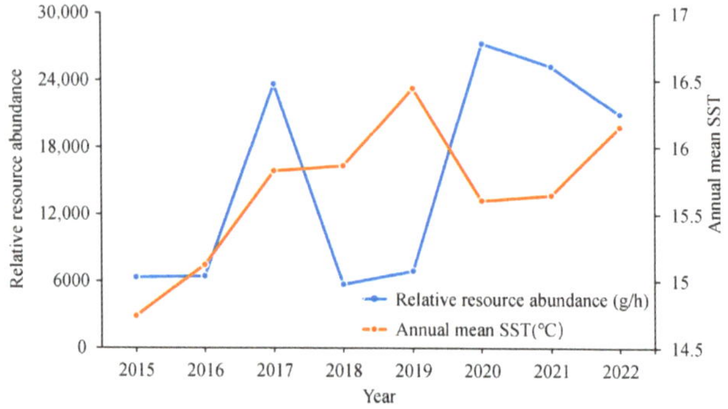

Figure 7. Resources of *S japonica* and sea surface temperature in Zhejiang's offshore area.

4.2. *Trends of the Change in the Resource Center of* S. japonica

Human activities and climate change may alter the distribution of a population and the changes in the location where a species is abundant [23]. Since the late 1970s, due to the trapping of spawning groups outside the spawning grounds, the mass fishing of overwintering cuttlefish, and the destruction of the spawning grounds, the *S. japonica* resources have decreased and their group composition and fishing status have changed (e.g., small central fishing ground, gorgonians, seaweed, and other spawning substrates are insufficient) [5,24,25]. Additionally, climate change is causing the geographic range of species to expand toward the poles [26]. Based on an analysis of the resource status of *S. japonica* from south to north (Nanji, Beiji, Pishan, Yushan, Jiushan, Zhongjieshan, and Shengsi Sea areas) in Zhejiang's offshore area from 1964 to 1984, Tang et al. [25] found that the fishing season in the spawning grounds of *S. japonica* in the offshore area of Zhejiang gradually disappeared in the late 1970s due to the decline of *S. japonica* stocks. Despite being a traditional spawning ground for *S. japonica*, the *S. japonica* fishing has not been prosperous in the Zhongjieshan fishing area, but the *S. japonica* production in Shengsi and the North Sea area has improved. The central fishery has tended to move northward from the small Zhoushan fishery [27]. Ni et al. [28] evaluated the *S. japonica* resources in Zhejiang's offshore area, and the results showed that the maximum sustainable yield in northern Zhejiang was 3.48 times that in southern Zhejiang, and the populations in southern Zhejiang and eastern Fujian could not form a fishing season. Chen et al. [29] showed that the density of cephalopod resources was high near 30.5° N in the coastal waters of Zhoushan and 28.5° N in the coastal waters of Taizhou, followed by the sea area off the Nanji Islands (121° E, 27°–27.5° N). In the large coastal area of Zhejiang province, the resource center of *S. japonica* shifted from the north to the central part, but this does not mean that the resource center of *S. japonica* shifted southward. Compared with the historical studies of 1964–1986, the resource center showed a northward trend. In general, the ecological environment of traditional spawning grounds has been deterioration, and the natural substrates used for attachment, such as gorgonian and *Sargassum fusiforme*, have been severely damaged, no longer meeting the reproduction and habitat needs of *S. japonica*, which has thus shifted its resource center in search of more suitable spawning grounds [24,30]. At the end of 1980s, the growth rate of cuttlefish accelerated, the sexual maturity advanced, and the fishing season gradually advanced. The accelerated individual growth and early sexual maturity, on the one hand, showed that the population was in a recession stage; on the other hand, it also showed that the time when cuttlefish enter the spawning ground also advanced. Under the influence of factors such as sea and meteorological conditions, the spawning ground of *S. japonica* has migrated. For example, in April every year, the surface water of the East China Sea enters the coastal waters of Zhejiang and converges with the runoff coastal waters of the Yangtze River and Qiantang River under the action of mixing. The water temperature is high in the south and low in the north, and high in the outside and low in the inside. After the *S. japonica* enters the spawning ground from the wintering ground, it quickly enters the breeding state under suitable water temperature and salinity conditions. At the same time, changes in habitat and the marine food web caused by global warming and extreme weather (El Niño and La Niña events) further affect the selection of spawning sites and changes in resource centers of *S. japonica* [31,32].

4.3. Effects of Environmental Factors on the Habitat and Distribution of S. japonica

Changes in the marine environment have a major impact on the habitat and reproduction of species, affecting their distribution and abundance [33]. Guerra et al. [34], regarding the spawning habitat of cuttlefish in northwestern Spain, showed that substrate type, depth, temperature, season, and spatial orientation (latitude and longitude) have significant effects on the distribution of cuttlefish eggs. Most of the exploited cephalopod species have short life cycles, typically only 1–2 years, and their short life cycles, high metabolic rates, and high growth rates are associated with a high degree of plasticity in life history traits and sensitivity to environmental change [35,36].

S. japonica is an offshore warm-water benthic animal. In spring, it migrates from the coastal deep-water overwintering ground (60~80 m) to the spawning ground in shallow coastal water, and where they spawn in the surface-to-50 m water layer [5]. Water depth, as an important factor affecting the distribution of the cuttlefish, has an important effect on its habitat, reproduction, and migration. Li et al. [24] showed that the water depth suitable for egg attachment in the Jungshan Islands was 10~20 m. The concentration of *S. japonica* in this study was at water depths of 15-25 m, which is consistent with the results of historical studies. Compared with deep water, the natural adhesion base in the shallow sea and near the islands and reefs is rich in resources, enabling easy spawning and attachment. In addition, shallow seas usually provide more biological food resources. For example, small fish and shrimp are mainly distributed in shallow waters and near islands and reefs. At the same time, *S. japonica* tends to be weak light, and afraid of strong light (preferring weak light), so it generally lives in the sea areas near coastal islands around rocks and seaweed during the breeding period [37].

The habitat of *S. japonica* is limited by salinity. In spring, the overwintering adults perform reproductive migration from the deep high-salt and low-temperature waters of the open sea to the shallow seas in coastal areas with higher temperature and lower salinity. In late autumn and early winter, the juvenile cuttlefish hatch after spawning and scatter in the high-temperature and low-salt waters along the coast to grow into overwintering adults. Then, they gradually swim toward high-salt sea areas [5] as they mature and forage. Salinity is an important ecological factor in marine ecosystems, and its fluctuation affects the behavior, immunity, and other physiological responses of marine organisms [38]. For cephalopods, salinity changes cause a series of physiological stress responses, including abnormalities in osmotic regulation, immunity, and hormone secretion [39]. *S. japonica* has a wide salt tolerance range and has a good low-salt tolerance. If the salinity exceeds the appropriate values, abnormal behavioral changes such as severe ink-jet phenomena occur, which result in energy consumption and serious damage to the tissues and organs that affect reproduction.

Studies have shown that ocean acidification has negative effects on larval development and calcification in mollusks [40]. For cephalopods, the binding of oxygen to the cephalopod respiratory protein, hemocyanin, is highly dependent on pH, and the increase in seawater

CO_2 concentration may further negatively affect cephalopod respiratory metabolism [41,42]. The pH tolerance range of *S. japonica* is narrow, between 7.6 and 8.3, and the oxygen consumption rate of *S. japonica* increases in more acidic environment, which is also not conducive to bivalve growth. In addition, gorgonian is an important substrate for *S. japonica* spawning, accounting for about 60% of *S. japonica* spawning attachment in natural waters. Corals are very sensitive to changes in the marine environment [43]. The negative effects of ocean acidification on corals are multifaceted, affecting not only the growth and survival of the corals but also causing significant damage to the entire marine ecosystem.

Due to their fast growth, cephalopods have a high demand for oxygen [36]. All activities of cephalopods require oxygen consumption, and the oxygen consumption rate of *S. japonica* is positively proportional to the water temperature. In areas with low dissolved oxygen levels, the growth of *S. japonica* slows, which reduces its metabolic rate and affects individual activities [44–48]. As the seasons change, the water temperature gradually increases, which is beneficial for the growth and development of *S. japonica* in the areas of the sea with higher dissolved oxygen concentrations. In the short life of most cephalopods (except Nautilus), they lay eggs later in their lives, die soon after the end of parental reproduction, and the spawning grounds turn into nurseries for the offspring. Sea areas with a higher dissolved oxygen concentration not only provide a more suitable ecological environment for the cuttlefish parents but also sustain the hatchlings of a large number of cuttlefish eggs and ensure the survival of the larvae.

Wang et al. found that sea surface temperature had a significant effect on cuttlefish fishing [49]. As an important marine environmental factor, sea water temperature directly affects the time when cuttlefish enters the spawning ground and the rate of reproduction, thus affecting the distribution of this resource. There is a certain correlation between the concentration of chlorophyll a and the distribution of cephalopods in the water. For example, the distributions of *Loligo edulis*, *Sepia kobiensis*, *Sepia esculenta*, and *Octopus variabilis* fishing grounds are influenced by the concentration of chlorophyll a. Fish and shrimp (which feed on algae and other organic matter) are important food for cephalopods. Chlorophyll a, as an important indicator of phytoplankton abundance, can provide favorable growth conditions for planktonic organisms [50]. During the reproductive period, *S. japonica* shows decreased feeding intensity but still requires a large amount of food to promote gonad development and ensure sufficient energy for reproduction.

5. Conclusions

Cephalopods actively respond to changes in the marine environment, allowing for both active migration and passive utilization. As one of the most promising species for fisheries, as increasingly emphasized by fishery managers around the world, and the increase in global warming, fishery productivity and the distribution of fishery resources will change [36]. In this study, we investigated the interannual variation in and spatial distribution characteristics of the abundance of *S. japonica* and its response to environmental factors in the offshore waters of Zhejiang province. The results showed that *S. japonica* concentrated in the Dachen Sea area, and the spawning ground had shifted compared with the historical locations. The environmental factors affecting the distribution of *S. japonica* were water depth, salinity, pH, dissolved oxygen, water temperature, and Chl-a in descending order of importance. Among these factors, water depth, water temperature, dissolved oxygen, and pH were significantly correlated. The results of this study can be used to characterize the resource quantity and spawning area locations of *S. japonica* to a certain extent, providing theoretical guidance for resource conservation, spawning habitat protection and restoration, and the scientific management of *S. japonica* resources within the context of global climate change. There are many factors influencing changes in species distribution, and the factors included in this study may not be comprehensive. Therefore, in future work, we will continue to identify more influencing factors and comprehensively consider the effects of these additional factors on the spatial and temporal distribution of *S. japonica*. This will further improve the prediction accuracy and provide theoretical support for the scientific conservation, propagation, and release of *S. japonica*.

Author Contributions: T.W. and J.L.: methodology, formal analysis, and writing—original draft preparation. W.X., G.F., Y.Z. (Yazhou Zhang) and F.C.: software and data curation. J.L. and Y.Z. (Yongdong Zhou): writing—review and editing, visualization, supervision, and project administration. All authors have read and agreed to the published version of the manuscript.

Funding: This study was supported financially by funds from the National Key R&D Program of China (2019YFD0901204, 2020YFD0900804, 2019YFD0901205) and National Natural Science Foundation of China Youth Science Foundation Project (31702346).

Institutional Review Board Statement: Not applicable.

Data Availability Statement: The datasets that support the findings of this study are available from the corresponding author upon reasonable request.

Acknowledgments: The authors express their gratitude to the staff of the Zhejiang Marine Fisheries Research Institute and Key Laboratory of Sustainable Utilization Research for Fishery Resources of Zhejiang Province for help with conducting the experiments.

Conflicts of Interest: The authors declare no conflicts of interest.

References

1. Pauly, D.; Christensen, V.; Dalsgaard, J.; Froese, R.; Torres, F.S. Fishing Down Marine Food Webs. *Science* **1998**, *279*, 860–863. [CrossRef] [PubMed]
2. Pyke, C.R. Habitat loss confounds climate change impacts. *Front. Ecol. Environ.* **2004**, *2*, 178–182. [CrossRef]
3. Lotze, H.K.; Lenihan, H.S.; Bourque, B.J.; Bradbury, R.; Cooke, R.G.; Kay, M.C.; Kidwell, S.M.; Kirby, M.X.; Peterson, C.; Jackson, J.B. Depletion, degradation, and recovery potential of estuaries and coastal seas. *Science* **2006**, *312*, 1806–1809. [CrossRef] [PubMed]
4. Mantyka-Pringle, C.; Martin, T.G.; Rhodes, J.M. Interactions between climate and habitat loss effects on biodiversity: A systematic review and meta-analysis. *Glob. Change Biol.* **2011**, *18*, 1239–1252. [CrossRef]
5. Chen, X.J.; Wang, G.Y.; Qian, W.G. *Important Cephalopod Resources and Fisheries in the Coast of China*; Science Press: Beijing, China, 2013; pp. 17–40.
6. Zheng, X.D.; Lin, X.Z.; Liu, Z.S.; Wang, Z.K.; Zheng, W.B. *Sepiella Japonica*; Iglesias, J., Fuentes, L., Villanueva, R., Eds.; Cephalopod Culture; Springer: Dordrecht, NL, USA, 2014; pp. 241–252. [CrossRef]
7. Qiu, J.Y.; Zheng, L.B.; Chi, C.F. Identification, Characterization, and Expression of a PRQFVamide-Related Peptide in Cephalopod *Sepiella japonica*. *Front. Mar. Sci.* **2022**, *9*, 805209. [CrossRef]
8. Wada, T.; Takegaki, T.; Mori, T.; Natsukari, Y. Reproductive Behavior of the Japanese Spineless Cuttlefish *Sepiella japonica*. *Venus (J. Malacol. Soc. Jpn.)* **2006**, *65*, 221–228. [CrossRef]
9. Yan, L.P.; Li, S.F.; Ling, J.Z.; Zheng, Y.J. Study on the resource alteration of commercial cuttlefish in the East China Sea. *Mar. Sci.* **2007**, *4*, 27–31. [CrossRef]
10. Zheng, B.; Chen, X.J.; Li, G. Relationship between the resource and fishing ground of mackerel and environmental factors based on GAM and GLM models in the East China Sea and Yellow Sea. *J. Fish. China* **2008**, *32*, 379–386. [CrossRef]
11. Zhang, X.M.; Wang, L.L.; Zhang, Y.Y.; Luo, G.; Gao, H.Y. Strategy optimization of stock enhancement of golden cuttlefish, (*Sepia esculenta*) based on structural characteristics of reproductive and recruitment populations. *J. Fish. China* **2019**, *43*, 1890–1899. [CrossRef]
12. *GB/T 12763. 6-2007*; Specifications for Marine Surveys-Part 6: Marine Biological Surveys. General Administration of Quality Supervision, Inspection and Quarantine of the People 's Republic of China. Standardization Committee of the People 's Republic of China: Beijing, China, 2008.
13. Breiman, L. Random Forests. *Mach. Learn.* **2001**, *45*, 5–32. [CrossRef]
14. Xu, Z.; Huang, X.; Lin, L.; Wang, Q.; Liu, J.; Yu, K.; Chen, C. BP neural networks and random forest models to detect damage by Dendrolimus punctatus Walker. *J. For. Res.* **2018**, *31*, 107–121. [CrossRef]
15. Zhang, Y.; Xu, B.; Ji, Y.; Zhang, C.; Ren, Y.; Xue, Y. Comparison of habitat models in quantifying the spatio-temporal distribution of small yellow croaker (*Larimichthys polyactis*) in Haizhou Bay, China. *Estuar. Coast. Shelf Sci.* **2021**, *261*, 107512. [CrossRef]
16. Genuer, R.; Poggi, J.; Tuleau-Malot, C. Variable selection using random forests. *Pattern Recognit. Lett.* **2010**, *31*, 2225–2236. [CrossRef]
17. Valavi, R.; Guillera-Arroita, G.; Lahoz-Monfort, J.J.; Elith, J. Predictive performance of presence-only species distribution models: A benchmark study with reproducible code. *Ecol. Monogr.* **2022**, *92*, e01486. [CrossRef]
18. Walsh, W.A.; Kleiber, P. Generalized additive model and regression tree analyses of blue shark (*Prionace glauca*) catch rates by the Hawaii-based commercial longline fishery. *Fish. Res.* **2001**, *53*, 115–131. [CrossRef]

19. Cui, Y.H.; Liu, S.D.; Xu, B.D.; Ji, Y.P.; Zhang, C.L.; Ren, Y.P.; Xue, Y. Potential Effects of Climate warming on Spatial Distribution of *Octopus ocellatus* inhabiting Haizhou Bay in spring. *Period. Ocean. Univ. China* **2023**, *53*, 42–48. [CrossRef]
20. Forsythe, J.W. Accounting for the effect of temperature on squid growth in nature: From hypothesis to practice. *Mar. Freshw. Res.* **2004**, *55*, 331–339. [CrossRef]
21. Borges, F.O.; Sampaio, E.; Santos, C.P.; Rosa, R. Climate-change impacts on Cephalopods: A meta-analysis. Integrative and comparative biology. *Integr. Comp. Biol.* **2023**, *63*, icad102. [CrossRef]
22. Kang, B.; Pecl, G.T.; Lin, L.; Sun, P.; Zhang, P.; Li, Y.; Zhao, L.; Peng, X.; Yan, Y.; Shen, C.; et al. Climate change impacts on China's marine ecosystems. *Rev. Fish Biol. Fish.* **2021**, *31*, 599–629. [CrossRef]
23. Pecl, G.T.; Jackson, G.D. The potential impacts of climate change on inshore squid: Biology, ecology and fisheries. *Rev. Fish Biol. Fish.* **2008**, *18*, 373–385. [CrossRef]
24. Li, J.J.; Guo, B.Y.; Wu, C.W. A Review of the Resource Evolvement and the Way of Restoration of *Sepiella maindroni* in Coastal Waters of Zhejiang Province. *J. Zhejiang Ocean Univ. (Nat. Sci.)* **2011**, *30*, 381–385. [CrossRef]
25. Tang, Y.M.; Wu, C.W. Biological characteristics of *Sepiella maindroni* de rochebrune and changes of the clam-fishing grounds. *J. Zhejiang Coll. Fish.* **1986**, *2*, 165–170.
26. Hughes, L. Biological consequences of global warming: Is the signal already apparent? *Trends Ecol. Evol.* **2000**, *15*, 56–61. [CrossRef] [PubMed]
27. Wu, C.W.; Zhao, S.J.; Xu, D.N. Biology and resource of cuttlefish *Sepia andreana* in zhoushan fishing area. *Oceanol. Limnol. Sin.* **2006**, *37*, 231–237.
28. Ni, Z.Y.; Xu, H.X. The preliminary study of the population of the cuttlefish *Sepiella maindroni* de rochebrune along the east china sea. *Mar. Sci.* **1985**, *4*, 41–45.
29. Chen, F.; Qu, J.Y.; Fang, Z.; Zhang, H.L.; Zhou, Y.D.; Liang, J. Variation of community structure of Cephalopods in spring and autumn along the coast of Zhejiang Province. *J. Fish. China* **2020**, *44*, 1317–1328. [CrossRef]
30. Li, X.J.; Dai, J.S.; Tong, Z.Y. An ecological environment Investigation on the egg-attached area of *Sepiella maindroni* de rochebrune. *J. Zhejiang Coll. Fish.* **1986**, *5*, 121–124.
31. Boavida-Portugal, J.; Guilhaumon, F.; Rosa, R.; Araújo, M.B. Global Patterns of Coastal Cephalopod Diversity Under Climate Change. *Front. Mar. Sci.* **2022**, *8*, 740781. [CrossRef]
32. Pierce, G.J.; Valavanis, V.D.; Guerra, Á.; Jereb, P.; Orsi-Relini, L.; Bellido, J.M.; Katara, I.; Piatkowski, U.; Pereira, J.S.; Balguerías, E.; et al. A review of cephalopod–environment interactions in European Seas. *Hydrobiologia* **2008**, *612*, 49–70. [CrossRef]
33. Anderson, C.I.; Rodhouse, P.G. Life cycles, oceanography and variability: Ommastrephid squid in variable oceanographic environments. *Fish. Res.* **2001**, *54*, 133–143. [CrossRef]
34. Guerra, Á.; Hernández-Urcera, J.; Garci, M.E.; Sestelo, M.; Regueira, M.; Gilcoto, M.; Gonzáleza, Á.F. Spawning habitat selection by the common cuttlefish *Sepia officinalis* in the Cíes Islands (Northwest Spain). *Fish. Res.* **2016**, *183*, 44–54. [CrossRef]
35. Wang, R.; Zhang, R.; Song, P.; Liu, S.; Li, Y.; Li, H. Diversity and Distribution of 18 Cephalopod Species, and Their Link with Some Environmental Factors in the NW Pacific. *Diversity* **2023**, *15*, 694. [CrossRef]
36. Rodhouse, P.G.; Pierce, G.J.; Nichols, O.C.; Sauer, W.H.; Arkhipkin, A.I.; Laptikhovsky, V.V.; Lipiński, M.R.; Ramos, J.E.; Gras, M.; Kidokoro, H.; et al. Environmental effects on cephalopod population dynamics: Implications for management of fisheries. *Adv. Mar. Biol.* **2014**, *67*, 99–233. [CrossRef] [PubMed]
37. Zheng, M.L.; Xiao, Z.H.; Zheng, W.Y.; Chai, M.J.; He, D.R. Studies on the Phototactic behaviour of cuttlefish (*Sepiella maindroni* de Rochebrune). *J. Xiamen Univ. (Nat. Sci.)* **1980**, *3*, 91–99.
38. Luz, R.K.; Martínez-Álvarez, R.M.; Pedro, N.D.; Delgado, M.J. Growth, food intake regulation and metabolic adaptations in goldfish (*Carassius auratus*) exposed to different salinities. *Aquaculture* **2008**, *276*, 171–178. [CrossRef]
39. Zhu, L.Q.; Zhang, X.M.; Zhang, Z.H.; Shi, H.L.; Guo, H.Y. Behavioral and physiological responses of *Sepiella japonica* exposed to salinity changes. *Acta Ecol. Sin.* **2023**, *19*, 1–12. [CrossRef]
40. Parker, L.M.; Ross, P.M.; O'Connor, W.A.; Pörtner, H.O.; Scanes, E.; Wright, J.M. Predicting the response of molluscs to the impact of ocean acidification. *Biology* **2013**, *2*, 651–692. [CrossRef]
41. Pörtner, H.O.; Langenbuch, M.; Reipschläger, A. Biological Impact of Elevated Ocean CO_2 Concentrations: Lessons from Animal Physiology and Earth History. *J. Oceanogr.* **2004**, *60*, 705–718. [CrossRef]
42. Kaplan, M.B.; Kaplan, M.B.; Mooney, T.A.; McCorkle, D.C.; Cohen, A.L. Adverse Effects of Ocean Acidification on Early Development of Squid (*Doryteuthis pealeii*). *PLoS ONE* **2013**, *8*, e63714. [CrossRef]
43. Allemand, D.; Osborn, D.I. Ocean acidification impacts on coral reefs: From sciences to solutions. *Reg. Stud. Mar. Sci.* **2019**, *28*, 100558. [CrossRef]
44. Guerra, Á. Ecology of *Sepia officinalis*. *Vie Milieu* **2006**, *56*, 97–107.
45. Buentello, J.A.; Gatlin, D.M.; Neill, W.H. Effects of water temperature and dissolved oxygen on daily feed consumption, feed utilization and growth of channel catfish (*Ictalurus punctatus*). *Aquaculture* **2000**, *182*, 339–352. [CrossRef]
46. Valverde, J.C.; García, B.G. Suitable dissolved oxygen levels for common octopus (*Octopus vulgaris* cuvier, 1797) at different weights and temperatures: Analysis of respiratory behaviour. *Aquaculture* **2005**, *244*, 303–314. [CrossRef]
47. Wang, C.L.; Wu, D.H.; Dong, T.Y.; Jiang, X.M. Oxygen consumption rate and effects of hypoxia stress on enzyme activities of *Sepiella maindron*. *Chin. J. Appl. Ecol.* **2008**, *19*, 2420–2427.

48. Qian, S.J.; Zhang, Y.; Wang, Y.; Yin, F.; Song, W.H. Respiratory Metabolism, Energy Utilization and Biochemical Responses of Juvenile Common Cuttlefish (*Sepiella maindroni*) to Hypoxia. *Indian J. Anim. Res.* **2019**, *54*, 31–35. [CrossRef]
49. Wang, T.Z.; Han, Q.; Luo, N.J.; Zhang, H.; Wang, Y.B. Catches of several major demersal fish species inhabiting Zhejiang sea area and their relationship with main influencing factors. *Trans. Oceanol. Limnol.* **2021**, *43*, 77–85. [CrossRef]
50. Chen, W.F.; Ye, S.; Yu, Y.; Fan, Q.S.; Qin, S.; Peng, X.; Chen, S.B. Analysis on species composition and ecological niche of cephalopod in the coastal area of south zhejiang. *Acta Hydrobiol. Sin.* **2021**, *45*, 428–435. [CrossRef]

fishes

MDPI

Article

Evaluation Performance of Three Standardization Models to Estimate Catch-per-Unit-Effort: A Case Study on Pacific Sardine (*Sardinops sagax*) in the Northwest Pacific Ocean

Yongchuang Shi [1,2], Haibin Han [3], Fenghua Tang [1,2], Shengmao Zhang [1,2], Wei Fan [1,2], Heng Zhang [1,2,*] and Zuli Wu [1,2,*]

1 East China Sea Fisheries Research Institute, Chinese Academy of Fishery Sciences, Shanghai 200090, China; syc13052326091@163.com (Y.S.)
2 Key Laboratory of Fisheries Remote Sensing, Ministry of Agriculture and Rural Affairs, Shanghai 200090, China
3 College of Marine Sciences, Shanghai Ocean University, Shanghai 201306, China; hhbfishery@163.com
* Correspondence: zhangziqian0601@163.com (H.Z.); wuzl@ecsf.ac.cn (Z.W.)

Abstract: Catch-per-unit-effort (CPUE) standardization in fisheries is a critical foundation for conducting stock assessment and fishery conservation. The Pacific sardine (*Sardinops sagax*) is one of the economically important fish species in the Northwest Pacific Ocean (NPO). Hence, the importance of choosing an appropriate CPUE standardization model cannot be overstated when it comes to achieving a precise relative abundance index for the efficient management of Pacific sardine fishery. This study's main aim was to assess and compare the efficacy of three models, specifically the General Linear Model (GLM), the Generalized Linear Mixed Model (GLMM), and the spatio-temporal GLMM (VAST), in the CPUE standardization for Pacific sardine fishery in the NPO, with the ultimate goal of identifying the most appropriate model. An influence analysis was applied to analyze the impact of individual variables on the disparity among standardized and nominal CPUE, and the main explanatory variables influencing standardized CPUE were identified. A coefficient–distribution–influence (CDI) plot was generated to analyze the impact of the different models on the annual standardized CPUE. Additionally, a simulation testing framework was developed to evaluate the estimated accuracy of the three models. The results indicated that the standardized CPUE and the nominal CPUE exhibited similar trends between 2014 and 2021 for the three models. Compared to the GLM and the GLMM, the VAST demonstrates larger conditional R^2 and smaller conditional AIC, indicating a better performance in standardizing the CPUE for Pacific sardines due to its consideration of spatial and temporal variations. The interaction terms within the three models exert significant influences on the annual standardized CPUE, necessitating their inclusion in the model construction. CDI plots indicate that the spatio-temporal influence of the VAST model exhibits a smaller variation trend, suggesting that the VAST is more robust when standardizing the CPUE for Pacific sardines. Simulation testing additionally demonstrated that the VAST model displays smaller model root mean squared error (RMSE) and bias, establishing it as the superior performer for standardizing CPUE. Our results provide a theoretical basis for the scientific management of Pacific sardines in the NPO and can be extended to CPUE standardization for other small pelagic fish species worldwide.

Keywords: *Sardinops sagax*; CPUE standardization; VAST model; model evaluation; Northwest Pacific Ocean

Key Contribution: This study evaluates and compares the performance of three different models in standardizing the CPUE for Pacific sardines, obtaining accurate and biologically significant indices of resource abundance, establishing a foundation for assessing the status of this fishery resource. Simultaneously, it aims to explore the influence of each explanatory variable on standardized CPUE and identify its primary drivers.

check for updates

Citation: Shi, Y.; Han, H.; Tang, F.; Zhang, S.; Fan, W.; Zhang, H.; Wu, Z. Evaluation Performance of Three Standardization Models to Estimate Catch-per-Unit-Effort: A Case Study on Pacific Sardine (*Sardinops sagax*) in the Northwest Pacific Ocean. *Fishes* **2023**, *8*, 606. https://doi.org/10.3390/fishes8120606

Academic Editor: Susana Franca

Received: 30 October 2023
Revised: 25 November 2023
Accepted: 5 December 2023
Published: 11 December 2023

1. Introduction

Abundance indices are the bedrock of global fishery stock assessments and fishery conservation [1]. The catch-per-unit-effort (CPUE) serves as a commonly utilized metric for revealing relative fluctuations in stock abundance [2]. Currently, the stock assessment of most fisheries in the world relies heavily on fisheries-dependent data [3]. However, the nominal CPUE from fisheries-dependent data is likely to be biased and limited, which may be influenced by factors such as catchability, fishing effort, spatial heterogeneity, and environmental changes [4]. For instance, fishers may increase their efforts in response to declining fish populations, leading to higher CPUE values which may not accurately reflect the true fish abundance. Therefore, the assumption of a proportional relationship between the CPUE and stock abundance is often criticized in fishery stock assessment [5]. CPUE standardization is a critical step in fisheries' stock assessment that involves removing or adjusting for factors that may influence catchability [6]. By standardizing the CPUE, fisheries' managers can obtain more accurate information on fish stock status, which has significant implications for scientific stock assessment and management.

In recent years, many authors have employed various statistical models other than the traditional General Linear Model (GLM) and Generalized Additive Model (GAM) for CPUE standardization research [7]. For example, two recruitment indices were constructed by Hashimoto et al. [8] utilizing pelagic trawl survey data for the Chub mackerel (*Scomber japonicus*) employing a delta-GLM model. Hazin et al. [9] standardized the CPUE of swordfish in the equatorial and southwestern Atlantic Ocean using a Generalized Linear Mixed Model (GLMM). Thorson et al. [10] introduced a spatio-temporal GLMM (referred to as VAST) for the purpose of estimating abundance indices for West Coast groundfish species and found that the model could improve the accuracy of CPUE standardization. The GLM, GLMM, and VAST models were commonly used methods for CPUE standardization in the past and have been successfully applied to various fish species [11]. However, these models have their own advantages and limitations, and there is relatively little research on the comparative estimation performance between the models [12]. Therefore, selecting CPUE standardization models with a higher estimation accuracy by comparing their evaluation performance is crucial for fishery stock assessment and management.

The Pacific sardine (*Sardinops sagax*) is one of the essential target species for commercial fishing in the Northwest Pacific Ocean (NPO) [13,14]. Pacific sardines are a short-lived species, typically living for 6–7 years, and they reach sexual maturity at around one year of age and spawn in large schools near the surface of the water [15,16]. The Pacific sardine is a small pelagic fish, feeding primarily on zooplankton such as copepods, krill, and small fish larvae [17], and it is also a critical prey for larger fish, seabirds, and marine mammals and plays a vital ecological role in the marine ecosystem. Therefore, sustainable management and conservation of this species can positively impact the entire marine food web. Currently, the primary harvesters of the Pacific sardine population in the NPO are China (including Chinese Taipei), Japan, and Russia. In 2021, China reported an annual catch of approximately 237,301 tons of Pacific sardines, representing 22.20% of the global production [18,19]. In addition, the catch percentage of Pacific sardines for China in the NPO has been increasing year by year [20]. With increasing attention from global researchers, the North Pacific Fisheries Commission (NPFC) has officially recognized the Pacific sardine as a priority species, and preliminary fishery stock assessment and management have been conducted [21].

Recently, the potential impact of climate and ocean environmental changes on Pacific sardine populations has been gathered increasing attention. Several researchers have suggested that the distribution and population size of Pacific sardines is highly impacted by environmental conditions. Shi et al. [22] utilized the ensemble distribution model to examine the population variation for Pacific sardine in the NPO, and they pointed out that sea surface height (SSH) and sea surface temperature (SST) were critical environmental variables. Takasuka et al. [23] studied the suitable temperature for the Pacific sardine and found that 16.2 °C is the optimal growth temperature for Pacific sardines. Ito [24] stated that the spawning grounds' temperature of Pacific sardines in the northern and southern Pacific coast of Japan were 14–17 °C and 17–19 °C, respectively. Wada et al. [25] found that the population size of Pacific sardine has experienced drastic fluctuations, closely related to the climate and the oceanic environment Therefore, it is necessary to consider environmental factors when standardizing the CPUE for Pacific sardines. However, currently, research on Pacific sardines mainly focuses on biology [26,27] and potential habitat distribution [28,29], and there are few reports on the CPUE standardization for Pacific sardine fishery. Furthermore, it is crucial to accurately grasp the impact of explanatory variables in a model on the standardized CPUE in CPUE standardization studies [30]. However, previous research has rarely taken this into account.

Due to the urgent need for stock assessment and management of Pacific sardine fishery, we used the GLM, GLMM, and VAST models to analyze the CPUE for the Pacific sardine, gauging the effectiveness of each model to select the most suitable model for standardizing the CPUE for Pacific sardine fishery. This can obtain an accurate abundance index of Pacific sardine resources. Additionally, we conducted an influence analysis to appraise the effect of each variable on standardized indices. Finally, simulation tests were used to assess the effectiveness of the various models in CPUE standardization. To our knowledge, this study marks the inaugural endeavor in this field to standardize Pacific sardine CPUE within the NPO of China, employing an array of standardization models. Our study aims to achieve the following three primary goals: (1) to assess and compare the performance of three models in CPUE standardization; (2) to derive accurate CPUE data and provide support for stock assessment of the Pacific sardine fishery; and (3) to examine how the inclusion of each explanatory variable affects the standardized CPUE and determine its main drivers. The results of this investigation can provide technical support for the stock assessment and management of Pacific sardines in the NPO region.

2. Materials and Methods

2.1. Data Sources

Figure 1 shows the research region, encompassing the geographical range between 30° N and 45° N latitude and between 143° E and 165° E longitude, and the study focuses on the primary fishing season, which occurs from May to November, from 2014 to 2021. The fishery logbook data, acquired from the Technical Group for Trawl-purse seine Fishery within the Distant-water Fishery Society of China, included information on date, longitude, latitude, catch, hauls, and vessel length, with the location data having a spatial resolution of 0.25°. No zero catches were included in the dataset (removed approximately 1% of zero catch data), and the daily vessel catch defined the nominal CPUE.

Based on the references, we selected SST(°C), SSH (m), and SSTG (°C/km) as the environmental factors when constructing the model for Pacific sardines. The SST and SSH, featuring a monthly temporal resolution and a spatial resolution of 0.25°, were acquired from the Copernicus Marine Service. This access took place on the 23rd of November 2022. The SST and SSH are the monthly average values for this grid. Regarding the SSTG, which signifies the temperature change across the ocean surface, it can influence the distribution and behavior of fish species. Additionally, it may affect the distribution of planktonic organisms, thereby impacting the food sources of fish and other marine organisms. Therefore, this environmental factor needs to be taken into account in CPUE standardization. In this study, it represents temperature variations within a 0.25° × 0.25° grid [31]. The variance inflation factor (VIF) measures the degree of multicollinearity among explanatory variables in a regression analysis, which assesses how much the variance of an estimated regression coefficient is inflated due to multicollinearity in the model. The VIF numbers represent the extent to which multicollinearity inflates the variance of the estimated regression coefficients: the larger the value, the higher the degree. According to the VIF test (Table 1), we found that there is no multicollinearity among the explanatory variables [32,33].

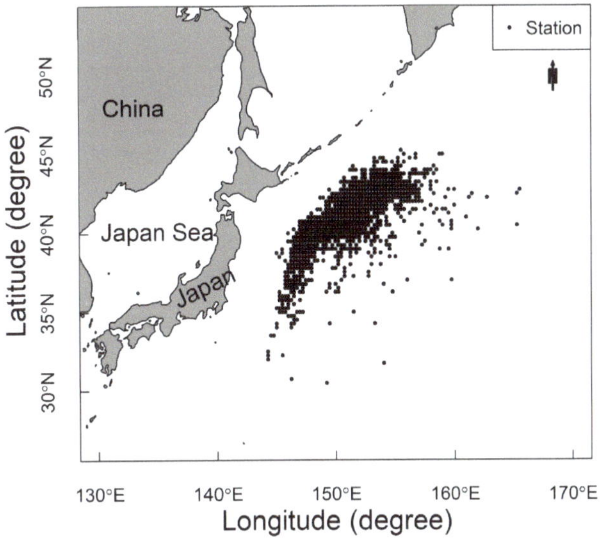

Figure 1. Sampling locations (black dots) of Pacific sardines in the Northwest Pacific Ocean.

In this study, all data processing and model construction were conducted using R (V4.0.3). "Spatial 1", introduced by Hsu et al. [34], was selected as the area stratification approach for the GLM and the GLMM. Using spatial proximity and average CPUE as the criteria, Spatial 1 employed the *k*-medoids algorithm to divide the Pacific sardine CPUE grids into five area strata. (Figure 2) [35]. Briefly, the *k*-medoids algorithm clusters dataset observations into *k* groups, each led by a representative observation (called a "medoid"). It selects these medoids by minimizing distances within clusters, ensuring every data point connects to its closest medoid. This method, using actual observations as the cluster centers, enhances its resilience to outliers. For more details on the method of partitioning area strata, please refer to Hsu et al. [34] and Ono et al. [35].

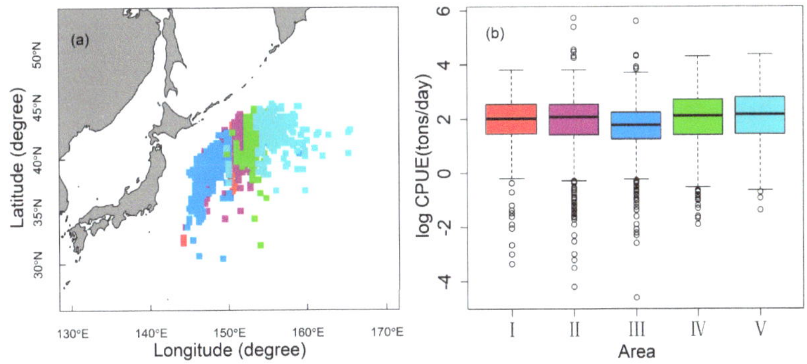

Figure 2. The distribution (**a**) and the boxplots of the CPUE data (**b**) for each area strata. The colors of each area in Figure (**a**) are consistent with Figure (**b**), i.e., the blue represents the area three. The boxes in Figure (**b**) represents the middle 50% of the data; the median line indicates the median of the data, the small circles is very big or small data in CPUE, and the lines extending from the box represent the data outside the box.

Table 1. Explanatory variables test.

Explanatory Variables	Year	Area	Vessel	SST	SSTG	SSH
VIF	1.02	1.01	1.02	1.18	1.06	1.16

2.2. Methods

2.2.1. General Linear Models

The GLM is the most common modeling approach for CPUE standardization, assuming a linear relationship between the response variable and the explanatory factors [36]. In this study, in addition to the explanatory variables shown in Table 1, interaction terms for year and spatial location were also considered for inclusion in the GLM. *Ln(CPUE)* is the response variable; so, assuming that the response variable follows a normal distribution, the GLM model is as follows:

$$Ln(CPUE_i) = \alpha_{Year(i)} + \alpha_{Area(i)} + \alpha_{Vessel(i)} + \alpha_{SST(i)}SST_i + \alpha_{SSTG(i)}SSTG_i + \alpha_{SSH(i)}SSH_i + \alpha_{Year(i)\times Area(i)} \quad (1)$$

where $CPUE_i$ is the predicted CPUE for the *i*th data, and α denotes the estimated coefficient associated with its respective subscript. In the GLM, the year, area, vessel, and year × area were treated as discrete variables, and the *SST*, *SSTG*, and *SSH* were continuous.

2.2.2. Generalized Linear Mixed Models

The GLMM is a commonly employed model that integrates the advantages of both the GLM and the LMM, finding extensive application in CPUE standardization studies [37]. The GLMM is an extension of the GLM model that allows for the inclusion of random variables in the linear predictor for CPUE standardization [38,39]. We fit the GLMM for Pacific sardine CPUE standardization as defined by the equation presented below:

$$Ln(CPUE_i) = \beta_{Year(i)} + \beta_{Area(i)} + \beta_{Vessel(i)} + \beta_{SST(i)}SST_i + \beta_{SSTG(i)}SSTG_i + \beta_{SSH(i)}SSH_i + \beta_{Year(i)\times Area(i)}Year_i \times Area_i \quad (2)$$

where $CPUE_i$ represents the predicted CPUE for the *i*th data, and β denotes the estimated coefficient associated with its respective subscript. We consider year, area, *SST*, *SSTG*, and *SSH* as fixed effects, and treat the other variables as random effects in the GLMM.

2.2.3. Spatio-Temporal GLMM (VAST)

For our spatio-temporal modeling, the VAST R package (version 3.4.0) was applied. VAST's default model framework is founded on a delta-generalized linear mixed model, which separates the catch probability distribution into the following two discrete elements: the probability of encounter and the positive catch rate. Since there were no instances of zero CPUE data in the Pacific sardine dataset, we exclusively considered the positive catch rate component and assigned an encounter probability of one for all the years. We employed a lognormal GLMM to estimate the positive catch rate, utilizing linear predictors and a log-link function. The model also incorporated Gaussian Markov random fields (GMRFs) to effectively capture and consider both spatial and spatio-temporal effects [40].

It is necessary to define spatial knots in advance for computational convenience in the VAST. To evaluate correlations for both the spatial and spatio-temporal effects, we utilized the *K*-means algorithm to partition all the grid cells into 100 spatial knots, which served as reference points (Figure 3). Furthermore, the spatial and spatio-temporal random effects of each grid cell were assumed to originate from the cell's nearest spatial knot [41], and the formula of the CPUE for Pacific sardine using logarithms is detailed as follows:

$$\log(p_i) = \beta(t_i) + \omega(s_i) + \varepsilon(s_i, t_i) + \delta(v_i) + \sum_{j=1}^{n} \gamma(j) X(s_i, t_i, j) \quad (3)$$

where p_i is the predicted CPUE of the *i*th data; the intercept for year t_i, denoted as $\beta(t_i)$, is considered a fixed effect and is assumed to be independent across various years; $\omega(s_i)$ and

$\varepsilon(s_i, t_i)$ are time-invariant spatial variances in location s_i and time-varying spatio-temporal variances for location s_i in year t_i, respectively. $\delta(v_i)$ represents the influence of vessel v_i, and $\delta(v_i) \sim$ Normal$(0, 1)$, $\gamma(j)$ is the jth catchability covariate $X(s_i, t_i, j)$ on location s_i during year t_i, and the variable n represents the total number of catchability covariates.

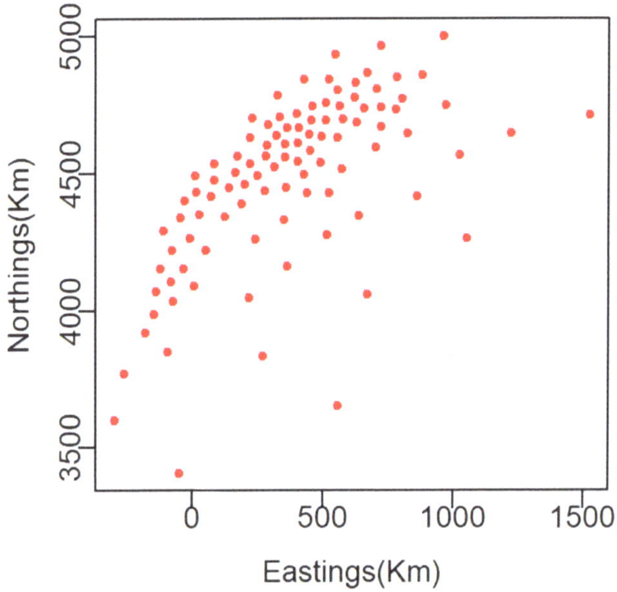

Figure 3. Distribution of the 100 core knots in the VAST. The northing is the "distance traveled or measured northward, especially at sea", and the easting is the distance traveled or measured to the east.

We applied the Laplace approximation as implemented in the Template Model Builder for parameter estimation [42]. This technique facilitated the integration of random-effect parameters and the derivation of the marginal likelihood of fixed-effect parameters. These fixed-effect parameters were subsequently estimated using a maximum likelihood estimation within the R computing environment.

2.2.4. Model Evaluation

In the process of identifying the most suitable models among the GLM, GLMM, and VAST, we employed a forward approach by incrementally incorporating explanatory variables into each of the three models. This led to the creation of models with varying sets of explanatory variables. Subsequently, we utilized the Akaike Information Criterion (AIC) to make the final selections for the optimal GLM, GLMM, and VAST. AIC values are indicative of a model's fit, with smaller values suggesting superior-fitting models.

To compare the performance of different models, the conditional R^2 and conditional AIC (*CAIC*) were employed, since those metrics provide a more accurate measure of model complexity [43,44]. An enhanced model performance in a standardized context is indicated with a higher conditional R^2 and a lower *CAIC* value. The *CAIC* formula is given below:

$$CAIC = -2ln(L) + kln(n) \tag{4}$$

where L denotes the value of likelihood function. In addition, we assessed the normality of the GLM, GLMM, and VAST using histograms of the residuals and quantile–quantile normal probability plots (Q-Q plots).

2.2.5. Standardized CPUE Value

Using the optimal models chosen for the three models, we computed the standardized abundance indices, and we employed the "predict" function to estimate the CPUE for the GLM and GLMM models. Following that, we utilized the aggregate function to calculate the CPUE for each year and area ($CPUE_{Year,Area}$). Lastly, we determined the area-weighted annual CPUE using the formula below:

$$\widehat{CPUE}_{Year} = \sum_{Area=1}^{n} SA_{Area} \times CPUE_{Year,Area} \tag{5}$$

where n represents the total number of areas, which amounts to five. The term SA_{Area} denotes the proportion of the study domain's surface area allocated to a particular area.

Utilizing the VAST model, which includes an explicit spatial correlation component, we estimated the Pacific sardine density in every spatial cell across the study region. The description of the standardized abundance indices for the study area in year t is as follows:

$$\widehat{CPUE}_t = \sum_{s=1}^{n} SA_s \times exp\left(\beta(t) + \omega(s) + \varepsilon(s,t) + \delta(v) + \sum_{j=1}^{n} \gamma(j) X(s_i, t_i, j)\right) \tag{6}$$

where n refers to the number of knots denoted as s; $\beta(t)$ is the year-specific effect at time t; $\omega(s)$ signifies the spatial effect at knot s, and $\varepsilon(s,t)$ stands for the spatio-temporal effect at knot s during time t. Additionally, SA_s represents the surface area associated with the mesh corresponding to knot s.

2.2.6. The Impact of Predictor Variables on Standardized CPUE

In this study, we calculated the influence value for each variable. Subsequently, we generated coefficient–distribution–influence (CDI) plots, which encompass coefficient values, changes in distributions, and annual impacts [30]. These CDI plots served as a valuable resource for gaining a deeper understanding of the annual influence patterns.

To measure the extent of influence, we determined the normalized coefficient (ρ) associated with each explanatory variable using the formula below:

$$\rho = \sum_{i=1}^{n} a_i / n \tag{7}$$

where a_i signifies the anticipated coefficient associated with the ith observation, and n is the overall number of CPUE observations.

Calculating the influence value (AI_y) for a variable in a specific year involves finding the average difference between the coefficients linked to all the observations within that year and the normalized coefficient in a multiplicative GLMM.

$$\delta_y = \frac{\sum_{i=1}^{m} a_i - \rho}{m} \tag{8}$$

where m represents the number of observations during year y. Given that the log-link function was utilized, the AI_y was calculated using the subsequent formula:

$$AI_y = \exp(\delta_y) \tag{9}$$

As referenced in Bentley et al. [30], when the value of AI_y for a variable is larger than one, it signifies the variable's contribution to an increase in the nominal CPUE for that year. Conversely, if the value is below one, it implies the opposite, and an AI_y of one reveals that the variable had no discernible impact on the difference between the nominal and standardized CPUE.

The formula for computing the average influence metric $\overline{AI}$ for a variable across all years is as follows:

$$\overline{AI} = \exp\left(\frac{\sum_{y=1}^{m} \delta_y}{m}\right) - 1 \quad (10)$$

where m signifies the number of years under consideration in the analysis.

We employed the year × spatial interaction random-effect CDI plot format, following the guidance of Bentley et al. [30], for creating CDI plots in both the GLM and GLMM. For the VAST, the initial set of 100 spatial knots was initially classified into 20 groups, referred to as "grouped" knots, based on the coefficients of the spatio-temporal random effect, and for more details, please refer to Hsu et al. [34].

2.2.7. Simulation Testing for Three Models

Simulation testing is a valuable tool for assessing the effectiveness of CPUE standardization models in predicting "true" abundance trends [45,46]. We established a simulation testing framework under a preferential sampling scenario [47]. Utilizing simulated data for Pacific sardines, we applied the three models to standardized CPUE data. Following that, we conducted a comparison between the models' estimated CPUE values and the "true" values as part of the assessment of the models' performance. For a more comprehensive understanding of the way to calculate the "true" values of CPUE and the preferential sampling pattern, we recommend consulting the research conducted by Hsu et al. [34] and Thorson et al. [10].

To assess the models' performance, we employed a comparison of the computed indices with area weightings against the "true" indices. The evaluation criteria included the use of the relative error metric in year y (RE_y), the root mean squared error (RMSE), and the bias metric (β).

The formula for calculating RE_y is as follows:

$$RE_y = \frac{\left(\widehat{I_y} - T_y\right)}{T_y} \quad (11)$$

where $\widehat{I_y}$ represents the estimated CPUE during year y, and T_y represents the true CPUE during year y.

The calculation of the $RMSE$ was performed according to the formula provided by Ducharme-Barth et al. [47], as follows:

$$RMSE = \sqrt{\frac{\sum_{y=1}^{n}\left(\widehat{I_y} - T_y\right)^2}{n}} \quad (12)$$

where n is the total count of years.

Based on the research by Thorson et al. [10] and Ducharme-Barth et al. [47], the bias metric, denoted as coefficient β, was applied in the linear model as follows:

$$\widehat{I_y} = \alpha + \beta \times T_y + \varepsilon_y \quad (13)$$

$$\varepsilon_y \sim Normal\left(0, \sigma_\varepsilon^2\right) \quad (14)$$

where α is the intercept of linear model, and β is the slope to the slope parameter that relates the "true" CPUE to the estimated CPUE [48].

3. Results

3.1. *Diagnostic and Selection of Three Model*

Tables 2–4 illustrate the model decision process for the three models based on the AIC criterion. The values of the AIC suggest that the most effective models for the GLM and the GLMM incorporate the following explanatory factors: year, area, vessel, SST, SSTG, and Year × Area (Tables 2 and 3). As for the VAST, the favored model includes only the covariate SST (Table 4). Compared to the GLM and the GLMM, the VAST exhibits a higher conditional R^2 and smaller CAIC values (Table 5). Additionally, the histogram of the residuals and Q-Q plots of the three models, which are based on lognormal distributions, demonstrate normality, affirming the appropriateness of the assumption regarding the error distribution for the CPUE standardization in the three models (Figure 4).

Table 2. Selection process of the optimal GLM.

No.	GLM Model	AIC
1	Ln(CPUE)~Year + Area + Year: Area	9755.40
2	Ln(CPUE)~Year + Area + Year:Area + Vessel	9720.78
3	Ln(CPUE)~Year + Area + Year:Area + Vessel + SST	9722.72
4	Ln(CPUE)~Year + Area + Year:Area + Vessel + SST + SSTG	9708.86
5	Ln(CPUE)~Year + Area + Year:Area + Vessel + SST + SSTG + SSH	9710.58

Table 3. Selection process of the optimal GLMM.

No.	GLMM Model	AIC
1	Ln(CPUE)~Year + Area + Year: Area	9795.28
2	Ln(CPUE)~Year + Area + Year:Area + Vessel	9765.51
3	Ln(CPUE)~Year + Area + Year:Area + Vessel + SST	9776.61
4	Ln(CPUE)~Year + Area + Year:Area + Vessel + SST + SSTG	9761.08
5	Ln(CPUE)~Year + Area + Year:Area + Vessel + SST + SSTG + SSH	9765.23

Table 4. Selection process of the optimal VAST.

No.	Covariates	AIC
1	None	8789.27
2	SST	8782.16
3	SST + SSTG	8785.38
4	SST + SSTG + SSH	8787.45

Table 5. Summary of the three models fitted to Pacific sardine CPUE from 2014 to 2021.

Model	Conditional R^2	CAIC
GLM	0.27	9709.18
GLMM	0.32	9690.90
VAST	0.36	8776.27

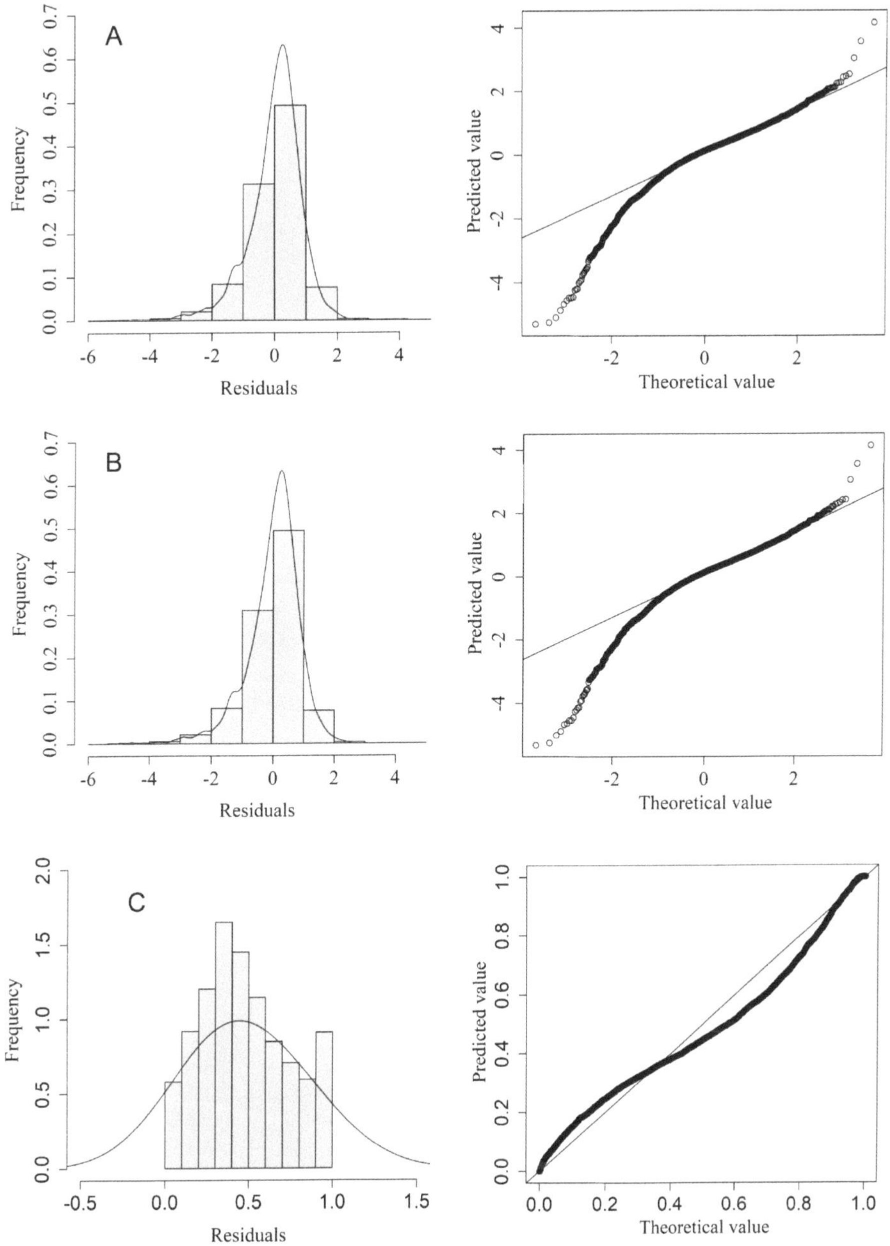

Figure 4. Diagnostic plots for the optimal GLM, GLMM, and VSAT. The histogram of the residuals (**left**) and Q-Q plot (**right**) from (**A**) the GLM, (**B**) the GLMM, and (**C**) the VAST. The thick black lines in Q-Q plot represent the consistency between theoretical CPUE and predicted CPUE, and the circles represent abnormal data.

3.2. The Nominal and Standardized CPUE

An overall upward pattern was noted in both the nominal and the standardized CPUE for Pacific sardine from 2014 to 2021 (Figure 5). Except for 2015 and 2016, the relative standardized CPUE with area weightings of the three models exhibited a similar trend compared to the nominal CPUE. Specifically, the standardized CPUE showed a slight decrease from 2014 to 2015. Subsequently, from 2015 to 2018, it increased annually, followed by a decrease in 2019, and, from 2019 to 2021, there was a substantial growth in the standardized CPUE (Figure 5). Furthermore, regarding the CPUE values, except for 2020 and 2021, the standardized CPUE values of all three models were higher than the nominal CPUE. Notably, in 2021, the nominal CPUE (3.79) was significantly higher than the standardized CPUE values of the three models (GLM: 2.05, GLMM: 2.06, and VAST: 1.94).

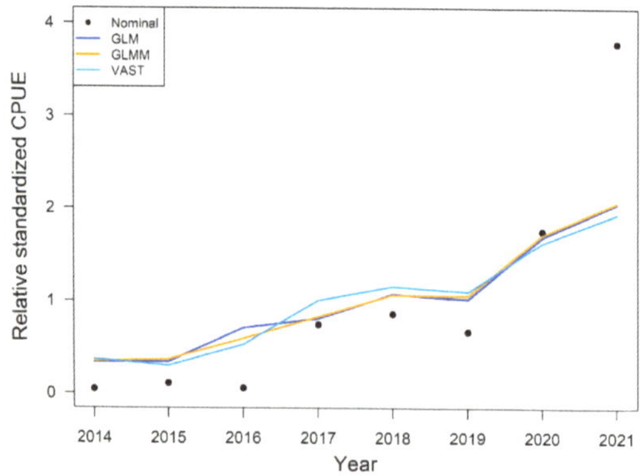

Figure 5. Yearly relative nominal and standardized CPUE for the three models, all of which are calculated with area weightings.

3.3. *Influence of Explanatory Variables*

Figure 6 depicts the annual influence plots of the explanatory variables. The spatial, year × spatial, and vessel annual influence values exhibit diverse patterns across the various models (Figure 6A–C), whereas the SST's values consistently follow a similar trend across all the models (Figure 6D). This highlights the crucial role of model selection in CPUE standardization. In contrast to the GLMM and the VAST, the GLM demonstrates a wider spectrum of variation in annual influence, indicating that the GLMM and the VAST are more stable than the GLM in CPUE standardization. The annual relative standardized CPUE of Pacific sardine was higher than the nominal CPUE from 2014 to 2019 (Figure 5), which can be attributed to the fact that the annual influence values of the year × spatial and vessel were mainly below one in the corresponding years. However, the annual influence of the explanatory variables consistently exceeded a value of one from 2019 to 2021, resulting in standardized CPUE values lower than the nominal CPUE. The annual influence of the SST in all three models was concentrated around one, indicating that the effect of the SST on the difference between the standardized and nominal CPUE is relatively minor. Notably, the year × spatial overall influence in the GLM and the GLMM, as well as the spatio-temporal variable in the VAST, surpassed that of other variables (Table 6).

Table 6. Overall influence of the three models.

Model	Variable	Overall Influence
GLM	Year	0.258
	Area	0.173
	Year × Area	0.287
	SST	0.015
GLMM	Year	0.196
	Area	0.204
	Year × Area	0.303
	SST	0.015

Table 6. *Cont.*

Model	Variable	Overall Influence
VAST	Year	0.217
	Area	0.209
	spatio-temporal	0.312
	SST	0.015

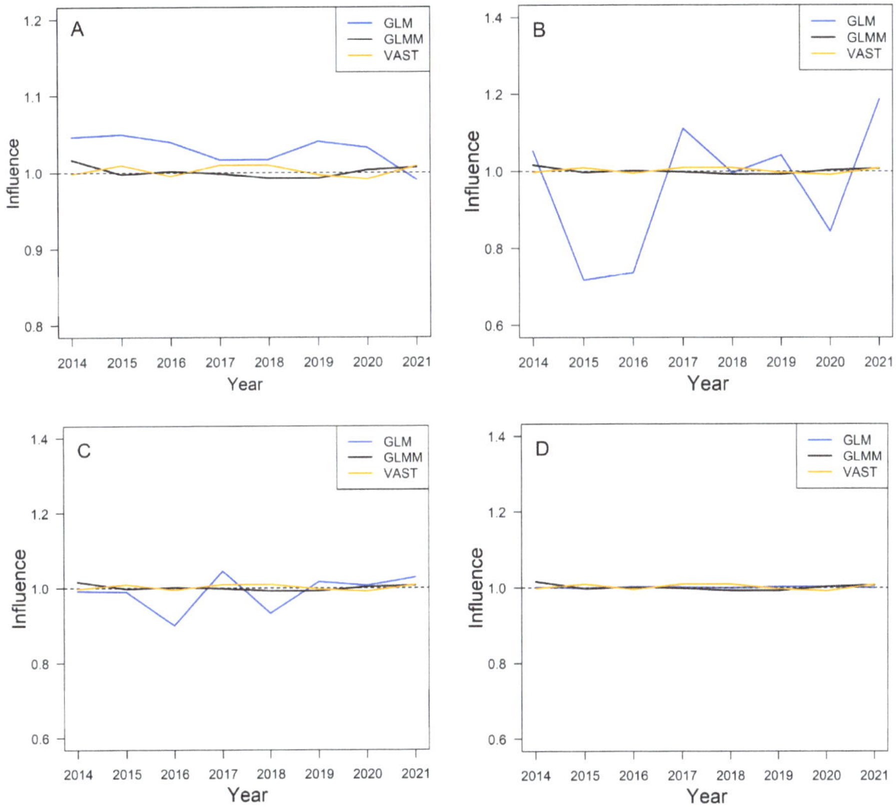

Figure 6. Yearly impact of explanatory factors on the standardized CPUE of Pacific sardine for the three models. (**A**) corresponds to the spatial effect; (**B**) denotes the interaction between the year and the spatial/spatio-temporal random effect; (**C**) represents the vessel effect, and (**D**) signifies the SST's effect. The dotted lines is the reference line for each explanatory factors to judge the influence on the standardized CPUE.

3.4. The Influence of Various Models on Yearly Standardized CPUE

Using the same data stratification method (Spatial 1), this study employed three models for CPUE standardization, allowing us to assess their impact on the standardized CPUE by examining the CDI plot of the explanatory variables. In Figure 7, the CDI plot for the year × spatial effect in the GLM is presented, and it demonstrates notable variations in the annual influence of the year × spatial (right bottom panel). More precisely, during 2015 and 2016, the values of the yearly influence were notably subdued, registering at only 0.72 and 0.74, respectively. This can be ascribed to the concentration of data primarily within areas 1, 3, and 5, where the associated coefficients are relatively modest (Figure 7). Conversely, the yearly influence value for 2021 surged to 1.19, marking a significant increase. This notable upswing is linked to the greater quantity of accessible data, primarily distributed in regions with elevated coefficients. The CDI plot illustrating the year × spatial interaction random effect in the GLMM is shown in Figure 8. Although the data distribution across the areas aligns with the GLM, the impact of the year × spatial in the GLMM model exhibits minimal variation. The lowest influence value was observed in 2015 (0.98), while the highest was observed in 2019 (1.02). This variation could be ascribed to differences in model structure between the GLM and the GLMM. Figure 9 displays the CDI plot for the spatio-temporal random effect in the VAST. In 2014 and 2017, the annual influence values remained relatively modest, primarily due to the concentration of data within knots with lower coefficients. In contrast, a higher annual influence value in 2015 was associated with the year's elevated coefficients.

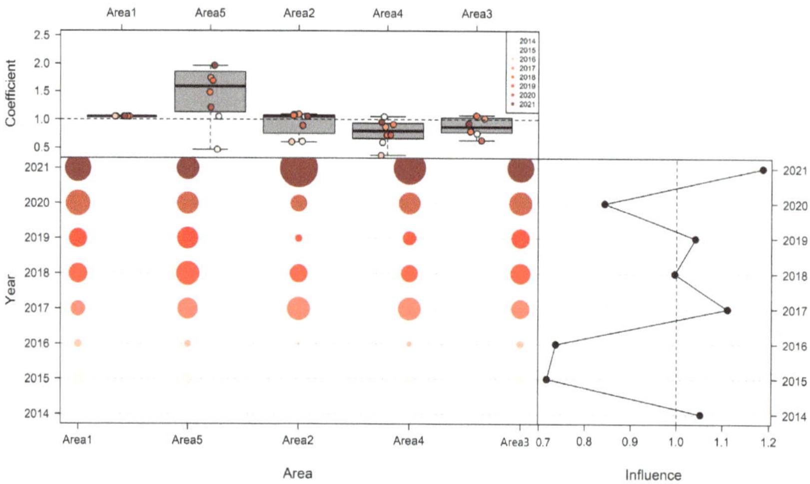

Figure 7. The CDI plot of the year × spatial interaction in the GLM analysis. The top section of the picture displays the normalized coefficients corresponding to each geographical area. Each year is represented by a solid color point indicating its coefficient. To enhance graphical visualization, a small amount of random noise was added to the area coefficient for each year. The lower left portion of picture displays bubbles that represent the yearly distribution of observed CPUE values across each area stratum. The size of each bubble corresponds to the number of data records it represents. The lower right portion of picture displays the yearly influence value of the year × spatial effect.

3.5. Model Evaluation Using a Simulation Test

From Figure 10, it can be observed that the VAST exhibits smaller relative errors compared to the GLM and the GLMM, and it fluctuates around the zero value, which indicates that the VAST estimates CPUE values closer to the "true" value. Among the three models, the GLM model has the most significant relative errors, followed by the GLMM model (Figure 10). Regarding the RMSE, the VAST model has the smallest RMSE value (0.093), while the GLM model has the largest value (0.143). The RMSE value for the GLMM model is 0.126 (Figure 11A). Similarly, the bias metrics values for the three models, from largest to smallest, were the GLM, the GLMM, and the VAST (Figure 11B), which suggested that the VAST model demonstrates a superior evaluation performance.

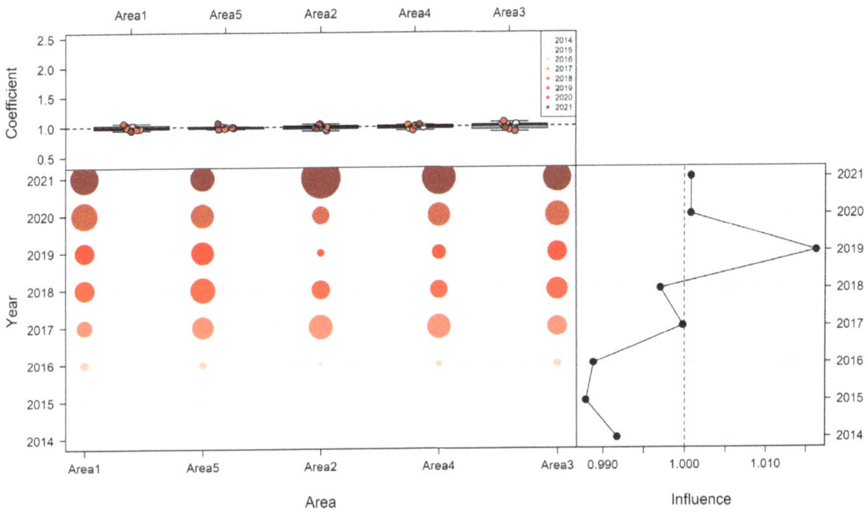

Figure 8. The CDI plot of the year × spatial interaction random effect in the GLMM. The top section of the picture displays the normalized coefficients corresponding to each geographical area. Each year is represented by a solid color point indicating its coefficient. To enhance graphical visualization, a small amount of random noise was added to the area coefficient for each year. The lower left portion of picture displays bubbles that represent the yearly distribution of observed CPUE values across each area stratum. The size of each bubble corresponds to the number of data records it represents. The lower right portion of picture displays the yearly influence value of the year × spatial effect.

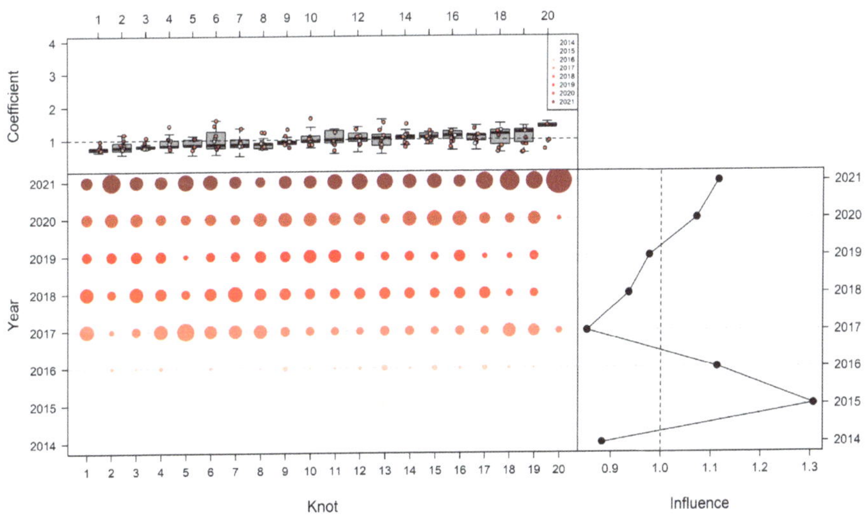

Figure 9. The CDI plot of the spatio-temporal random effect in the VAST. The top section of the picture displays the normalized coefficients corresponding to each geographical area. Each year is represented by a solid color point indicating its coefficient. To enhance graphical visualization, a small amount of random noise was added to the area coefficient for each year. The lower left portion of picture displays bubbles that represent the yearly distribution of observed CPUE values across each area stratum. The size of each bubble corresponds to the number of data records it represents. The lower right portion of picture displays the yearly influence value of the spatio-temporal effect.

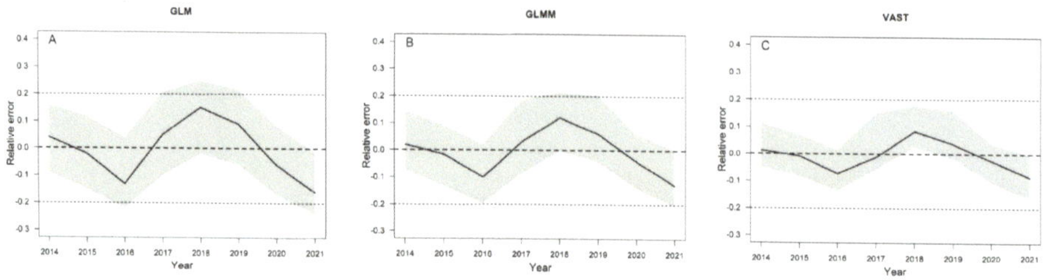

Figure 10. The relative error trends for the three models, with the 95% confidence interval depicted as a light grey polygon.

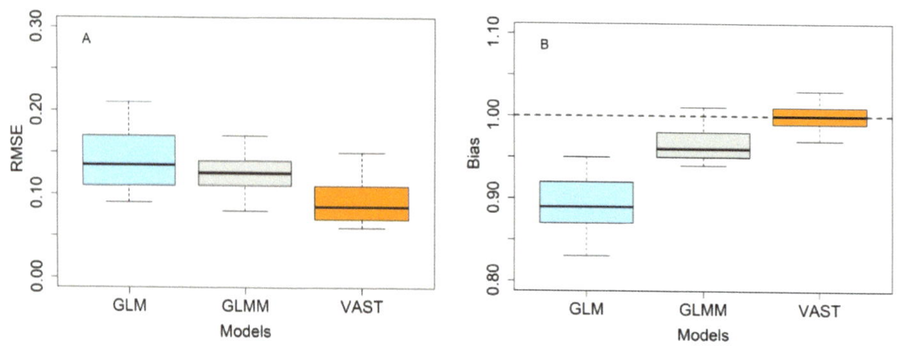

Figure 11. RMSE and bias measures for the GLM, GLMM, and VAST models. In Figure (**A**) the boxes represent the RMSE of each model. In Figure (**B**), a horizontal dashed line serves as the reference for the bias metrics, showing that there is no bias present.

4. Discussion

Pacific sardine fishery is a traditional fishery in the NPO [49]. Since the 1990s, the catch of Pacific sardines has experienced significant fluctuations. However, after 2010, the population of Pacific sardines gradually recovered, leading to an increase in the catch. In 2021, the catch of Pacific sardines exceeded one million tons [50]. With the increase in fishing scale and economic worth, there has been a growing focus on the stock assessment and governance of Pacific sardine resources, and the NPFC has prioritized its stock assessment and management [21]. Therefore, obtaining reliable relative abundance indices is of great significance for the management of Pacific sardines. The Pacific sardine, known for its short life cycle and highly migratory behavior, demonstrates continuous fluctuations in abundance and availability that are influenced by biotic and abiotic environmental factors, exhibiting a spatial structure [51]. Hence, including spatial autocorrelation as a continuous covariate in the CPUE standardization model for Pacific sardines appears fitting. This will improve the model's capacity to precisely grasp the spatial distinctions in the fish occurrence patterns, thus reflecting the variations more reasonably [52].

This study assessed the effectiveness of three models—the GLM, the GLMM, and the VAST—using Chinese Pacific sardine fishery data and marine environmental data including the SST, the SSTG, and the SSH. The reason for selecting the SST, the SSTG, and the SSH is that previous research has indicated that the habitat distribution and abundance of small pelagic fish species, such as the Pacific sardine, are highly sensitive to marine environmental factors, particularly the SST, the SSTG, and the SSH [53,54]. The SST has long been considered the most important influencing factor for Pacific sardine resources, and the CAIC results revealed that the best-performing GLM, GLMM, and VAST models all included the SST, confirming its significance. Nevertheless, the findings from the influence analysis revealed that the annual influence values of the SST were distributed around one

(Figure 6D), suggesting that the factors affecting resource distribution might not always make a substantial contribution to the disparities between the nominal and standardized CPUE, aligning with the conclusions of Hsu et al. [34].

Based on the CAIC and the Conditional R^2, the VAST demonstrates a better fitting performance compared to the GLM and the GLMM (Table 5), which was consistent with the results of Kai [55]. Meanwhile, according to the simulation tests, the VAST model showed smaller RMSE and model bias, indicating that the bias values were closer to one. This suggests that the VAST model can better accommodate the spatial variations in the CPUE data and has a higher assessment accuracy. Ducharme-Barth et al. [47] conducted simulation tests on the VAST model for CPUE standardization and reached conclusions consistent with our study. This can be ascribed to three primary factors. To begin with, the VAST model encompasses both spatial and temporal data variations, accounting for spatial correlations between sampling locations and temporal correlations across different time intervals, and allows for a more accurate representation of the complex spatio-temporal patterns in the data [56]. Additionally, by introducing spatial and temporal random effects, the VAST can effectively handle the inherent correlations and disparities within the CPUE data [57]. Lastly, the VAST provides the flexibility to incorporate supplementary covariates that elucidate the fluctuations in CPUE. This helps in capturing the impacts of pertinent environmental factors and other significant variables on the standardized CPUE, facilitating a thorough and precise analysis [58].

From Figure 5, it can be observed that the values of the standardized CPUE surpassed the nominal CPUE between 2014 and 2019. However, in 2020 and 2021, the opposite trend was observed. This may be due to the fact that the values of the influence of the year × spatial or spatio-temporal variables were mostly below one from 2014 to 2019, whereas, in the following two years, their influence values exceeded one. This also demonstrates the importance of this explanatory variable in influencing the disparities between the nominal and standardized CPUE [30]. The yearly relative standardized CPUE shows a progressive upward trend between 2014 and 2021 (Figure 5), indicating the gradual recovery of Pacific sardine resources. This pattern corresponds to the noted surge in Pacific sardine catches in China during recent years. Additionally, these research findings align with those of Yang et al. [14] and can offer valuable scientific support for the establishment or adjustment of management regulations in Pacific sardine fisheries. The objective is to strike a balance between ecological preservation and the concerns of fishers.

CPUE standardization models are applied to mitigate the confounding influences of external factors and derive an indicator that accurately reflects fish biomass [59]. Therefore, it is imperative to thoroughly scrutinize the results, instead of merely adopting the CPUE data generated using the standardization model, in order to grasp the impact of including each explanatory factor in the standardized CPUE. We conducted an influence analysis for each explanatory variable of the three models, and the results indicated that, in contrast to other explanatory variables, the influence of the SST on the disparities among the standardized and nominal CPUE is relatively minor (Figure 6). The overall influence of the year × spatial or spatio-temporal variable is the highest (Table 6), which is consistent with the results of Hsu et al. [34]. From Figure 6, it can be observed that the influence values of the explanatory variables in the VAST model fluctuate less over time, while, in the GLM and GLMM models, the influence values of explanatory variables exhibit larger fluctuations over time. This also demonstrates the robustness of CPUE standardization in the VAST model.

In this study, the spatial stratification method "Spatial 1" was employed to stratify the fishery data of Pacific sardines in the NPO [34]. The influence values of the year × spatial random effect or spatio-temporal random effect in the GLMM and VAST models were consistent with the influence criteria of the relationship between the nominal and standardized CPUE (Figures 8 and 9), whereas the year × spatial effect of the GLM did not exhibit this trend (Figure 7). This discrepancy highlights the differences in the data handling capabilities among the models [60]. We also discovered that the distribution of data had a significant impact on the standardized CPUE. In instances where the data clustered in regions with high coefficients, the influence value for the respective year was relatively elevated, while it was diminished when the data dispersed across areas with low coefficients (Figures 7–9). Based on the existing literature, several approaches, such as the ad hoc approach [61] and the binary recursive partitioning approach [62], can be used to determine area stratification in CPUE standardization. Therefore, it is advisable to undertake future studies to examine the influence of various fishery data stratification approaches on CPUE standardization. This will impart valuable insights, delivering scientific management recommendations to a wide audience which includes fishers, managers, and stakeholders.

Our study has provided crucial insights for the stock assessment of Pacific sardine. It has also established a spatio-temporal model framework for the CPUE standardization for other small pelagic fish species worldwide, thereby supporting the conservation and sustainable use of other fish stocks. Importantly, our study once again demonstrated the impact of the SST on the distribution and abundance of Pacific sardines. However, we also discovered that explanatory variables with a high explanatory power in CPUE standardization models may not necessarily have a significant influence on the disparities between the nominal CPUE and the standardized CPUE. Therefore, this aspect should be clarified in future research. Our study revealed that the variables of year × spatial or spatio-temporal exerted the most significant overall impact concerning the standardized CPUE, indicating their significant role in explaining the disparities between the nominal and standardized CPUE. Furthermore, it highlighted the advantages of the VAST in standardizing the CPUE for highly migratory small pelagic fish species, suggesting its incorporation as a routine CPUE standardization tool. This research will facilitate the application of accurate biomass indices in stock assessment and ultimately promote the scientific management and conservation of Pacific sardines.

5. Conclusions

We effectively developed multiple models for Pacific sardine CPUE standardization and assessed the effectiveness of three models (GLM, GLMM, and VAST). While all three models displayed parallel patterns in the standardized CPUE when compared to the nominal CPUE, the VAST method exhibited a lower CAIC and a higher conditional R^2, indicating a better performance in fitting and prediction. This is because the VAST has the capability to consider and address temporal and spatial correlations and disparities within the CPUE data, leading to a more precise estimation of spatial and temporal effects. The VAST with only the SST as a covariate had a smaller CAIC. Influence analysis is an effective method for comprehending how each explanatory factor affects the disparities between the nominal and standardized CPUE. We performed a quantitative assessment on how each explanatory factor influenced the optimal models for the three models. The results demonstrated that the interaction terms of the three models exert a more significant influence on the standardized CPUE. This underscores the importance of thoroughly incorporating these elements in future research on Pacific sardines in the NPO. Subsequently, we produced CDI plots representing the year × spatial interaction effect or the spatio-temporal random effect. These plots illustrated that the VAST model displays an enhanced reliability when standardizing the CPUE for Pacific sardines. Simulation testing also confirmed that the VAST model exhibits the highest accuracy (with the smallest RMSE and model bias) in CPUE standardization. This suggests that the VAST model can be considered a practical approach for CPUE standardization for this fish species. The findings presented in this study represent a significant initial stride towards the balanced conservation and eco-friendly utilization of Pacific sardine resources.

Author Contributions: Y.S.: writing—original draft preparation; conceived the project, and designed the experiments. H.H.: investigation and resources. F.T.: writing—review and editing. S.Z.: data processing, review, and comment. W.F.: conceptualization. H.Z.: review and comment. Z.W.: manuscript revision and improvement. All authors have read and agreed to the published version of the manuscript.

Funding: This study was sponsored by the Laoshan Laboratory (No. LSKJ202201804), the Shanghai Sailing Program (22YF1459900), the Shanghai Science and Technology Committee (20dz1206400), the Central Public-Interest Scientific Institution Basal Research Fund, the ECSFR, the CAFS (2021T04, 2021M06), and the Zhejiang ocean fishery resources exploration and capture project (CTZB-2022080076).

Institutional Review Board Statement: The fishery logbook data is acquired from the Technical Group for Trawl-purse seine Fishery within the Distant-water Fishery Society of China, so the ethical approval is not applicable.

Data Availability Statement: The datasets used and/or analyzed during the current study are available from the corresponding author upon reasonable request.

Conflicts of Interest: The authors declare no conflict of interest.

References

1. Campbell, R.A. CPUE standardisation and the construction of indices of stock abundance in a spatially varying fishery using general linear models. *Fish. Res.* **2004**, *70*, 209–227. [CrossRef]
2. Hilborn, R.; Walters, C.J. *Quantitative Fisheries Stock Assessment and Management: Choice, Dynamics and Uncertainty*; Chapman & Hall: New York, NY, USA, 1992; p. 570.
3. Callwood, K.A. Examining the development of a parrotfish fishery in The Bahamas: Social considerations & management implications. *Glob. Ecol. Conserv.* **2021**, *28*, e01677.
4. Harley, S.J.; Myers, R.A.; Dunn, A. Is catch-per-unit-effort proportional to abundance? *Can. J. Fish. Aquat. Sci.* **2001**, *58*, 1760–1772. [CrossRef]
5. Cao, J.; Thorson, J.T.; Richards, R.A.; Chen, Y. Spatiotemporal index standardization improves the stock assessment of northern shrimp in the Gulf of Maine. *Can. J. Fsih. Aquat. Sci.* **2017**, *74*, 1781–1793. [CrossRef]
6. Walters, C. Folly and fantasy in the analysis of spatial catch rate data. *Can. J. Fish. Aquat. Sci.* **2003**, *60*, 1433–1436. [CrossRef]
7. Li, G.; Lu, Z.W.; Cao, Y.M.; Zou, L.J.; Chen, X.J. CPUE Estimation and Standardization Based on VMS: A Case Study for Squid-Jigging Fishery in the Equatorial of Eastern Pacific Ocean. *Fish* **2023**, *8*, 2. [CrossRef]
8. Hashimoto, M.; Nishijima, S.; Yukami, R.; Watanabe, C.; Kamimura, Y.; Furuichi, S.; Ichinokawa, M.; Okamura, H. Spatiotemporal dynamics of the Pacific chub mackerel revealed by standardized abundance indices. *Fish. Res.* **2019**, *219*, 105315. [CrossRef]
9. Hazin, H.G.; Hazin, F.; Mourato, B.; Carvalho, F.; Frédou, T. Standardized catch rates of swordfish (*Xiphias gladius*) caught by the Brazilian fleet (1978–2012) using generalized linear mixed models (GLMM) using delta log approach. *Collect. Vol. Sci. Pap. ICCAT.* **2014**, *70*, 1875–1884.
10. Thorson, J.T.; Shelton, A.O.; Ward, E.J.; Skaug, H.J. Geostatistical delta-generalized linear mixed models improve precision for estimated abundance indices for West Coast groundfishes. *ICES J. Mar. Sci.* **2015**, *72*, 1297–1310. [CrossRef]
11. Maunder, M.N.; Punt, A.E. Standardizing catch and effort data: A review of recent approaches. *Fish. Res.* **2004**, *70*, 141–159. [CrossRef]
12. Rodriguez, M.E.; Arrizabalaga, H.; Ortiz, M.; Rodriguez, C.C.; Moreno, G.; Kell, L.T. Standardization of bluefin tuna (*Thurmus thynnus*) catch per unit effort in the bait boat fishery of the Bay of Biscay (Eastern Atlantic). *ICES J. Mar. Sci.* **2003**, *60*, 1216–1231. [CrossRef]
13. Cortes, G.G.; Montanez, J.A.D.A.; Sánchez, F.A.; Salas, S.; Balart, E.F. How do environmental factors affect the stock–recruitment relationship? The case of the Pacific sardine (*Sardinops sagax*) of the northeastern Pacific Ocean. *Fish. Res.* **2010**, *102*, 173–183. [CrossRef]
14. Yang, C.; Han, H.B.; Zhang, H.; Shi, Y.C.; Su, B.; Jiang, P.W.; Xiang, D.L.; Sun, Y.Y.; Li, Y. Assessment and management recommendations for the status of Japanese sardine *Sardinops melanostictus* population in the Northwest Pacific. *Ecol. Indic.* **2023**, *148*, 110111. [CrossRef]
15. Tameishi, H.; Shinomiya, H.; Aoki, I.; Sugimoto, T. Understanding Japanese sardine migrations using acoustic and other aids. *ICES J. Mar. Sci.* **1996**, *53*, 167–171. [CrossRef]
16. Sarr, O.; Kindong, R.; Tian, S.Q. Knowledge on the Biological and Fisheries Aspects of the Japanese Sardine, *Sardinops melanostictus* (Schlegel, 1846). *J. Mar. Sci. Eng.* **2021**, *9*, 1403. [CrossRef]
17. Furuichi, S.; Niino, Y.; Kamimura, Y.; Yukami, R. Time-varying relationships between early growth rate and recruitment in Japanese sardine. *Fish. Res.* **2020**, *232*, 105723. [CrossRef]
18. FAO. *FAO Yearbook*; Fishery and Aquaculture Statistics: Rome, Italy, 2021; p. 13.
19. Shi, Y.C.; Zhang, X.M.; He, Y.R.; Fan, W.; Tang, F.H. Stock Assessment Using Length-Based Bayesian Evaluation Method for Three Small Pelagic Species in the Northwest Pacific Ocean. *Front. Mar. Sci.* **2022**, *9*, 775180. [CrossRef]
20. Zhao, G.Q.; Shi, Y.C.; Fan, W.; Cui, X.S.; Tang, F.H. Study on main catch composition and fishing ground change of light purse seine in Northwest Pacific. *S. China Fish. Sci.* **2022**, *18*, 33–42, (In Chinese with English Abstract).
21. North Pacific Fisheries Commission. *NPFC Yearbook 2017*; North Pacific Fisheries Commission: Tokyo, Japan, 2017; p. 385. Available online: www.npfc.int (accessed on 21 November 2017).

22. Shi, Y.C.; Kang, B.; Fan, W.; Xu, L.L.; Zhang, S.M.; Cui, X.S.; Dai, Y. Spatio-Temporal Variations in the Potential Habitat Distribution of Pacific Sardine (*Sardinops sagax*) in the Northwest Pacific Ocean. *Fish* **2023**, *8*, 86. [CrossRef]
23. Takasuka, A.; Oozeki, Y.; Aoki, I. Optimal growth temperature hypothesis: Why do anchovy flourish and sardine collapse or vice versa under the same ocean regime? *Can. J. Fish. Aquat. Sci.* **2007**, *64*, 768–776. [CrossRef]
24. Ito, S. Fishery biology of the sardine, *Sardinops melanosticta* (T. & S.), in the waters around Japan. *Bull. Jpn. Sea Reg. Fish. Res. Lab.* **1961**, *9*, 1–227, (In Japanese with English abstract).
25. Wada, T.; Jacobson, L.D. Regimes and stock-recruitment relationships in Japanese sardine (*Sardinops melanostictus*), 1951–1995. *Can. J. Fish. Aquat. Sci.* **1998**, *55*, 2455–2463. [CrossRef]
26. Matsuyama, M.; Adachi, S.; Nagahama, Y.; Kitajima, C.; Matsuura, M.S. Annual reproductive cycle of the captive female Japanese sardine (*Sardinops melanostictus*): Relationship to ovarian development and serum levels of gonadal steroid hormones. *J. Mar. Biol.* **1991**, *108*, 21–29. [CrossRef]
27. Nyuji, M.; Hongo, Y.; Yoneda, M.; Nakamura, M. Transcriptome characterization of BPG axis and expression profiles of ovarian steroidogenesis-related genes in the Japanese sardine. *BMC Genom.* **2020**, *21*, 668. [CrossRef] [PubMed]
28. Muko, S.; Ohshimo, S.; Kurota, H.; Yasuda, T.; Fukuwaka, M.A. Long-term distribution changes in distribution of Japanese sardine in the Sea of Japan during the stock fluctuations. *Mar. Ecol. Prog. Ser.* **2018**, *593*, 141–154. [CrossRef]
29. Sakamoto, T.; Komatsu, K.; Shirai, K.; Higuchi, T.; Ishimura, T.; Setou, T.; Kamimura, Y.; Watanabe, C.; Kawabata, A. Combining microvolume isotope analysis and numerical simulation to reproduce fish migration history. *Methods Ecol. Evol.* **2019**, *10*, 59–69. [CrossRef]
30. Bentley, N.; Kendrick, T.H.; Starr, P.J.; Breen, P.A. Influence plots and metrics: Tools for better understanding fisheries catch-per-unit-effort standardizations. *ICES J. Mar. Sci.* **2012**, *69*, 84–88. [CrossRef]
31. Howell, E.A.; Kobayashi, D.R. El Niño effects in the Palmyra Atoll region: Oceanographic changes and bigeye tuna (*Thunnusobesus*) catch rate variability. *Fish. Oceanogr.* **2006**, *15*, 477–489. [CrossRef]
32. Graham, M.H. Confronting multicollinearity in ecological multiple regression. *Ecology* **2003**, *84*, 2809–2815. [CrossRef]
33. Tien, B.D.; Lofman, O.; Revhaug, I.; Dick, O. Landslide susceptibility analysis in the Hoa Binh province of Vietnam using statistical index and logistic regression. *Nat. Hazards* **2011**, *59*, 1413–1444.
34. Hsu, J.; Chang, Y.J.; Ducharme-Barth, N.D. Evaluation of the influence of spatial treatments on catch-per-unit-effort standardization: A fishery application and simulation study of Pacific saury in the Northwestern Pacific Ocean. *Fish. Res.* **2022**, *255*, 106440. [CrossRef]
35. Ono, K.; Punt, A.E.; Hilborn, R. Think outside the grids: An objective approach to define spatial strata for catch and effort analysis. *Fish. Res.* **2015**, *170*, 89–101. [CrossRef]
36. Tian, S.Q.; Chen, X.J. Impacts of different calculating methods for nominal CPUE on CPUE standardization. *J. Shanghai Ocean Univ.* **2010**, *19*, 240–245.
37. Pinheiro, J.C.; Bates, D.M. Linear mixed-effects models: Basic concepts and examples. In *Mixed-Effects Models in S and S-Plus;* Springer: Berlin/Heidelberg, Germany, 2000; pp. 3–56.
38. Mikkonen, S.; Rahikainen, M.; Virtanen, J.; Lehtonen, R.; Kuikka, S.; Ahvonen, A. A linear mixed model with temporal covariance structures in modelling catch per unit effort of Baltic herring. *ICES J. Mar. Sci.* **2008**, *65*, 1645–1654. [CrossRef]
39. Forrestal, F.C.; Schirripa, M.; Goodyear, C.P.; Arrizabalaga, H.; Babcock, E.A.; Coelho, R.; Ingram, W.; Lauretta, M.; Ortiz, M.; Sharma, R. Testing robustness of CPUE standardization and inclusion of environmental variables with simulated longline catch datasets. *Fish. Res.* **2019**, *210*, 1–13. [CrossRef]
40. Sculley, M.L.; Brodziak, J. Quantifying the distribution of swordfish (*Xiphias gladius*) density in the Hawaii-based longline fishery. *Fish. Res.* **2020**, *230*, 105638. [CrossRef]
41. Grüss, A.; Walter III, J.F.; Babcock, E.A.; Forrestal, F.C.; Thorson, J.T.; Lauretta, M.V.; Schirripa, M.J. Evaluation of the impacts of different treatments of spatio-temporal variation in catch-per-unit-effort standardization models. *Fish. Res.* **2019**, *213*, 75–93. [CrossRef]
42. Kristensen, K.; Nielsen, A.; Berg, C.W.; Skaug, H.; Bell, B. TMB: Automatic Differentiation and Laplace Approximation. *J. Stat. Softw.* **2016**, *70*, 1–21. [CrossRef]
43. Greven, S.; Kneib, T. On the behaviour of marginal and conditional AIC in linear mixed models. *Biometrika* **2010**, *97*, 773–789. [CrossRef]
44. Nakagawa, S.; Schielzeth, H. A general and simple method for obtaining R^2 from generalized linear mixed-effects models. *Methods Ecol. Evol.* **2013**, *4*, 133–142. [CrossRef]
45. Carruthers, T.R.; Li, C.J.; McAllister, M.K. Assessing methods for detecting marine ecosystem regime shifts: A simulation test. *Fish. Res.* **2004**, *67*, 173–187.
46. Methratta, E.T.; Link, J.S.; Dow, D.M. Simulation testing of stock assessment models. *Fish. Res.* **2003**, *65*, 253–264.
47. Ducharme-Barth, N.D.; Grüss, A.; Vincent, M.T.; Kiyofuji, H.; Aoki, Y.; Pilling, G.; Hampton, J.; Thorson, J.T. Impacts of fisheries-dependent spatial sampling patterns on catch-per-unit-effort standardization: A simulation study and fishery application. *Fish. Res.* **2022**, *246*, 106169. [CrossRef]
48. Wilberg, M.J.; Thorson, J.T.; Linton, B.C.; Berkson, J. Incorporating time-varying catchability into population dynamic stock assessment models. *Rev. Fish. Sci.* **2010**, *18*, 7–24. [CrossRef]

49. Nakai, Z. Studies relevant to mechanisms underlying the fluctuation in the catch of the Japanese sardine, *Sardinops melanosticta*. *Jpn. J. Ichthyol.* **1962**, *9*, 1–115.
50. Saito, R.; Oozeki, Y.; Takasaki, K.; Saito, T.; Uehara, S.; Miyahara, M. Transshipment activities in the North Pacific Fisheries Commission Convention Area. *Mar. Policy* **2022**, *146*, 105299. [CrossRef]
51. Watanabe, Y.; Zenitani, H.; Kimura, R. Population decline of the Japanese sardine Sardinops melanostictus owing to recruitment failures. *Can. J. Fish. Aquat. Sci.* **1995**, *52*, 1609–1616. [CrossRef]
52. Thorson, J.T.; Barnett, L.A. Comparing estimates of abundance trends and distribution shifts using single-and multispecies models of fishes and biogenic habitat. *ICES J. Mar. Sci.* **2017**, *74*, 1311–1321. [CrossRef]
53. Song, H.; Miller, A.J.; McClatchie, S.; Weber, E.D.; Nieto, K.M.; Checkley, D.M.J. Application of a data-assimilation model to variability of Pacific sardine spawning and survivor habitats with ENSO in the California Current System. *J. Geophys. Res.* **2012**, *117*, C03009. [CrossRef]
54. Vander, L.C.D.; Castro, L.; Drapeau, L.; Checkley, D., Jr. (Eds.) *Report of a GLOBEC-SPACC Workshop on Characterizing and Comparing the Spawning Habitats of Small Pelagic Fish*; GLOBEC Report 21; GLOBEC: Swindon, UK, 2005; Volume 7, p. 33.
55. Kai, M. Spatio-temporal changes in catch rates of pelagic sharks caught by Japanese research and training vessels in the western and central North Pacific. *Fish. Res.* **2019**, *216*, 177–195. [CrossRef]
56. Thorson, J.T. Guidance for decisions using the Vector Autoregressive Spatio-Temporal (VAST) package in stock, ecosystem, habitat and climate assessments. *Fish. Res.* **2019**, *210*, 143–161. [CrossRef]
57. Han, Q.P.; Shan, X.J.; Jin, X.S.; Gorfine, H. Overcoming gaps in a seasonal time series of Japanese anchovy abundance to analyse interannual trends. *Ecol. Indic.* **2023**, *149*, 110189. [CrossRef]
58. Thorson, J.T.; Adams, C.F.; Brooks, E.N.; Eisner, L.B.; Kimmel, D.G.; Legault, C.M.; Rogers, L.A.; Yasumiishi, E.M. Seasonal and interannual variation in spatio-temporal models for index standardization and phenology studies. *ICES J. Mar. Sci.* **2020**, *77*, 1879–1892. [CrossRef]
59. Quirijns, F.; Poos, J.; Rijnsdorp, A. Standardizing commercial CPUE data in monitoring stock dynamics: Accounting for targeting behaviour in mixed fisheries. *Fish. Res.* **2008**, *89*, 1–8. [CrossRef]
60. Campbell, R.A. A new spatial framework incorporating uncertain stock and fleet dynamics for estimating fish abundance. *Fish Fish. Oxf.* **2016**, *17*, 56–77. [CrossRef]
61. Nakano, H. Stock status of Pacific swordfish, *Xiphias gladius*, inferred from CPUE of the Japanese longline fleet standardized using general linear models. *US Nat. Mar. Fish. Serv. NOAA Tech. Rep. NMFS* **1998**, *142*, 195–209.
62. Ichinokawa, M.; Brodziak, J. Using adaptive area stratification to standardize catch rates with application to North Pacific swordfish (*Xiphias gladius*). *Fish. Res.* **2010**, *106*, 249–260. [CrossRef]

MDPI AG
Grosspeteranlage 5
4052 Basel
Switzerland
Tel.: +41 61 683 77 34

Fishes Editorial Office
E-mail: fishes@mdpi.com
www.mdpi.com/journal/fishes

www.ingramcontent.com/pod-product-compliance
Lightning Source LLC
Chambersburg PA
CBHW041400210926
51850CB00013B/3137
* 9 7 8 3 7 2 5 8 2 8 0 6 7 *